YIELD MODELLING AND DEFECT TOLERANCE IN VLSI

Yield Modelling and Defect Tolerance in VLSI

PAPERS PRESENTED AT THE INTERNATIONAL WORKSHOP ON DESIGNING FOR YIELD, OXFORD, 1–3 JULY 1987

EDITED BY

Will Moore

DEPARTMENT OF ENGINEERING SCIENCE, UNIVERSITY OF OXFORD
FELLOW OF JESUS COLLEGE

Wojciech Maly and **Andrzej Strojwas**

DEPARTMENT OF ELECTRICAL AND COMPUTER ENGINEERING
CARNEGIE-MELLON UNIVERSITY

Adam Hilger, Bristol and Philadelphia

British Library Cataloguing in Publication Data

International Workshop on Designing for
Yield *(1987: Oxford, England).*
Yield modelling and defect tolerance in
VLSI.
1. Electronic equipment. Very large scale
integrated circuits. Design
I. Title II. Moore, Will III. Maly,
Wojciech IV. Strojwas, Andrzej
621.381′73

ISBN 0-85274-398-X

Library of Congress Cataloging-in-Publication Data are available

Consultant Editor: **Professor M H Rogers**, University of Bristol

Published under the Adam Hilger imprint by IOP Publishing Ltd
Techno House, Redcliffe Way, Bristol BS1 6NX
242 Cherry Street, Philadelphia, PA 19106, USA

Printed in Great Britain by J W Arrowsmith Ltd, Bristol

Contents

Preface

This book is an edited selection of the papers presented at the International Workshop on Designing for Yield which was held at Oxford University in July 1987. Our thanks go to all of the contributors and especially to the Programme Committee for all their hard work. Thanks are also due to the IEE and the IEEE Computer Society for publicising the event and to Oxford University for their active support. We are particularly grateful to David Cawley and Pat Tollestrup for coping with the administrative problems.

Will Moore
Wojciech Maly
Andrzej Strojwas

January 1988

Programme Committee

Simon Jones, Bangor (UK)
Israel Koren, Technion (Israel)
Andrew McCabe, GEC (UK)
Wojciech Maly, CMU (USA)
John McCanny, QUB (UK)
John MacDonald, RPI (USA)
Tulin Mangir, TRW (USA)
Will Moore, Oxford (UK)
Peter Schvan, Northern Telecom (Canada)
Charles Stapper, IBM (USA)
Andrzej Strojwas, CMU (USA)
Geoff Sumerling, Plessey (UK)
Anthony Walton, Edinburgh (UK)

List of Contributors

V K Agarwal, McGill (Canada)
H S Bolouri, Brunel (UK)
M Breuer, USC (USA)
D M Campbell, CMU (USA)
J J Charlot, Matra-Harris (France)
M F Chang, Illinois (USA)
J D Chinn, CMU (USA)
M Cutler, SUNY (USA)
J G Delgado-Frias, Oxford (UK)
W K Fuchs, Illinois (USA)
S Gandemer, ENST (France)
E Gautrin, INRIA (France)
S Gupta, RPI (USA)
M C Howells, McGill (Canada)
S R Jones, Bangor (UK)
J M Kim, Iowa (USA)
I Koren, Technion (Israel)
S Y Kuo, Illinois (USA)
R M Lea, Brunel (UK)
A K Lewis, GEC (UK)
H T Lin, RPI (USA)
R Logan, Motorola (FRG)
A P H McCabe, GEC (UK)
A MacConaill, Ferranti (UK)
J F McDonald, RPI (USA)
R D McKirdy, Brunel (UK)
W Maly, CMU (USA)
W P Marnane, Oxford (UK)
P Mazumder, Illinois (USA)
W R Moore, Oxford (UK)
P K Mozumder, CMU (USA)
U Ramacher, Siemens (FRG)
S M Reddy, Iowa (USA)
M Rivier, IBM (France)
S Sastry, USC (USA)
R Selvaraj, RPI (USA)
C H Stapper, IBM (USA)
A J Strojwas, CMU (USA)
C B Stunkel, Illinois (USA)
S Y H Su, SUNY (USA)
M E Thomas, CMU (USA)
B Tremintin, Matra-Harris (France)
J A Trotter, Oxford (UK)
A J Walton, Edinburgh (UK)
M Wang, SUNY (USA)
A Wild, Motorola (FRG)

Prologue

Traditionally the microelectronics engineer implements a custom VLSI circuit with fixed design rules mostly defining minimum device dimensions and spacings on the silicon layout. These design rules are typically derived from test chips used by the process engineer to characterize the process. Modern VLSI circuits are enormously complex devices with perhaps a million devices and minimum dimensions of the order of one micron, and circuit fabrication involves many processing steps requiring precise control of chemical and physical parameters. Because of this it is impossible to fabricate VLSI circuits without introducing some defects.

The best available fabrication lines may typically still introduce around two defects per square centimetre. Other lines with new processes, aggressive designs or poorer control may have many more defects than this. The chip yield is (to a first approximation) an exponentially decreasing function of the defect density. Therefore, the presence of defects limits the maximum chip size that a designer can use and still expect an economic yield level. As a result, very few commercial chips exceed 1 cm^2 in area.

Design rules are generally formulated as a compromise between yield, performance and circuit area. They will typically be oversimplified in order to be usable by the designer or his CAD tools, but often also because the available data do not justify more sophisticated rules. There is therefore nothing magical about a particular set of rules and the best circuit yields can be expected when the designer has a good understanding of the defect mechanisms and can adapt his design in response to the most critical geometries.

Circuit designers are hungry for circuit area and have also employed defect-tolerance strategies through the addition of redundant circuits. The prime example of this is the spare rows and/or columns added to commercial memory chips. In such cases the circuit designer requires an even clearer picture of the defect mechanisms in order to implement defect tolerance successfully.

The papers in this book address this often neglected (and certainly poorly developed) area of 'designing for yield'. It is quite clear that the better the information the designer has about yield loss mechanisms, the greater is the likelihood of the designer and the process engineer together producing large chips (with or without defect tolerance) that work in sufficient quantity.

Another important contributor to successful chip production is the testing of the circuit. It is already widely recognised that circuit design and circuit test should be integrated as closely as possible and it is also true that the test engineer needs the best available information about defect mechanisms in order to test the chip with confidence. For defect-tolerant designs, tests are also needed in order to diagnose which parts of a chip are faulty and this places even more importance on effective testing.

Chapter 1 reviews our current understanding of these issues. It describes the manufacturing process and physical mechanisms causing defects, the ways they may be modelled to predict circuit yields, and the kind of fault-tolerance techniques that may be useful for ULSI. **Chapter 2** contains papers which examine the problem from four alternative perspectives. It discusses the redesign of circuits for improved yield, the planning of redundancy in RAMs, and the reliability and repairability of ULSI circuits. **Chapter 3** looks at catastrophic yield loss mechanisms which are primarily due to random spot defects and suggests how the circuit layout together with data on the size and distribution of defects can be used as a model for yield prediction. **Chapter 4** reviews traditional aspects of process monitoring and the problems of parametric yield loss. It shows the importance of good statistical process control. **Chapter 5** gives examples of typical defect-tolerant architectures for ULSI and **Chapter 6** looks at the additional problems posed for the yield modelling by defect-tolerant designs. Finally **Chapter 7** contains three papers which illustrate the typical level at which defect models currently influence production testing of ICs.

1 YIELD LOSS MECHANISMS AND DEFECT TOLERANCE

Wojciech Maly, Will Moore and Andrzej Strojwas

INTRODUCTION

Wafer Scale Integration (WSI) is a term that can be seen in many conference proceedings and is still discussed in a number of conferences and workshops. But sessions and panels with WSI in the title are not very popular - they are still shaded by the unfortunate and spectacular failure of Trilogy. On the other hand an Ultra Large Scale of Integration (ULSI) must be the next step in the microelectronics evolution and of course such a step will be possible, if and only if, it is supported by a further decrease in the minimum feature size and an increase in the die size. Further drastic decreases in the minimum feature size seem, however, to be rather unlikely due to basic physical limitations of IC devices. Therefore the trend towards larger area circuits, and ultimately towards WSI, is the obvious direction in the evolution of modern microelectronics.

For this reason, and despite of some disappointing past experiences with WSI, large area VLSI circuits should be investigated and various options leading to better design and manufacturing strategies of ULSI-WSI systems should be examined. In particular, one should ask whether new advanced processing techniques and new CAD tools and design techniques are adequate to successfully design and manufacture WSI-ULSI circuits. (Such questions should be answered before another Trilogy-like failure.) The goal of this paper is to examine one aspect of the above question dealing with models and tools that are needed to predict manufacturing yield. Such models are very important because yield is a central issue in the design and manufacture of large WSI systems.

This paper is organized in the following way. First, we use information presented by Maly et al (1986) and we discuss manufacturing process and physical reasons for failures of fabricated ICs. In this discussion we indicate the main reasons for yield losses and we briefly describe the main mechanisms causing IC faults. Then we describe yield modeling techniques with the emphasis on defect models. Finally, a number of fault-tolerance techniques and their impact on manufacturing yield are presented. At the end of this paper we list a number of issues that have to be solved to answer the question about the feasibility of ULSI-WSI.

DISTURBANCES AND YIELD OF IC MANUFACTURING PROCESS

Due to the inherent fluctuations in any integrated circuit manufacturing process, the yield (which is defined as the ratio of the number of chips that perform correctly to the number of chips manufactured) is always less than 100%. As the complexity of VLSI chips increases, and the dimensions of VLSI devices decrease, the sensitivity of performance to process fluctuations increases, thus reducing the manufacturing yield. Since profitability of a manufacturing process is directly related to yield, the search for better design methodologies and strategies must include the yield issue as a central problem of investigation. Of course, this is especially true in the case of WSI defect-tolerant systems where yield is the most important problem.

Structure of the IC manufacturing process

The IC manufacturing process involves a sequence of basic processing steps which are performed on sets of wafers. The outcome of a manufacturing operation depends on three major factors: the *process recipe* which describes the sequence of fabrication steps and the process controlling parameters , the geometry of the fabricated IC, or *layout*, and some randomly changing environmental factors, called *disturbances*. The *control* of a manufacturing operation is the set of parameters which can be manipulated in order to achieve some desired change in the fabricated IC structure. Examples of control parameters are processing equipment settings which determine temperatures, gas pressures, step duration etc. The *layout* of an IC is described as a set of masks where each mask can be viewed as a collection of transparent regions within an opaque region. The *disturbances* are environmental factors that cause variations in outcome of a manufacturing operation. Since disturbances, by their nature are uncontrollable and inherent in any process, they should be well understood and characterized.

The process disturbances that occur in the IC manufacturing process can be characterized in terms of the physical nature of the disturbance or in terms of the effects they have on IC performance. Effects on performance can be described in terms of changes in the electrical characteristics of the IC or in terms of changes in circuit connectivity. In this section we describe and classify process disturbances and then classify the effects that these disturbances have on performance.

Sources of process disturbances

Sources of the random phenomena that occur in the IC fabrication process (see e.g. Sze 1983, Ghandhi 1983 , Wolf and Tauber 1986, and Maly 1987b) can be classified as:

- Human errors and equipment failures.

- Instabilities in the process conditions. These are random fluctuations in the actual environment that surrounds an IC chip and arise from phenomena such as a turbulent flow of gasses used for diffusion and oxidation, inaccuracies in the control of furnace temperature, etc. Because of these instabilities, each area of wafer is exposed to slightly different environmental conditions and, hence, no two manufactured chips can possibly perform identically.

- Material instabilities. These are variations in the physical parameters of the chemical compounds and other materials used in the manufacturing process. Typical examples of material instabilities are: fluctuations in the purity and physical characteristics of the chemical compounds, density and viscosity of photoresist, water and gas contamination etc.

- Substrate and surface inhomogeneities. These are local disturbances in the properties of substrate wafers which are of three types: *point defects*, *dislocations*, and *surface imperfections*. (They are discussed in more detail later in the paper.)

- Spots. These are mainly lithography related disturbances of IC geometry which may be caused by the mask fabrication process as well as by the lithography process itself. (They are also discussed in more detail in the following sections of this paper.)

Not all of the disturbances described above have the same influence on IC performance. Thus, we need to consider what electrical and geometrical effects result from these disturbances.

Process related deformations in the actual IC

Disturbances in the process can be viewed as affecting the geometry of an IC device and/or the electrical characteristics of an IC device.

Geometrical Deformations. The geometry of an IC is determined by the two dimensional (2D) IC layout and the physics of the manufacturing process. Geometrical deformations can be of three types: *lateral effects*, *vertical effects*, and *spot defects*. Lateral effects include all processing effects which cause the location of the boundary of the region in an actual IC to differ from the corresponding boundary in the ideal IC. A lateral effect can be composed of a *lateral edge displacement* and a *mask misalignment* . Lateral edge displacements are caused by all processing steps with anisotropic characteristics, such as:

- Lateral diffusion;

- Lateral oxidation i.e., bird's beak formation;

- Over- and under-etching of metal, polysilicon, oxide and nitride layers;

- Over- and under-exposures of the photoresist and photoemulsions applied in all lithography steps;

- UV light or e-beam diffraction on the mask edges.

Lateral edge displacements may cause drastic deformations in the topology of the actual IC by merging regions which are to be disjoint. In terms of electrical behavior such deformations manifest themselves as shorts between, or breaks in, conducting paths.

Mask misalignment which is an error in the position of a lithography mask with respect to the features already engraved on the surface of a wafer, causes all edges defined by the mask to be shifted by the same amount with respect to the boundaries of the regions which already exist on the surface of the wafer. In fact, mask misalignment could be so significant that the geometry of an actual IC can be deformed from that of the corresponding ideal IC to the extent it results in changes of the electrical connectivity of the IC.

Vertical effects are deformations in the thickness of IC layers and include deformations which are due to the p-n junction depth variations and deformations in the thickness of the oxide and other deposited layers. Junction depth variations are a direct consequence of the fluctuations in the impurity concentrations while deformations in the thickness of the deposited or oxidized layers are due to process instabilities such as turbulent gas flow, temperature fluctuations, etc.

Spot defects are geometrical features, caused by the previously mentioned spots, that emerge during the manufacturing process and were not originally defined by the the IC layout.

Electrical Deformations. The electrical characteristics of an actual IC are determined by the three-dimensional impurity distributions in the conducting and semiconducting regions, and charge distribution in the insulating layers. As was the case for geometrical deformations, we can divide these phenomena into global and local phenomena. Global electrical deformations will affect various regions in different IC elements in the same way. Examples of global deformations include:

- Fixed positive oxide charges in the oxide due to the non-perfect (non-stechiometric) structure of the oxide;

- Emitter-push effect which, due to the enlarged diffusivity of boron in the active base of the n-p-n transistor caused by the emitter diffusion, induces damage to the crystal lattice;

- Increased junction depths under the oxide regions due to oxidation-enhanced diffusion phenomena.

These deformations are random and vary from wafer to wafer, and even across the wafer. Local electrical deformations are due to

- Spikes which are tiny conducting pipes, or paths, crossing junctions or even whole regions, e.g. the bases in the n-p-n transistors;

- Local impurity precipitates or silicides which affect the percentage of electrically active impurities in the silicon;

- Local contaminations of the SiO_2 - Si surface that affect surface recombination mechanisms in bipolar devices and disturb MOS transistor thresholds through local concentration of surface states.

General characteristics of process disturbances

From the above discussion we can observe that the disturbances can cause either global or local deformations. All deformations which affect all IC elements can be called *global*. Mask misalignment is an example of a disturbance that causes a global deformation of an ideal IC. We will refer to disturbances that cause global deformations as *global disturbances*. Similarly, disturbances that cause only local deformations (affected regions are small compared to the total area of an IC chip) will be called *local disturbances*. Pin-holes, spikes, spot defects are examples of local deformations.

It is also important to note that various process conditions interact very often in an indirect way. For instance, high temperature processes may cause an increase in the lithography errors due to the deformations in the shape of the wafer. Hence, in general, random fluctuations in process conditions of one processing step may result in changing process conditions of another step. Note also that the process induced deformations of any kind, local or global, geometrical or electrical, are random. For the case of global disturbances, the magnitude of process-related deformations is random and varies from one IC to another. For the case of local disturbances, both the magnitude (e.g. size of a defect) and location of the resulting deformations are random.

IC faults

We now turn to a discussion of the impact that the process disturbances have on IC device characteristics and indicate how these randomly changing characteristics affect IC performance. We then introduce a fault classification based upon the effects on performance.

We begin by noting that each process disturbance may cause a number of different changes in IC performance. We will call this change in behavior a *fault*. For instance, a spot defect introduced in a lithography step may cause a short between two adjacent conducting paths, a break in the path or may only alter effective device dimensions (e.g. affect the W/L ratio of a MOSFET). It is therefore convenient to classify faults as either *structural faults* or *functional faults*, which are due to the changes in the topology of the electrical circuit that represents an IC or *performance faults*, in which the IC topology remains unchanged, but in which the circuit performance (e.g., the speed or the dissipated power) falls outside of some set of allowable tolerance limits. Performance faults can be further divided into two classes: *soft-performance faults* and *hard-performance faults*. If an IC is functionally correct (e.g. in a digital circuit, all state transitions occur in the proper order and the output pattern is the correct response to an input pattern vector) but some performance measure (e.g. signal delay) lies outside of the specified range, we say a soft-performance fault has occurred. If an IC does not function properly (e.g. some state transitions do not occur) or some performances are orders of magnitude from the desired values, we say a hard-performance fault has occurred.

Structural faults require some further elaboration. While some deformations will not alter the structure of an IC under no bias, or very small bias conditions, for other values of DC bias a short between two conducting paths can occur. As an example, consider the lateral diffusion in two parallel interconnect lines. Due to the expansion of depletion layers of the p-n junctions, there might be an equivalent short between these two lines under some bias conditions. On the other hand, some geometrical variations may introduce new parasitic elements (such as overlap capacitances) which do not necessarily cause an IC to be defective. Therefore, structural faults are best defined in terms of changes in the equivalent DC circuit diagram under worst case bias conditions.

This fault classification scheme is illustrated in Figure 1. The bottom of the figure shows our classification of physical phenomena which cause yield loss while in the upper part of the figure, the classification is based upon effects (performance faults, structural faults). Note that each fault can be caused by the variety of phenomena. However, some of the relations (solid lines) depicted in Figure 1 dominate. For instance, all global effects will cause first of all soft performance faults while spot defects are mainly responsible for structural faults.

Process observability

To facilitate process observability and process control, one or more processing steps are usually separated by an evaluation or inspection, and selection step. An evaluation step is a measurement or sequence of measurements, called *in-line measurements*, performed on

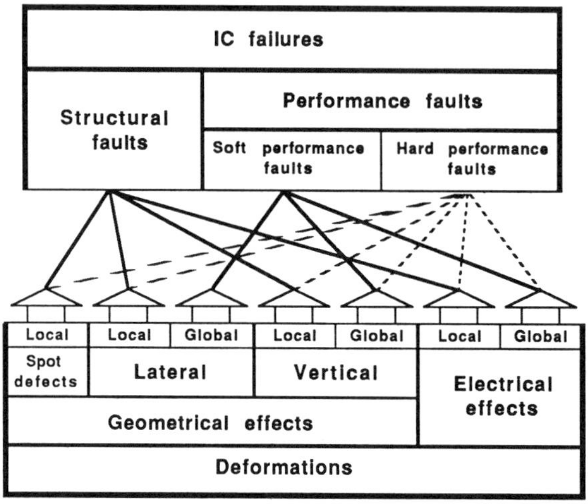

Figure 1: Classification of IC faults.

the wafer or, in some cases, a special test wafer that does not contain IC's. If one or more in-line measurements fall outside of some range defined by a set of *selection thresholds* , the wafers are considered defective and are discarded.

A few of the chips on the wafer are special in that they contain *test structures* and are called *test chips*. Test structures are special purpose devices which are designed so that their performance is sensitive to the quality of specific processing steps. Examples of test structures are long contact chains, large capacitors, and arrays of transistors of various geometries. In some cases, test structures are also placed on scribe lanes, i.e. on the spaces between IC chips.

The best assessment of the process quality is achieved via the probe tests that are performed at the end of the wafer manufacturing stage. Two types of tests are performed: parametric and functional. Parametric tests are used to eliminate the IC chips whose performance falls outside the tolerance limits. Functional tests are used to detect errors in the function performed by an IC.

Yield-disturbances relationships

Information presented above can be summarized in the following way.

- IC performance fluctuations as well as functional faults are due to inherent process disturbances of global and local nature. Thus manufacturing yield can be estimated if and only if the relationship between yield and process disturbances of both categories are well known and adequately modeled.

- IC parametric performance fluctuations can be observed through parametric performance measurement. It is likely that these observable performance fluctuations are caused by global disturbances of the process rather than by the local disturbances.

- IC functional faults are mainly due to structural changes in IC connectivity. Such changes are caused primarily by the local disturbances, i.e., spot defects.

The above observations have very important consequences. Yield losses that are due to global process disturbances have quite different characteristics from yield losses that are due to local disturbances.

In the case of global disturbances the communality of the disturbances of electrical and/or geometrical deformations of all elements of the IC allows for early detection of abnormally large disturbances and therefore for early correction of the process or rejection of the affected wafer. The same feature also allows the manipulation of design rules, to optimize the sizes of transistors and the structure of crucial functional blocks in order to minimize parametric yield losses. These losses, that we will call *parametric yield losses*, are also a weak function of the IC area.

Yield losses that are due to local disturbances have the opposite characteristic. They cannot be predicted until the end of the process because they manifest themselves through IC functional failures only. (Local defects are usually so small and their density is so low that probability of occurrence of local defects in small areas inspected in the process is very small.) In addition, for a uniform density of local defects yield drastically decreases with the increase of the area of the IC. Losses of this category we will call *functional yield losses*.

Consequently, IC design and manufacturing issues dealing with optimization of IC performance, yield oriented process control and diagnosis are mainly concerned with global process disturbances. On the other hand, IC design problems that deal with functional yield losses and optimization of the die size have to be related to the characteristics of local processing disturbances. Development of design strategies of VLSI large area fault-tolerant systems is therefore especially closely related to the development of appropriate modeling technique that relates functional yield losses and local process disturbances. Hence, it is essential to study the defect mechanisms, their impact on the IC failures and the resulting yield losses. In the next part of this chapter we will focus our attention on the spot defects and yield modeling issues.

LOCAL PROCESS DISTURBANCES

Local process disturbances manifest themselves as areas with extra, missing, or drastically changed material occurring in any IC layer forming spot defects or point defects of various kinds. Traditionally, they are assumed to be produced mainly by the lithography process and to cause shorts or breaks in IC connections. Actual spot defect and point defect mechanisms are rather complicated. They also causes failures in the IC functionality that are more complex than failures that can be modeled by simple shorts or opens. In this section we present a brief overview of mechanisms of spot defects and pinholes and models used for their characterization.

Spot defects

The summary of basic spot defect mechanisms (Doi et al 1987) is presented in Figures. 2 and 3. Figure 2 depicts basic processing steps that are used to pattern a segment of parallel metal lines. The first row of the table in Figure 2 shows the surface of the wafer during lithography (exposure) etching and the end of the process for the defect free process.

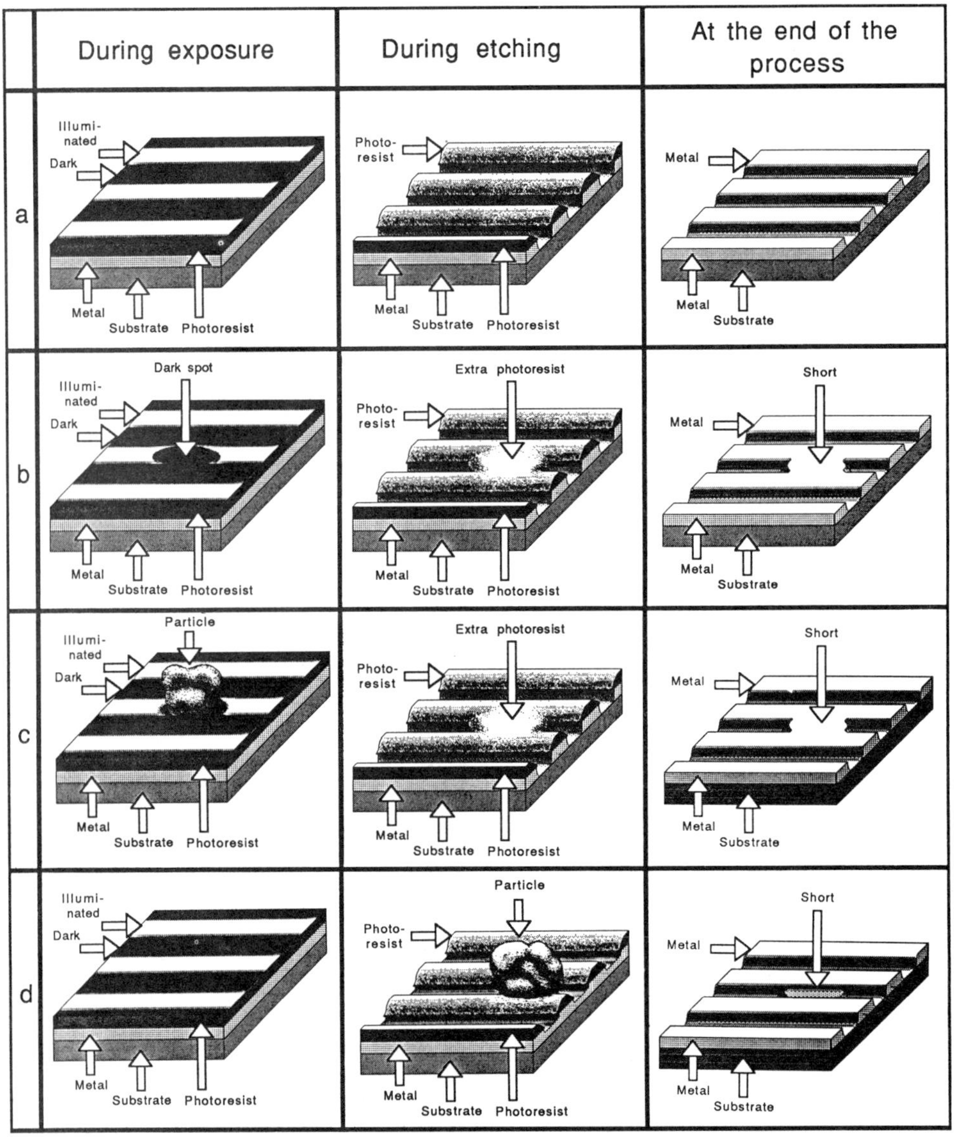

Figure 2: Examples of mechanisms that cause shorts between metal lines.

The second row (b) illustrates a typical, but in modern technologies not the most popular, defect mechanism, which is caused by an opaque spot in the mask. Such a spot shades some areas of the photoresist that should be exposed to the UV light and therefore causes a short. A very similar mechanism is shown in row (c) of Figure 2. In this case the

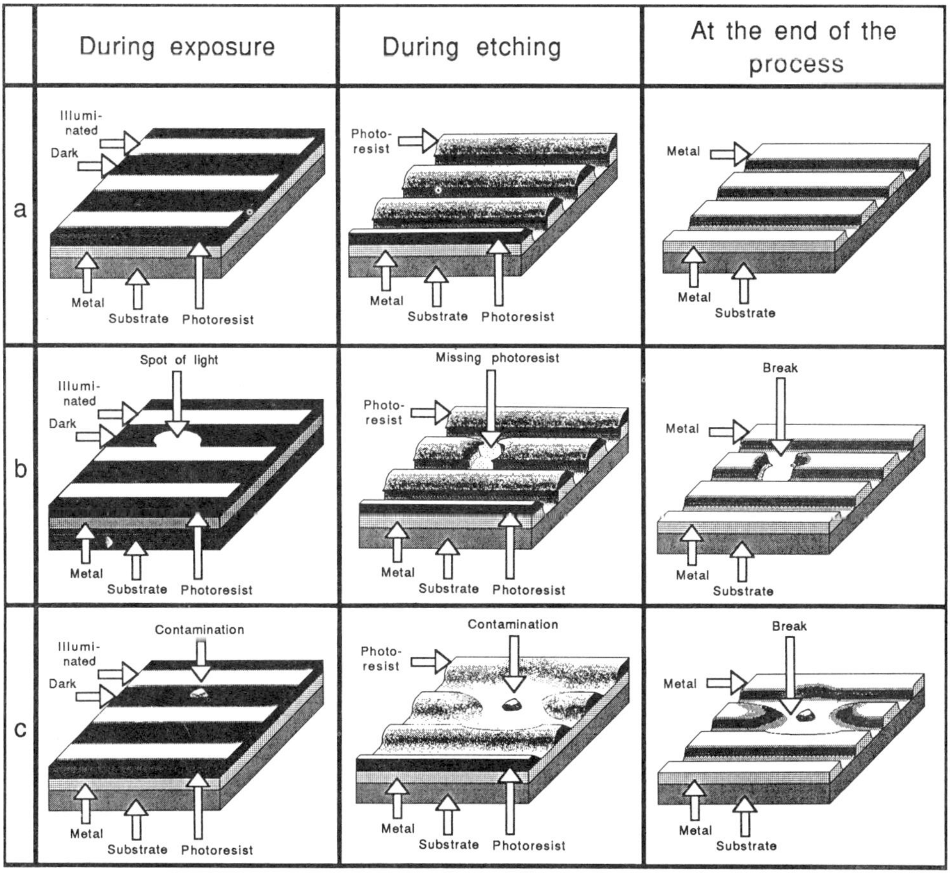

Figure 3: Examples of mechanisms that cause breaks of a metal line.

photoresist is shaded by a particle deposited directly on top of the photoresist during the exposition step. Such particles are among the most common reasons causing shorts. They are usually generated by the manufacturing equipment. They shade photoresist and cause a short. Row (d) of Figure 2 illustrates a defect mechanism that is caused by a particle deposited on the surface of the wafer during etching. The deposited particle protects the surface of the metal causing short. The above illustration describes defects in the metal layer, but of course, exactly the same mechanism can produce spots of extra material in other IC layer.

Figure 3 illustrates the basic mechanisms causing breaks, or in general, regions with missing material. Examples shown in this figure deal with breaks in metal connections caused by: mask defects (row b) and contamination in the photoresist (row c). The mask defect which is a transparent spot in the opaque mask region, (that actually should not occur in the modern process) damages photoresist that is supposed to protect metal connections during etching. A similar effect takes place if the surface of the photoresist is contaminated with the compound enhancing development of the unexposed photoresist (see Figure 3). If this effect is strong enough to cause complete development of unexposed photoresist then a small particle of contamination can cause etching of the metal in its neighborhood large enough to cause a break. Although the defect mechanisms depicted in Figure 3 describe breaks in the metal layer, they are general and illustrate phenomena causing breaks or regions with missing material in any layer of the IC.

Other local defects

There exist a large variety of physical reasons that may cause defects in the IC structure of a local nature and that are different to the defects described in previous paragraphs. To simplify their description we assume that they belong to one of two categories, namely *wafer mishandling defects* or substrate defects, that we will also call *point defects.*

Wafer mishandling defects are caused by physical contact of the surface of the wafer with the equipment or tools used to manipulate wafers, or with contaminants that were introduced to the process by mishandling processing equipment. Such contact damages the surface of the wafer causing faults. Typical examples of such defects are scratches. In the further part of this paper, we will disregard wafer mishandling defects assuming that in the automated IC factory wafer mishandling will be kept on a very low level.

Point defects which are local disorders in the lattice structure of a semiconductor material include: *vacancies* (a lack of the atom in the lattice site), *interstitial defects,* (an atom is located out of crystal site), and *vacancy-interstitial pairs,* called *Frenkel defects.* Point defects act as recombination centers and, therefore, affect carrier lifetime. The concentration of point defects may fluctuate and is affected by high temperature process operations. Dislocations, which are geometrical irregularities in the regular structure of the crystal lattice, are caused by crystal growth related thermal stresses and are of two different types: *edge dislocations* and *screw dislocations.* The most important characteristic of a dislocation is that it strongly interacts with point defects and impurities located in its neighborhood. As a consequence, a large increase of the impurity diffusion coefficient may occur which may lead to the formation of diffusion pipes that cause such problems as abnormal leakage currents or low breakdown voltage in p-n junctions.

The term "point defects" is also used to describe defects in thin gate oxide of MOS transistors. Oxide point defects, called also pin-holes, manifest themselves as a path of abnormally high conductivity that usually results in a short between transistor gate and channel.

Physical mechanisms causing point defects are very complicated and in many cases (e.g. oxide pin-holes) are not fully understood. In the majority of cases, however, they are caused by either surface or bulk contamination. They are usually very small and can be characterized in the same way as spot defects.

Spot defect and point defect basic characteristics

To estimate yield one has to know the probabilities of various IC failures. Faults and defects are very tightly related but in order to determine whether a particular defect causes a functional fault, one has to find out

1. Which layers of the IC are affected by this defect?

2. What is the location of the defect with respect to the IC topology?

3. What is the shape and size of the defect?

4. What are the electrical properties of the spot defect-generated device?

Hence, yield prediction oriented local spot defect characteristics should contain:

1. Density of defects in each IC layer, expressed in terms of number of defects per unit area. If distribution of defects is nonuniform, i.e., if some of the regions on the wafer have a tendency to collect more defects than others, then such tendencies have to be characterized as well.

2. Defect size distribution.

3. Defect electrical characteristics.

It is very important to see that the above characteristics have to be determined from the experimental data extracted from the manufacturing process.

YIELD MODELING

As was suggested in the second section of this paper and also as was formally demonstrated by Maly et al. (1986) the impact of local and global disturbances on manufacturing yield can be analyzed separately. In this section we use this observation and we discuss spot and point defect related yield losses and yield modeling techniques that can handle global process disturbances as two separate subjects.

Spot and point defect models

Modeling of spot defects involves two issues: determination of spot defect characteristics and modeling of basic features of spot defects.

Essentially, one can distinguish three categories of spot defect models.

- Models developed in the early sixties. In these models spot defects are any physical deformations that cause IC functional faults. They are characterized in terms of the density only, that is determined as a ratio of a number of faulty IC chips to the total area of fabricated silicon.

- The second generation of spot defect model, has been developed between the early seventies and early eighties. In these models data gathered through the analysis of memory chips is used to determine the density of "dimensionless" spot defects. Nonuniform distribution of defect densities and wafers, radial nonuniformities within wafers, and clustering effects are usually included in these models.

- In the third generation models, which are currently under development, spot defects are modeled as disks of missing or extra material that may occur in all IC layers. These models cover all previously mentioned features and also describe defect sizes. Parameters of such models are determined by using analysis of the memory test data and measurements from special purpose test structures.

The complexity of the spot defect model determines very strictly its accuracy and thus the area of application. We discuss this aspect of spot defect models in the following section.

Yield models

Yield models have been developed and investigated since the early sixties. They were used to determine the probability that a manufactured semiconductor device perform its function correctly. At the beginning of this period very simple models expressing probability of a failure of a single transistor (or later of a simple IC) as a function of the entire area of the fabricated device were developed (e.g. Wallmark 1960, Murphy 1964, Gupta 1970). Defects in these models were assumed to be dimensionless "points" uniformly distributed over the entire fabrication wafer. It was also assumed that each defect, regardless of its location or layer, causes functional failure of the device. These models use Poisson's statistics:

$$Y = \exp \{ - D_0 A \} \tag{1}$$

where Y is the functional yield, A is the device area and D_0 is the density of defects.

Later this basic model was modified in several ways. Stapper (1976) substituted in the formula (1) total device area with the *critical area* that represents this portion of the area of the IC that is prone to point defects (e.g. the area of thin oxide). It was also assumed that the total yield is a product of partial yields where each partial yield is associated with one distinct category of defects. Partial yield was described by the formula:

$$Y_k = \exp \{ - D_k A_k \} \tag{2}$$

where Y_k is the yield loss due to defects of the k-th category and D_k and A_k are density and critical area, respectively, determined for defects of the k-th category. Thus functional yield was computed using the following formula:

$$Y = \prod_{k=1}^{K} Y_k \tag{3}$$

where Y_k is defined by (2) and K is the number of defect categories. Typically each category of the defects represented all defects that occur in one layer of the IC, i.e., in metal, polysilicon, or thin oxide, etc.

An important modification of model (1) was developed by Murphy (1964) who assumed that D is not constant and that it is a random variable which describes changes in defect densities. With such an assumption he managed to obtain the following simple model that was successfully used by many researchers:

$$Y = \int_0^{+\infty} e^{-DA} f(D)\, dD \tag{4}$$

where f(D) describes distribution of defect density. Using equation (4) Murphy developed formulae for various distributions of D.

Seeds (1967a and 1967b) and Moore (1970) suggested models in which yield was expressed in terms of the ratio A/Ao where Ao is a *reference area*. Price (1970) uses a model which is a generalization of Seeds' model, in the form:

$$Y = \prod_k \frac{1}{(1 + D_k A_k)} \tag{5}$$

A similar model is suggested by Okabe (1972). He assumes, however, that critical areas and defect densities are the same for all layers. His model has the following form:

$$Y = \frac{1}{(1 + D_0 A/k)^k} \tag{6}$$

where D_0 is an average defect density and k is a number of processing steps.

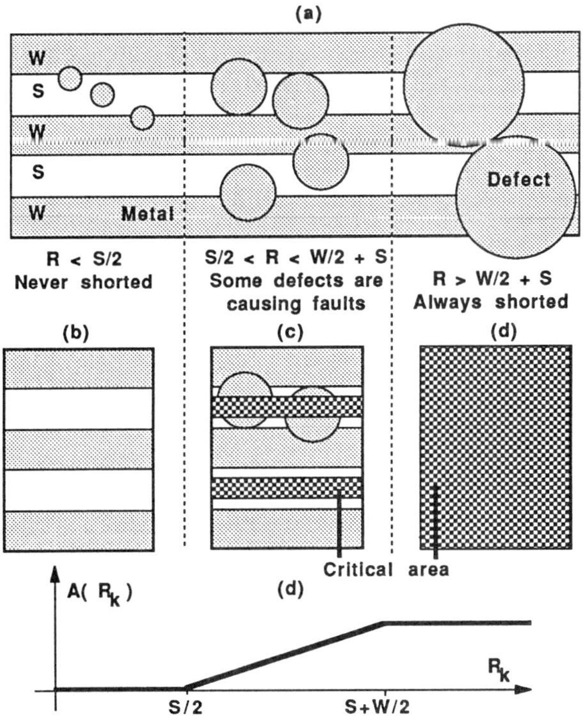

Figure 4: Critical area as a function of the defect size. (a) Segment of the metal bus having spacing S and connection width W, with spot defects of various sizes. Locations of defect centers that cause short (i.e. critical areas) (b) for small defects (c) medium defects, and (d) large defects. (e) Critical area as a function of defect size.

All of the above models, as well as other models developed in the same period, were successfully used to predict manufacturing yield as functions of the die size (see Stapper 1973, 1975, 1976, 1983a, 1983b, 1983c, 1986). They were found to be accurate enough when they were used to predict the yield of a product that was similar to the product from which such parameters as D_o, or reference areas were extracted. "Similar" means that they were valid within one family of products, such as different generations of memory chips whose layout styles, and therefore sensitivity to local defects, were similar. But they were not good enough to predict the yield of a new product designed with new design rules or new layout styles. The main reasons causing yield estimation errors in these models are due to the application of very simple defect models (dimensionless point defect) and an inadequate precision in the critical area estimation.

In the early eighties a new family of models that were supposed to solve these problems was discussed (Maly and Deszczka 1983, Stapper 1983b, Ferris-Prabhu 1985a and b). In these new models critical area is expressed as a function of the defect size, A_k, that takes into account the defect-layout relationship. The significance of this relationship is illustrated with a simple example shown in Figure 4. Similar relation can be obtained for more complicated layouts (Maly 1985, Ferris-Prabhu 1985a and 1985b).

The problem with the concept of the critical area as a function of the defect size is that the defect size is a random variable and therefore the critical area is a random variable, as well. In addition a typical distribution of the defect size is such as is shown in Figure 5 i.e.

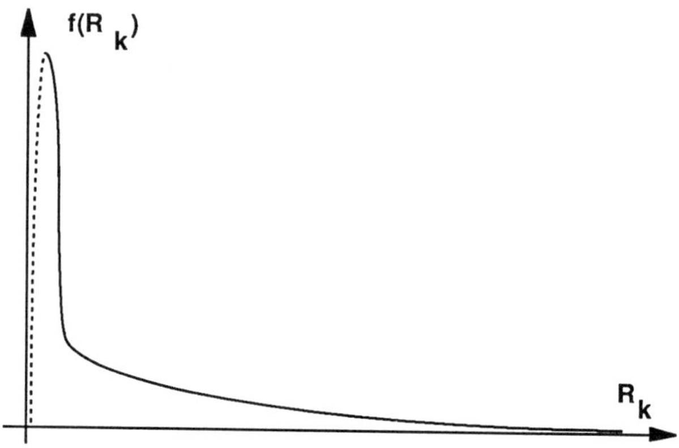

Figure 5: Defect size distribution.

the vast majority of defects have small sizes and the density of sharply increases with the decrease of the defect size. But on the other hand, as is illustrated in Figure 4 small defects may, or may not cause failure (small size defects have small critical areas). Consequently, inadequate estimation of the critical area that, disregards the random nature of the defect size and assumes that the critical area has a certain constant average value, may lead to a large error in yield estimation.

A modification of model (1) that takes into account critical area and defect size distribution is possible and can be done in the following way. Let us assume that A_k (R_k) describes critical area of defects of k-th category and the R_k is a random variable, with the distribution $f(R_k)$.

Let a sequence of segments $[R_k^i, R_k^{i+1}]$, for i = 1, 2, 3, ... , N, be arranged in such a way that they cover majority of actual values of R_k. For each of these segments we can estimate $p(R_k^i)$ which is the probability that R_k is contained in the segment $[R_k^i, R_k^{i+1}]$. Consequently, $D_k\, p(R_k^i)$ equals the density of defects whose radius belongs to the segment $[R_k^i, R_k^{i+1}]$. If now we assume that :

$$R_k^{*i} = (R_k^i + R_k^{i+1})/2 \tag{7}$$

and that $A_k(R_k^{*i})$ stands for the critical area computed for R_k^{*i} then from (1) we know that the probability of the event that the defect center is located within critical area is described by the formula:

$$\exp\{-A_k(R_k^{*i})\, p(R_k^i)\, D_k\}$$

Finally, using (1) one can show that :

$$Y_k = \prod_{i=1}^{N} \exp\{-D_k A_k(R_k^{*i})\, p(R_k^i)\} \tag{8}$$

where

$$p(R_k^i) = \int_{R_k^i}^{R_k^{i+1}} f(R_k)\, dR_k \tag{9}$$

Of course, (8) is valid for large enough N.

The problem with the above model, as with other models using $A_k(R_k)$, is that the computation of the critical area is not a trivial task and it may in some cases be prohibitively expensive. (Practically, critical area computations can be performed for regular layouts or can be approximated using the concept of virtual artwork (see Maly 1985)).

Problems with critical area computations can be avoided when yield estimation applies a Monte Carlo method (Walker and Director 1983 and 1986, Walker 1987). In this method random number generators are used to model spot defects and to "place" them on the layout of the IC. Each defect is then analyzed to determine the electrical consequences of its occurrence. Defects classified as causing structural failures are counted and in this way the yield can be estimated.

The Monte Carlo approach has a number of advantages. It allows for the best possible spot defect modeling. Classification of spot defect caused faults is also much better than in any other technique. Walker (1987) demonstrated that the Monte Carlo approach can be successfully implemented. His yield simulator VLASIC can handle nMOS technology but the simulation methodology implemented in VLASIC can be expanded to handle other technologies as well.

The Monte Carlo approaches also have important disadvantages. They are expensive to use and develop. These high costs are mainly due to the required large sample size and the complexity of the algorithms applied for the interpretation of the electrical behavior of the portion of the circuit affected by the defect. Thus, Monte Carlo approaches can be applied for the precise analysis of a crucial portion of the circuit rather than to predict the yield of a large VLSI circuit.

A new analytical yield estimation methodology that tends to combine the simplicity of the Monte Carlo methodology with the efficiency of the methods using analytical formulae was recently proposed by Chen and Strojwas (1987). In this approach the area prone to defects of a certain range of sizes is extracted from the IC layout. Such areas are usually topologically simple and therefore one can easily find analytical formulae describing the probability of occurrences of the defect in such an area. This approach is now in the phase of implementation and its final evaluation will be published soon.

Measurements of densities and defect size distributions

The accuracy of yield modeling depends on both accuracy of the yield models and accuracy of defect characterization used by these models. Unfortunately, for many reasons typical yield modeling approaches presented in the literature avoid discussion of the methodology and the interpretation of the measurements of basic defect characteristics. Consequently, relatively rich information about yield modeling methodologies cannot be effectively used or verified because of the inadequate information about defect densities, defect size distribution etc.

Hence it is very important to stress that an accurate yield estimation is possible if, and only if, an accurate, unbiased methodology is used to determine defect densities and defect size distributions for each layer in the fabricated IC. It was experimentally demonstrated by Maly et al. (1987a) that the defect size distribution is as important as defect densities. In the same experiment, in which a special purpose defect size sensitive test structure was used, it was determined that the defect size distribution is not a universal characteristic. It varies from one process to another because each manufacturing facility and each process has its own "personality". Therefore any yield prediction activity must be supported by the experimental data that provides local defect characteristics. An example of such experiment is described by Maly et al. (1987a).

Parametric yield modeling technique

As was suggested earlier each IC design is subjected to a number of constraints not only on the functionality but also on the IC performance measures such as delays and power dissipation. If the tolerance bounds on these performance measures are tight, the inherent process fluctuations can significantly reduce the fraction of circuits with the satisfactory performance. More precisely, if we define an *acceptable region* in the space of IC performance, the parametric yield can be defined as a ratio of the IC's with the performance indices within this acceptable region to the total number of IC's. To maximize this ratio, the IC designers attempt to perform the so-called design centering, i.e. to place the *nominal design* in the center of the acceptability region. This will result in a design which is immune to the process fluctuations. Naturally, centering might not be sufficient if the variations in the performance are large compared to the tolerance region. Hence, the designers aim at finding a solution at which the performance variations are minimal for given process fluctuations.

The ability to accurately predict parametric yield prior to IC manufacturing is of crucial importance in the design optimization. This is a very difficult task because the mapping of the process control parameters and IC layout into IC performances is very complicated. There is no chance of developing any analytical model to perform this mapping. Instead, a sequence of simulators must be used. Specifically, this sequence includes process simulators which generate IC device structures (e.g. layer thicknesses, impurity distributions), device simulators which generate device characteristics (e.g. I-V curves), and circuit simulators which evaluate IC performance. Most of the models in these simulators are described by partial differential equations and hence are computationally expensive. To evaluate the parametric yield, one needs to perform repeated simulations which may prove prohibitively expensive. To alleviate high computational costs, more efficient approaches to process/device/circuit simulations have been developed. In the area of process/device simulation a statistical simulator FABRICS (Maly 1982 and Nassif 1984) provides an attractive compromise between simulation accuracy and efficiency. (See Figure 6.)

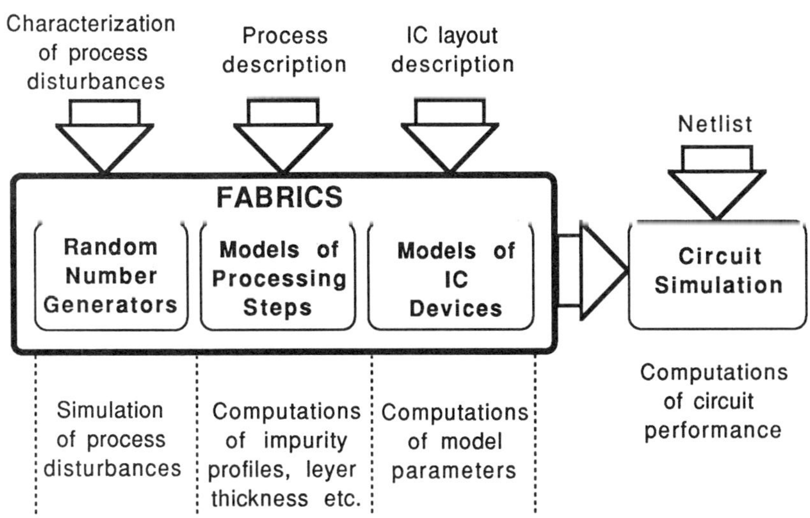

Figure 6: Statistical simulation using process simulator FABRICS.

It employs either efficient numerical or analytical models in which typically the boundary conditions are simplified to speed up the model evaluations. In the area of circuit simulation a number of efficient circuit simulators with voltage discretization schemes (e.g. E-logic) have been developed (e.g. CINNAMON, Vidigal 1986). If the performance evaluation can be limited to a subset such as timing then specialized simulators can be used such as a statistical timing simulator STAT! (Benkoski 1987)

Computation of the parametric yield can be significantly simplified if we can exclude local defects from the analysis. Then the estimation can be done for global disturbances only such as line width variations or mask misalignments. The most accurate but also the most expensive way to evaluate parametric yield is the Monte Carlo simulation. In this scheme, process fluctuations (disturbances) can be generated using pseudo-random number generators and for each set a full sequence of simulations has to be performed. To reduce the cost of this evaluations, more efficient Monte Carlo schemes such as *importance sampling* can be used. To drastically reduce this cost, the Monte Carlo scheme can be substituted by the worst-case analysis in which IC performances are evaluated for the worst-case combinations (e.g. 3σ) of the process disturbances (Nassif 1986). Hence, the performance limits, or corners, can be explored with some estimates of the probability of occurrence of the performances going beyond the acceptability region.

Yield estimation in IC design

In this section we have discussed basic aspects of yield estimation. Below we summarize this discussion and show how these yield estimation approaches may be used in the design of the large area defect-tolerant systems.

The goal of yield estimation in IC design is to determine whether a particular design decision affects manufacturing yield. There are two classes of decisions that have to be considered. Strategic decisions that are suppose to determine such quantities as the total area of the IC, or design rules, or circuit partitioning and redundancy, and low level decisions that deal with circuit level of abstraction and address such problems as timing, optimization of electrical and geometrical structure of the cell, transmission line or functional block. Decisions from the first class are affected by the estimation of functional yield losses. Decisions from the second class depend on circuit performances, so they must be supported by the parametric yield estimation.

This observation suggests that the design of the large area defect tolerant ICs should involve the following steps:

- Estimation of local defect characteristics that should be performed by gathering data in the manufacturing line.

- Estimation of functional yield losses for various design alternatives. Such estimation should be performed using both analytical yield models and Monte Carlo simulation.

- Statistical analysis and optimization of parametric performance of the design circuit. Such analysis should involve statistical process/device/design simulation applied in yield optimizing procedure.

Variables that can be manipulated during design include the full spectrum of parameters, starting with transistor dimensions up to the amount and architecture of the applied redundancy. Of course, higher level decisions are analyzed first to determine the structure of the designed system. Lower level yield optimization is performed later as a process-design final tuning step.

Since this paper is concerned with defect tolerance, which is determined by circuit architecture, in the remainder of this paper we do not discuss low level yield optimization. Instead in the next section we concentrate on the methodology of redundancy. This discussion will indicate various options that can be used to maximize functional yield of the defect tolerant IC.

LARGE MEMORY CIRCUITS

VLSI end users are greedy for ever larger circuits integrating more and more complex circuitry onto a single chip. The motives for this are to design more powerful systems and/or to reduce the number of chips and hence the system integration costs, volume and power dissipation. Nowhere is this demand more apparent than with memory circuits. Almost all computer systems resort to external storage (e.g. magnetic disc) not merely for longterm back-up but to supplement the otherwise inadequate amount of main VLSI memory during processing.

Because yield considerations have been the prime factor in limiting the size of VLSI circuits, leading RAM manufacturers have employed defect-tolerant techniques in order to obtain larger parts. This trend started around 1979 with 64K dynamic RAMs (DeSimone 1979, Cenker 1979) and is continuing now with 4Mbit RAMs and beyond.

The typical approach is to add spare rows and columns to the memory bit array (Fitzgerald 1980, Moore 1986). After fabrication the chips are probe-tested to determine the location of any defects and the circuit is configured by programming the decoders of the spare rows (columns) with the addresses of the defective rows they are to replace. The number, type and location of spare circuits is determined by consideration of the dominant defect mechanisms and yield predictions, along with other design factors such as layout and performance. A typical arrangement is shown in Figure 7.

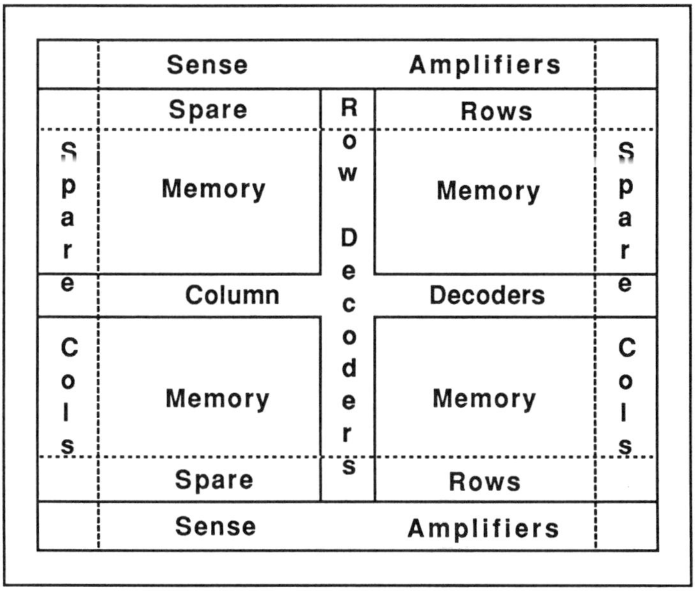

Figure 7: Typical arrangements of spares in a memory.

The configuration of defect-tolerant memory chips is fixed permanently by blowing fusible links to set the required address value. A number of techniques exist for this, the most common being to vaporize a polysilicon link by passing a high current through it or by focusing a laser beam on it (Smith et al. 1981).

Spare row and column techniques allow most local defects in the memory bit array to be tolerated. Effective schemes in commercial chips have added an overhead of between 2% and 5% to the chip area. The yield-enhancement that is obtained depends on the defects that occur but yield improvements of as much as 30-fold have been recorded on low yielding prototypes with perhaps 1.5-fold to 5-fold being more typical of mature devices with lower defect densities (Smith 1981, Sud and Hardee 1981).

Defect-tolerant RAMs have been introduced into the marketplace successfully because of the intense competition to produce larger parts and because of the insatiable demands of the end user. Also the extra development and chip design costs have been amortized by the large volumes of sales. Another important factor is the ease with which defect-tolerance can be introduced with the RAM architecture. Besides the spare row and column approaches, spare block, spare output bit, coding and hierarchical techniques have also been used (Moore 1986). The defect-tolerance is efficient in terms of the overhead of redundant circuits because a limited number of spares can serve the large number of repetitive circuit elements and the approach is easy to implement because there is no data communication between cells.

DEFECT-TOLERANT ARCHITECTURE

There is also a strong end-user demand to achieve higher levels of integration for other (non-memory) circuit architectures. Additional advantages in these cases are the potential for higher communication bandwidth through more interconnections and/or faster interconnections unconstrained by chip-to-chip bottlenecks and the potential to package a complete system or a substantial system component on a single integrated circuit.

Non-memory architectures have the additional complication of having to configure the internal communications around the defective parts, but architectures with extensive repetition of cells are increasingly common for efficient parallel computation and for ease of design. Thus it was to be expected that memory chips would be the first to employ defect-tolerant approaches but there is now a considerable potential to exploit that experience in a wider range of applications.

Non-memory architectures which have highly repetitive cells may be classified by the increasing complexity of their internal communications as: PLAs, one dimensional arrays (such as pipelines, bit sliced units and systolic arrays), two dimensional arrays (such as systolic arrays and SIMD processor arrays) and more general styles such as trees, hypercubes, etc. Increasing use is also being made of multiprocessing architectures with irregular resources but largely nearest neighbor communications (Delgado-Frias et al 1988). These are all clear targets for defect-tolerant designs and the variety of proposals is reviewed in Jesshope and Moore (1986) and Saucier and Trilhe (1987).

Typical proposals for one-dimensional pipelines center around starting with a longer pipeline and bypassing defective cells in it as in Figure 8 (a) and many of the proposals for two-dimensional arrays center around spare cells in one or both dimensions and a more flexible configuration scheme such as that illustrated by Figure 8 (b) (Evans et al 1986).

In broad terms, one-dimensional reconfiguration schemes appear to be appropriate to moderately sized VLSI chips of similar complexity to large memory chips which will be affected only by a small number of defects. The more complicated reconfiguration schemes are targeted at the design of large VLSI (or even WSI) chips where a larger number of defects may occur.

The development of defect-tolerant RAMs has demonstrated that such approaches are feasible and has provided appropriate technological advances such as the fusible links. Many parallel advances have also been made in the technology which will aid defect-tolerant design, for example the reconfiguration may be performed by reprogrammable devices (as in EPROMs) or by discretionary wiring (via E-beam direct write or even with laser-enhanced CVD (Saucier and Trilhe 1986)).

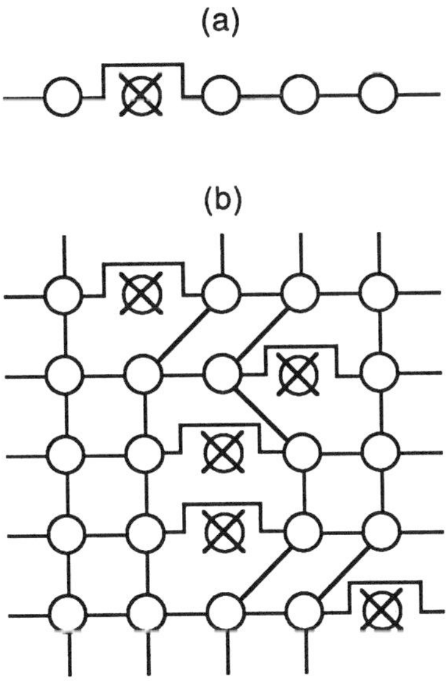

Figure 8: Typical defect-tolerant system for (a) one-dimensional pipeline and (b) two-dimensional array.

If effective design of large VLSI circuits without defect-tolerance is dependent on a good knowledge of potential defect mechanisms, the design of defect-tolerant circuits is even more dependent on such knowledge. Only a good model of these can permit an appropriate selection and placement of the redundant elements. On the other hand good defect and yield models based on the principles we have expounded above will enable designers to implement defect-tolerant designs with confidence.

YIELD-PREDICTION FOR DEFECT-TOLERANT DESIGNS

The designer of defect-tolerant circuits needs to be able to make good predictions of the yield of the circuit in order to make the proper allocation of redundancy. This can be done at the lowest level cells using the techniques already described for assessing the probability and consequences of VLSI yield loss mechanisms. The designer needs at this stage to see how best to avoid the risk of catastrophic defects (e.g. power to ground shorts) which may affect the whole circuit but the higher level defect-tolerance strategy will be designed to handle the failure of individual cells. In the event that some catastrophic faults do nevertheless occur, the possibility exists to eliminate their effects by techniques such as laser surgery, but this is best avoided if possible.

At the higher level, the choice of defect-tolerance approach and the amount of redundant circuitry can be decided by the predicted yields of the building blocks. The spatial distribution of the defects is also important as many defect-tolerant strategies are vulnerable to the worst case local grouping of faulty building blocks which will arise from defect clusters or large area defects. This will often dictate a strong preference for defect-tolerant schemes which are resistant to clustering. Finally the designer must consider the "hard core" circuitry such as the power and clock supplies and perhaps data busses. Such areas can be difficult to make defect-tolerant and the designer needs to examine the defect model carefully to see if these parts can be designed to avoid the more likely defect mechanisms, for example through sufficient spatial separation or by judicious use of different layers.

The yield of the low-level (non-defect tolerant) cells can be predicted using the techniques discussed above, but the defect-tolerant strategy must also be taken into account to predict the overall chip (or wafer) yield. In the case of a uniform defect density this can usually be done by simple binomial summation of the cell yields based on how many of each type are required to work. For example if in the linear array of Figure 8(a) the yield of a cell is Y_c, the yield of its bypass circuit is Y_b and the yield of the configuration control or other hardware is Y_h, then the yield of a circuit requiring n out of m cells to work is given by (Moore 1984(b)):

$$\sum_{i=1}^{m-n} \sum_{j=0}^{n} \frac{m!}{i!\,j!\,(m-i-j)!} Y_c^{m-i}(1-Y_c)^i \; Y_b^{m-j}(1-Y_b)^j \; Y_h \tag{10}$$

While this approach may be an adequate approximation for assessing the feasibility of a defect-tolerant proposal, detailed design will require clustering effects to be considered. The effects of clustering and the assessment of more complicated reconfiguration schemes are difficult to evaluate analytically and are probably best assessed by Monte Carlo defect simulation such as might be obtained by development of VLASIC (Walker 1987).

DESIGN FOR TESTABILITY

The concept offered here of designing for yield parallels the concept of designing for testability which is now recognized as being vital to good VLSI design. In fact designing for yield is even more pervasive not only because of its links with design which have been discussed above but because of the need for accurate defect models in order to determine effective test patterns for a chip.

Designing and assessing testability is done with defect models at a variety of levels of abstraction:

Functional models are easy to deal with but rapidly become impractical if it is necessary to instantiate every data value and sequence; *gate-level stuck-at fault models* are widely used by commercial test generators but as we have shown these do not correspond well to MOS circuit implementations. Realistic testing of VLSI circuits may therefore require test sequences as well to detect *stuck-open faults* and redesign of power supply monitors to detect *stuck-closed faults* (Liu and McClusky 1987). Tests should cover all the most likely faults (Maly 1987 d) rather than simply covering a high percentage of stuck-at faults as is often the case.

All these effects will become more important with the progress to sub-micron design rules and more complex chips. At the present time many designers use the higher level fault models and hope that a good test coverage for these faults will ensure a good coverage of the lower level faults. In the future we need hierarchical test generation tools which can build up efficiently from the lower levels.

Design for testability implies adapting the design to make it more testable and typically includes extra circuitry such as a scan path to improve the controllability and observability of the circuit. These measures should also use a realistic fault model (Maly 1987 d).

Much effort has been expended on built-in self-testing (BIST) designs typically through the use of pseudo random test generators and signature analyzers. The main advantages of this are to reduce test costs at the chip and at the board level, to permit real time (full speed) testing and to permit exhaustive test coverage. This test coverage is usually computed for stuck-at faults but should be computed for the realistic models.

For defect-tolerant circuits the test problem becomes more difficult and again requires the best available fault model. The additional problems include the need to diagnose the location of the defects, the need to continue testing after the first fault is found, the need to discriminate multiple faults and the need to test the redundancy circuits and the configuration logic. Finally the use of good diagnostic tests may also provide a good feedback to the defect monitoring activity.

SUMMARY AND CONCLUSIONS

In this paper yield loss mechanisms, yield estimation techniques, and redundancy were discussed as central issues in the design of defect-tolerant large area integrated circuits. Disturbances inherent in the manufacturing process, of both global and local nature, with emphasis on spot defects were indicated as main sources of yield losses. Then a full spectrum of yield modeling techniques was discussed. Finally, the impact of redundancy on the manufacturing yield was analyzed.

Such a comprehensive treatment of all yield related issues has been given in this paper in order to demonstrate that design and manufacturing of defect-tolerant ULSI systems must involve all levels of abstractions ranging from defect mechanisms up to optimization of the architecture of the designed system. None of them is less important than the other and higher level design decisions must be analyzed on the basis of the detailed understanding of the manufacturing process.

The discussion presented in this paper indicates also that in all three major areas dealing with yield of defect tolerant ULSI systems there still exist missing components that need to be investigated. In the area of defect modeling a better local defect model must be developed. Currently available spot defect models do not cover adequately enough the following aspects:

- Diversity of defect mechanisms;
- 3-D nature of some defects;
- Electrical properties of materials causing defects;
- Defect clustering mechanisms;
- Interaction between defects and process steps that take place after the defect occurrence;

In the area of yield modeling there still exists a need for yield estimation technique that

- Uses realistic spot defect model;

- Is much more efficient, i.e. can estimate yield for the circuits with a large number of elements at the same time can adequately handle interaction between small defects and any detail of the IC layout;

Analysis of the redundancy strategies must also be significantly improved. Weak points of currently existing redundancy analysis are due to:

- Application of very simple yield models;

- Lack of methodology that can handle:

    1. Yield losses in various non computing elements of the systems such as interconnect, power distribution, self testing functional blocks, etc.

    2. Optimization of redundancy that takes into account performance issues such as timing, power, global disturbances of the process, etc.

Information presented in this paper indicates also that the above listed shortcomings of the existing approaches should be, and can be solved through the development of a software package for yield estimation of large area defect-tolerant ICs. Such software should be built as a package that integrates CAD tools dealing with a complete spectrum of yield estimation and design optimization issues. This kind of software must be developed as soon as possible if we want to avoid further disappointments on the way towards ULSI-WSI systems.

References

Benkoski J and Strojwas A J 1987 A New Approach to Hierarchical and Statistical Timing Simulations IEEE Transactions on CAD of ICAS (CAD-6) 1039-1052

Cenker R P et al. 1979 A Fault Tolerant 64K Dynamic RAM, IEEE Transactions on Electronic Devices 26(6) 853-860

Chen I and Strojwas A J 1986 Realistic Yield Simulation for IC Structural Failures, IEEE International Conference on Computer-Aided Design (ICCAD) Digest of Technical Papers 220-223

Chen I and Strojwas A J 1987 Realistic Yield Simulation for VLSIC Structural Failures, IEEE Transactions on CAD of ICAS, (CAD-6) 965-980

Delgado-Frias J G et al., 1988 High Harvest Approaches for 2-D arrays, this proceedings.

DeSimone R R et al. 1979 FET RAMS, International Solid State Circuits Conference 154-155

Doi J A et al. 1987 Detection and Physical Characterization of Spot Defects in Metal IC Interconnections, 172 Meeting of The Electrocemical Society 637

Evans R A et al. 1986 Wafer Scale Integration Based on Self-Organization (in Jesshope and Moore 1986)

Ferris-Prabhu A V 1985a Defect Size Variations and Their Effect on the Critical Area of VLSI Devices, IEEE Journal of Solid-State Circuits SC-20(4) 874-878

Ferris-Prabhu A V 1985b Modeling of Critical Area in Yield Forecasts, IEEE Journal of Solid-State Circuits SC-20(4) 878-880

Fitzgerald B F and Thomas E P 1980 Circuit Implementation of Fusible Redundant Addresses on RAMs for Productivity Enhancement, IBM Journal of Research and Development 24(3) 291-298

Ghandhi S K 1983 VLSI Fabrication Principles, John Wiley & Sons

Gupta A and Lathrop J 1970 Comments on "Influence of Epitaxial Mounds on the Yield of Integrated Circuits", Proceedings of the IEEE 58(12) 1960-1961

Gupta A and Lathrop J 1972 Yield Analysis of Large Integrated-Circuit Chips, IEEE Journal of Solid-State Circuits SC-7(5) 389-395

Gupta A et al. 1974 Defect Analysis and Yield Degradation of Integrated Circuits, IEEE Journal of Solid-State Circuits SC-9(3) 96-103

Hofstein S and Heiman F 1963 The Silicon Insulated-Gate Field-Effect Transistor, Proceedings of the IEEE 51(9) 1190-1202

Hu S 1979 Some Considerations in the Formulation of IC Yield Statistics, Solid-State Electronics 22(2) 205-211

Jesshope C R and Moore W R (eds) 1986 Wafer Scale Integration, Adam Hilger

Liu D L and McClusky E J 1987 Designing CMOS Circuits for Switch-Level Testability, IEEE Design and Test 4(4) 42-49.

Maly W and Strojwas A J 1982 Statistical Simulation of the IC Manufacturing Process, IEEE Transactions on Computer-Aided Design of Integrated Circuits and Systems 1(3) 120-131

Maly W and Deszczka J 1983 Yield Estimation Model for VLSI Artwork Evaluation Electronics Letters 19(6) 226-227

Maly W et al. 1984 Systematic Characterization of Physical Defects for Fault Analysis of MOS IC cells, IEEE International Test Conference Proceedings 390-399

Maly W 1985 Modeling of Lithography Related Yield Losses for CAD of VLSI Circuits, IEEE Trans. on Computer-Aided Design of Integrated Circuits and Systems CAD-4(4) 166-177

Maly W et al. 1986 VLSI Yield Prediction and Estimation: A Unified Framework, IEEE Transactions on Computer-Aided Design of Integrated Circuits and Systems CAD-5(1) 114-130

Maly W et al. 1987a Double-Bridge Test Structure for the Evaluation of Type, Size and Density of Spot Defects, Research Report CMUCAD-87-2, SRC-CMU Center for Computer-Aided Design, Carnegie-Mellon University

Maly W 1987b Atlas of IC Technologies: An Introduction to VLSI Processes, The Benjamin/Cummings Publishing Company, Inc., Menlo Park, California

Maly W et al. 1987c Yield Diagnosis Through Interpretation of Tester Data, Proceedings of the International Test Conference 10-20

Maly W 1987d Tutorial: Realistic Fault Models for VLSI Testing, In Proc. of the 24th Design Automation Conference 173-180

Mangir T and Avizienis A 1982 Fault-Tolerant Design for VLSI: Effect of Interconnect Requirements on Yield Improvement of VLSI Designs, IEEE Transaction of Computers C-31(7) 609-615

Mangir T 1984 Sources of Failures and Yield Improvement for VLSI and Restructurable Interconnects for RVLSI and WSI: Part I - Sources of Failures and Yield Improvements for VLSI, Proceedings of the IEEE 72(6)690-708

Moore G 1970 What Level of LSI is Best for You?, Electronics 43(4) 126-130

Moore W R 1986 A Review of Fault Tolerant Techniques for the Enhancement of Integrated Circuit Yield, Proceedings of the IEEE 74(5) 684-698

Moore W R and Day M 1984a Yield-Enhancement of a Large Systolic Array Chip, Microelectronics and Reliability 24(3) 511-526

Moore W R 1984b Switching Circuits for Yield-Enhancement of an Array Chip, Electronics Letters 20(16) 667-669

Murphy B 1964 Cost-Size Optima of Monolithic Integrated Circuits, Proceedings of the IEEE 52(12) 1537-1545

Murphy B 1971 Comments on "A New Look at Yield of Integrated Circuits", Proceedings of the IEEE 59(8) 1128

Nassif S R et al. 1984 FABRICS II: A Statistically Based IC Fabrication Process Simulator, IEEE Transactions on Computer-Aided Design of Integrated Circuits and Systems CAD-3(1) 40-46

Okabe T et al. 1972 Analysis on Yield of Integrated Circuits and a New Expression for the Yield, Electrical Engineering in Japan 92:135-141

Paz O 1977 Modification of Poisson Statistics: Modeling Defects Induced by Diffusion, IEEE Journal of Solid-State Circuits SC-12(5) 540-546

Price J 1970 A New Look at Yield of Integrated Circuits, Proceedings of the IEEE 58(8) 1290-1291

Saucier G and Trilhe J (eds) 1987 Wafer Scale Integration, North Holland

Seeds R 1967a Yield, Economic, and Logistic Models for Complex Digital Arrays, In IEEE International Convention Record 60-61

Seeds R 1967b Yield and Cost Analysis of Bipolar LSI, In IEEE International Electron Devices Meeting 12

Smith R T 1981 Using a Laser Beam to Substitute Good Cells for Bad, Electronics July 28, 131-134

Smith R T et al. 1981 Laser Programmable Redundancy and Yield Improvement in a 64K DRAM, IEEE Journal of Solid-State Circuits 16(5) 506-513

Stapper C 1973 Defect Density Distribution for LSI Yield Calculations, IEEE Transactions on Electron Devices ED-20(7) 655-657

Stapper C 1975 On a Composite Model to the IC Yield Problem, IEEE Journal of Solid-State Circuits SC-10(6) 537-539

Stapper C 1976 LSI Yield Modeling and Process Monitoring, IBM Journal of

Research and Development 20(3) 228-234

Stapper C H 1983a Integrated Circuit Yield Statistics, Proc. of the IEEE 71(4):453-470

Stapper C H 1983b Modeling of Integrated Circuit Defect Sensitivities, IBM Journal of Research and Development 27(6) 549-557

Stapper C 1983c Modeling Redundancy in 64K to 16Mb DRAMs, In IEEE International Solid-State Circuits Conference Digest of Technical Papers, 86-87

Stapper C 1984 Modeling of Defects in Integrated Circuit Photolithographic Patterns, IBM Journal of Research and Development 28(4) 461-475

Stapper C 1984b Yield Model for Fault Clusters Within Integrated Circuits, IBM Journal of Research and Development 28(5) 636-639

Stapper C 1985 The Effects of Wafer to Wafer Defect Density Variations on Integrated Circuit Defect and Fault Distributions, IBM Journal of Research and Development 29(1) 87-97

Stapper C 1986 Yield Statistics for Large Area ICs, In IEEE International Solid-State Circuits Conference Digest of Technical Papers, 168-169

Sud R and Hardee K C 1981 Designing Static RAMS for Yield as Well as Speed, Electronics July 28 121-126

Sze S M 1983 VLSI Technology, McGraw-Hill Book Company, New York

Vidigal L. M. et al 1986 CINNAMON: Coupled Integration and Nodal Analysis of MOS Networks The 23rd Design Automation Conference

Walker H and Director S W 1983 Yield Simulation for Integrated Circuits In IEEE International Conference on Computer-Aided Design Digest of Technical Papers 256-257

Walker H and Director S W 1986 VLASIC: A Catastrophic Fault Yield Simulator for Integrated Circuits, IEEE Transactions on Computer-Aided Design of Integrated Circuits and Systems CAD 5(4) 541-556

Walker H 1987 Yield Simulation for Integrated Circuits, Kluwer Academic Publishers, Boston

Warner R M 1974 Applying a Composite Model to the IC Yield Problem, IEEE Journal of Solid-State Circuits SC-9(3) 86-95

Wolf S and Tauber R N 1986 Silicon Processing for the VLSI Era, Lattice Press

Yanagawa 1972 Yield Degradation of Integrated Circuits Due to Spot Defects IEEE Transactions on Electron Devices ED-19(2) 190-197

2 ALTERNATIVE PERSPECTIVES

In contrast with the broad overview of Chapter 1, this chapter contains four rather specific perspectives on 'designing for yield' and on the future of very large integrated circuits.

In the first section, Wild and Logan (§2.1) of Motorola examine ICs whose popularity in the market place has given them a long enough lifetime to warrant a re-evaluation of their design. The authors present a number of criteria which can be used for assessing the efficiency of a redesign. These include: yield, die size, process standardization and circuit performance. They address both circuit and layout redesign and the need for good CAD tools. As well as describing typical improvements which may be possible, they promote the need for good management of such an exercise.

In §2.2, Stapper of IBM discusses the effect of redundancy in dynamic RAMs on the yield of such chips. Various redundancy schemes are discussed in the context of yield, or productivity. The importance of defect learning for a particular fabrication line is stressed. A simplified yield model has been used successfully to design several generations of memory chips.

In the next section, McKirdy and Lea (§2.3) discuss the potential of a WSI technology from the reliability viewpoint. A number of comparisons are made between VLSI, ULSI and WSI technologies. Image processing systems implemented in these three technologies are being used as a research vehicle. Various reliability measures such as relative component failure rates, package, PCB and connector reliabilities are employed. A cost comparison is presented as well. The authors conclude that the reliability of ULSI and WSI systems should significantly exceed the reliability of VLSI systems.

The last viewpoint by McDonald *et al.* (§2.4) presents applications of the focused ion beam technique in VLSI/WSI manufacturing. Specifically, it demonstrates the applications in inspection, characterization and repair of defects. The authors give a brief description of the FIB technique and the equipment used in the actual manufacturing. They discuss a number of the most typical defects in VLSI/WSI interconnections and show how they can be repaired by using FIB. Also, a via verification method is described in this section. The ion beam technique may also have application in the repair of prototype circuits and another paper presented at the Workshop described a lower powered, high precision milling tool which has been employed for such work (R Clampitt, R Watkins and J Whitaker, Custom IC repairs using focused ion beams, *Microcircuit Engineering*, Sept. 1987, Paris).

2.1 A Design Strategy for ICs

Andreas Wild and Robert Logan

INTRODUCTION

In the semiconductor industry, new, high performance integrated circuits appear at an incredible rate. At the time of their introduction, the top priority is usually given to the timely execution of the marketing plan: indeed, any delay in bringing a product on the market will result in a loss of market share. The manufacturers will find out later that some of the ICs are there to stay for several years longer than the large majority. As time goes by, the number of competitors increases and the obtainable cost decreases (the effect of the "learning curve"). A moment will come at which all the shortcuts taken during the first development have to be eliminated in order to upgrade the technical performance, to optimize the manufacturing and to incorporate in the devices the state of the art technology: a redesign will be considered. The "redesign" conditions are fundamentally different. The resources are limited (the main efforts are still oriented toward new products), the glamour is missing. The design objectives are emphasizing the manufacturability, quality and reliability as well as the cost reduction. The candidates to be redesigned have to be chosen carefully and the activities prioritized to get the maximum impact with the smallest investments. Last but not least, the program must be given adequate support and recognition.

SELECTING REDESIGN CANDIDATES

Several criteria have to be taken into account in the process of selecting and prioritizing the candidates.

Yield

A plot of the yield versus die size will allow the identification of the offenders, i.e. of the devices not falling on the regular curve (Figure 1). The reasons for this behaviour have to be thoroughly investigated.
There are various answers:

- Improvement may be needed within the operating circuitry. A device may be marginal to an AC or DC limit.
- The layout may be sensitive to process variation (e.g. notching at photolithography).
- Poor manufacturability, e.g. difficult bonding due to pad layout (an example is given in Figure 2).
- Non-standard fabrication flow.

Die size

Again, there are many areas that should be considered:
- Is the layout too loose?
- Can the present internal circuitry of the device be simplified, therefore making it smaller?
- Does the layout use the latest design rules (e.g. spacings)?
- Are minimum feature sizes used for transistors etc. whenever possible?
- Does the circuit utilise the latest process innovations, tighter tolerances for etching, diffusion etc.?
- Are bond pads larger than need be? (due to increased usage of automated equipment which presently gives greater accuracy).

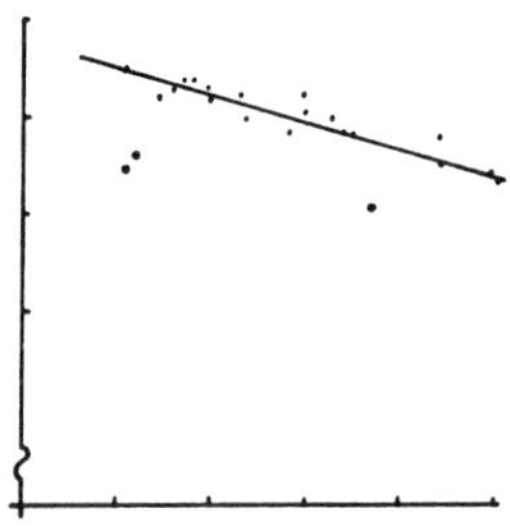

Figure 1. A typical yield-versus-size curve. Three offenders are easily recognisable.

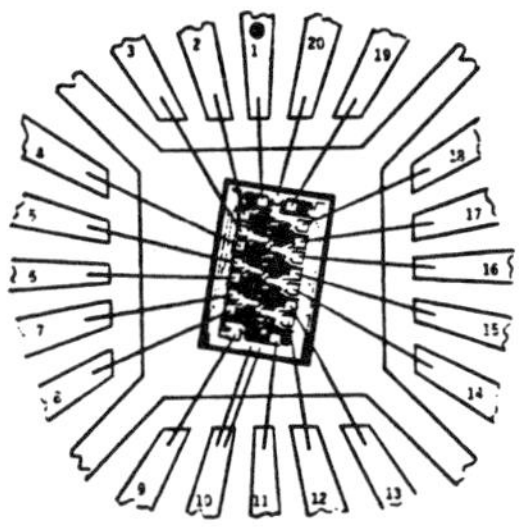

Figure 2. Inconvenient pad layout resulting in bad bondability.

Process

Various alternatives may be considered, taking into account also developments that occurred since the devices have been released to production. The target must be to standardize the processing as much as possible. However, a careful analysis of the cost involved in different solutions has to be performed. One example for standard logic devices is the use of a single metal layer versus a double metal layer. It is obvious that a double metal layer offers in principle a possibility to reduce the die size, however, this advantage has to outweigh the disadvantages of the additional processing steps.

Known problems

There may be some devices in the family which do not perform well in specific applications or with lower than standard quality and reliability. A few problems that may typically plague a bipolar logic family are listed below:

- Input threshold or output voltage drop at high temperature.
- Setup and hold times (AC parameters).

- Propagation delays (in general).
- Oscillation problems.
- Current consumption.
- Speed versus power considerations.
- Output drive current capability.

Some possible ways to achieve improvements are (respectively):

- Rethink the input structure (diode sizes, diode combination etc.).
- Redesign critical paths, improve voltage levels on internal transistors.
- Resize transistors, use speed up diodes.
- Centre the input thresholds.
- Increase internal resistors.
- Increase or decrease resistor values to particular needs.
- Increase base drive.

Production Quantity

In a product line there are "high runners", i.e. devices manufactured in considerably higher quantities than the average. It seems reasonable to concentrate on such a device first in order to get the biggest impact in the shortest time. Of course, a known problem device will override any volume considerations

The outcome of the evaluation process will be a comprehensive list of devices and the design objectives for each of them.

CIRCUIT ANALYSIS/SYNTHESIS

Circuit simulation is performed to prove that a circuit has similar functionality when it has been internally altered, the goal being to simplify or improve it, but not to change its logical operation. This is an important part of the operation as it contributes significantly to a first pass success. It is essential that the correlation between measured values and simulation is consistent. Accurate simulation models are required. Therefore an extensive correlation has to be performed first, to collect the newest process data and extract the parameters to feed into the simulation program. There are many simulation programs available, such as SPICE, STIME, MTIME etc., each one having its own strengths and weaknesses.

AC simulations for marginal values seen, whether at high temperature or room temperature, will be performed by capturing the circuit schematics and specifying the nodes to perform measurements on as well as the waveforms to be used. Alternative solutions will be evaluated to find out the most cost effective solutions in terms of silicon real estate.

Some examples (for TTL type devices):

- The elimination of internal level shifting by adequate sizing of the internal interface devices to adjust thresholds (Figure 3).

- Simulation of a squaring network versus a resistor on the bottom transistor in internal block outputs. Also optimization of the resistor value.
- Simulation of an internal block output built as a totem pole versus a pulled up open collector.
- Simulation of a transistor with the collector and base short-circuited versus a diode, to evaluate the silicon real estate consumption.

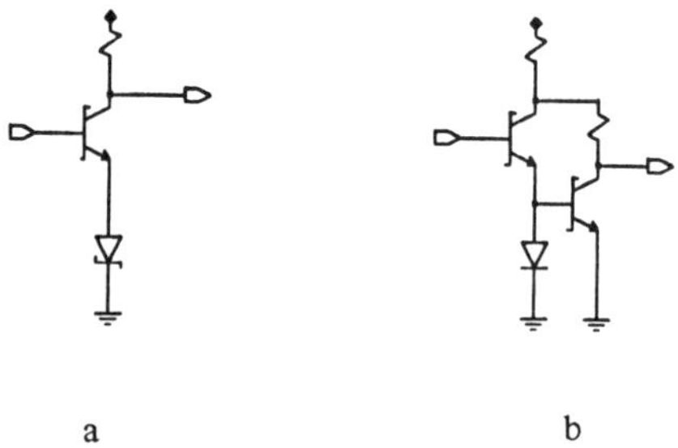

Figure 3. A properly sized Schottky diode in the emitter of npn transistor (a) can produce an acceptable "0"-level at the block output, making a supplementary level shifting (b) redundant.

The results of the simulation will constitute the basis for the redesign. Values and measurements taken will reflect the true circuit once it has been fabricated. Many important decisions are taken on the strength of the simulation as this is the only pointer to the circuit performance before putting the design to silicon. Therefore it must be right. This is the only guide the engineer has from the theoretical design to the actual manufactured device. Decisions amounting to tens of thousands of dollars may rest on the outcome of the simulation.

LAYOUTING

The modern CAD packages offer automatic layouting as a feature for at least some design techniques (e.g. standard cells). A designer can run a program and have the layout produced automatically. This dramatically increases the development rate and eliminates human errors. The problem still left is the optimum usage of silicon. The place and route programs are still satisfactory as long as the eventual size of the circuit is not a premium. At this stage it is important above all to maintain total flexibility, i.e. also the ability to adapt geometries to the available spaces, continuously taking into account the design rules, the process data, as well as the possible influence of the chosen geometry on the electrical features. It is essential that the designer and the layouter work closely together. One of the first decisions to be made is the process to be used, if several alternatives exist. If sufficient size reduction/cost gains cannot be realized in the highest volume process, then another proved alternative may be considered. However, the benefits must then compensate for the disadvantage of departing from the basic flow.

Whatever the decision of what process to use, it is at this stage that a layout can be 'moulded' together avoiding future problems that could occur, e.g. open circuit metal paths due to minimum sized metal lines for a critical geometry that could be increased. Similarly, metal lines can be rerouted to eliminate possible step coverage problems. The layout sensitivity to mask misalignment or misregistration can be further reduced by using insensitive structures (Figure 4), proper placement, selectively increasing size or altering the rotation of the geometries.

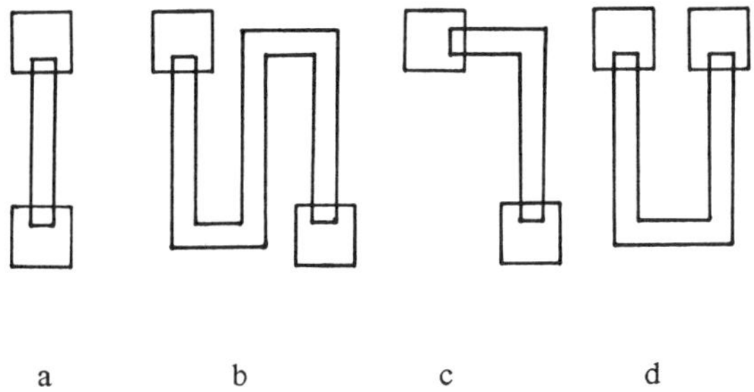

Figure 4. Four possible resistor geometries with ion implanted body. (a) and (b) are insensitive to misalignment, (c) is affected by both vertical and horizontal misalignment, (d) only by vertical misalignment one, but twice as much as (c).

Also to be considered during the layouting stage would be the manufacturability aspect, e.g. the placement of the bond pads in the most suitable position for bondability making the task of assembly as error-free as possible. This can also be achieved by increasing their size so as to completely eliminate bond misplacements.

CAD TOOLS

CAD tools to ensure first pass success in the design cycle will have to be extensively used. The need to repeatedly tweak a redesign is in strong contradiction with the objectives of the program. The layout has to be thoroughly checked to eliminate any possibilities of layout errors once the electrical circuit has been proven. This is performed in a two stage process.

- First, a schematic of the circuit has to be captured on a CAD system. Programs which can address a library of basic devices are used to support the sophisticated graphic generation. The complete circuit is then verified by an automatic routine able to recognise and highlight any drawing errors. Once this is clean then the circuit is ready to have a netlist generated (another background job for the CAD system). This netlist is an alphanumeric file of the drawn circuit with all the associated nodes and will represent the circuit as it should be from the electrical standpoint.

- The second stage is an output of the actual layout created for the particular device being designed, a database dump. This is then loaded into a computer, along with the particular netlist file which has been created.

Even at this stage of the design no corners are cut to speed up the design process, everything must be meticulously checked down to the last detail, no compromise is made. Once these two required data files are loaded into the computer, other sophisticated software is used to enable logical verification to take place. This includes information such as the position of the bond pads relative to the die, to enable the program to output co-ordinates when an error occurs to allow the easy location of that error. The verification program can compare the two files present for any logical errors that may exist and print these out as error messages. It will detect wrong devices, incorrect structures, wrongly positioned and wrongly connected structures. The Logic Verification program is run over and over until the circuit is clear from errors. When it is clear, the database dump of the layout is then assumed to have no logical errors and is therefore ready to be Design Rule Checked (DRC'd). The DRC is run on the layout to highlight any design rule violation to ensure that nothing slips through that could present problems, e.g. diffusion spacings, metal widths, minimum geometries etc.

Before commencing with these activities it is highly recommended to check the software for both logical verification and DRC as to whether they comply with the latest processing and tolerance specifications. Often, the rules incorporated in these programs have not been updated for years, remaining the same as the original design, while the processing was continuously improved. An update is then mandatory.

EVALUATION

When the device is considered to be free from all errors, it can be released for processing. When the first silicon hits the probe area, the complex evaluation cycle can be started. This involves probing and also encapsulating some samples to ensure that a demanding evaluation is performed. The evaluation strategy for redesigned devices is basically the same as for newly designed ones. The data collection has to go far beyond the data book specifications.

It is extremely helpful to have access to a highly automated evaluation environment, in which both data collection and data reduction are computer assisted and trained personnel perform routine manual testing for the features that are not (yet) automated.

The DC/AC operation including functionality at supply voltage and temperature extremes that go well beyond normal operating conditions, as well as parametric testing again, are gathered on automatic data acquisition equipment to improve standardisation of results, and also to improve the rate at which measurements can be taken, without human error. Any device characteristics that need to be looked at such as input/output current characteristics, input thresholds etc. are measured on the bench. Also parameters such as ESD (Electrostatic Discharge) and latch-up (formation of parasitic transistors which turn on under certain conditions and damage a device) are checked. All the parameters mentioned must measure up to the limits set for that device, or in the case of ESD and latch-up it must equal or better the standard for the particular logic family being redesigned. There is no compromise at this stage. If a device does not meet a particular parameter then it will not be qualified until the reason is found and corrected, whether this is by another modification, relayout, or redesign.

Reliability testing must also be performed to qualify a new process that is being used, to verify that a known problem has been fixed or that the device operating lifetime under normal environment or under high humidity has been improved. Also to be checked is the fact that no adverse results are obtained that would degrade the operation of the redesign versus the old design, even though in both case the minimum or maximum limit specified is met. At the end of the evaluation process it has to be demonstrated that the redesigned devices have all the parameters properly centred with reasonable margins to the specified limits and that the design objectives have been consistently achieved. There are the conditions for releasing the redesigned parts for mass production.

It is also to be kept in mind that the extensive verifications and evaluations both during the redesign phase and on the finished products aim to make sure that the quality is inbuilt into every circuit that is redesigned.

GENERAL CONSIDERATIONS

A redesign program is a team effort, in which all engineering departments must participate. All the inputs have to be gathered at the very beginning from design, process, device, product, quality and reliability departments as well as from the field. The goal setting and the priorities have to be agreed upon by every organization participating in the effort. No easy compromises can be admitted, e.g. a solution that improves cost but impedes quality, or the other way around. Solutions have to be found that result in simultaneous improvements of all features so that there are no "losers" among the participants.

Performing such a program requires, besides dedicated and ambitious engineering resources, a lot of drive and coordination to be guaranteed by the involvement of a reasonably high management level.

CONCLUSIONS

The necessity may occur to re-evaluate the design solutions for existing product families in order to improve their competitivity. This is a complex activity, requiring a reasonable amount of technical cleverness and creativity, and a strong organization.

The results that can be obtained have to be carefully evaluated in advance, and solutions must be chosen to simultaneously improve cost, manufacturability and quality.

Well organised, it will be a rewarding program. For instance, within a redesign program for a logic family the number of good dies per wafer for selected devices could be increased by as much as 50%, resulting in a significant reduction of the number of wafers to be started for the same output. This was achieved due to both die size reduction and yield increase. As a bonus, some other features more difficult to quantify have been considerably improved, such as aspect ratio, pad placement and susceptibility to fail at overstress.

2.2 Productivity Optimization of VLSI DRAM Chips with Redundant Circuits

Charles Stapper

INTRODUCTION

The addition of redundant circuits in memory chips has made it possible to significantly increase the yield of such chips. The redundant circuits, however, increase the complexity and size of such chips. The proper amount of redundancy, therefore, depends on the defect levels associated with the manufacturing process.

An example is given of the yield strategy for a dynamic RAM which was to be introduced into a mature manufacturing line.

CHIP LAYOUT

A yield model described by Stapper et al has been used for more than a decade to optimize the productivity of dynamic and static random access memory chips. Such chips consist primarily of an array of storage cells. Data is written into this array by means of decoders and drivers. There are two groups of such circuits. One set of these circuits are the word decoders and drivers. They provide access to the word lines. The other set connects to the bit lines. Data is read from an array by raising the potential of a word line. This results in data signals on the bit lines. These signals are amplified with sense amplifiers to bring them up to usable output levels.

A relatively small section of memory chips is set aside for voltage generators, input output circuits and generators for timing pulses. A simplified schematic diagram of these circuit groups is shown in Figure 1. Actual chip layouts are much more complex than this. Also shown are additional circuits for redundancy. This includes extra cells in the array, additional word decoders, bit decoders, and sense amplifiers. In addition, a set of steering circuits is needed to steer signals away from the defective word or bit lines. The steering circuits are activated by blowing a set of fuses. This makes it possible to replace the word lines, and bit lines that are defective.

YIELD AND PRODUCTIVITY FOR DIFFERENT REDUNDANCY SCHEMES

Increasing the amount of redundancy usually results a higher yield. An example of this is shown in Figure 2. The four bars represent DRAM yields calculated with the yield model of reference (1). Each bar consists of two parts. The left part represents the yield of each chip without activating the redundancy. This is referred to as the perfect chip yield. The right part represents the gain in yield obtained by replacing defective circuits with the redundant ones. It is therefore known as the fixed chip yield. Since yield plans are proprietary information, the bars indicate relative yields. The bottom bar, designated as D,

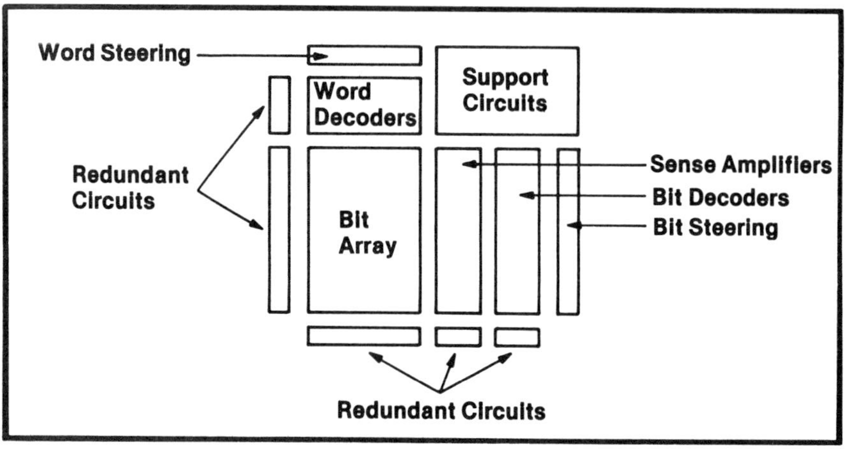

Figure 1. Circuits on a memory chip. This is a simplified diagram of the circuitry that makes up a memory chip. On actual chips some of these circuits are often intermixed.

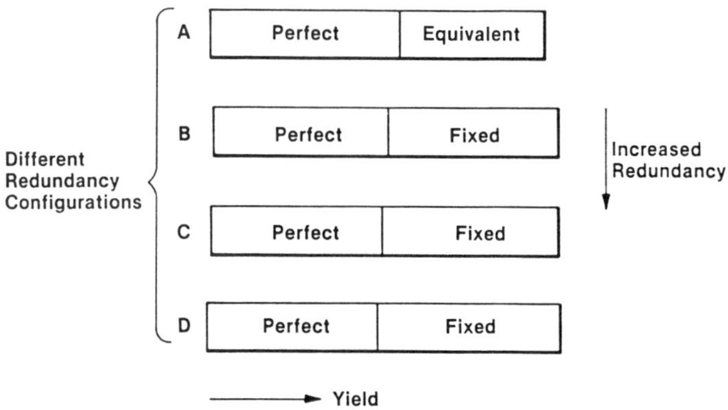

Figure 2. Yields for different redundancy schemes. These relative yields are the result of a partially good product scheme, two word lines of redundancy, four word lines of redundancy, and an arrangement of two redundant word lines and two redundant bit lines.

represents the case with the most redundancy. This consists of two redundant word lines and two redundant bit lines. The perfect chip yield in the bottom bar is the lowest in the group. This is caused by the increased defect sensitivities from the redundant circuits. The fixed chip yield, however, exceeds that of all the other ones, resulting in the highest overall yield.

The redundancy configurations for case C consisted of four redundant word lines and no redundant bit lines. The chip represented by bar B contained only two redundant word lines. The yields depicted by bar A in Figure 2 result from no redundancy at all. Defective chips, however, could be used if the defects were located in one of the eight word addresses. Such chips are referred to, in general, as partially good chips. In this specific case such chips are known as being 7/8 good. Since not all the bits in a partially good chip are usable, the value is less than that of perfect or even fixed chips. To obtain a yield comparison, the percentage of these partially good chips is, therefore, multiplied by 7/8 to obtain an equivalent yield. This yield is represented by the right side of bar A in Figure 2.

The results in Figure 2 show that the chip with the two word and bit lines redundancy has the highest yield. We have, however, at this point to consider the bit productivity. This quantity is defined as the product of the yield, the number of chips per wafer, and the number of bits per chip. The units of bit productivity are "number of usable bits per wafer." Bit productivity corresponding to the four sets of yields in Figure 2, are shown in Figure 3. The actual productivities are proprietary information. The results are, therefore, again shown as relative quantities. The chip without any redundancy turns out to be the winner.

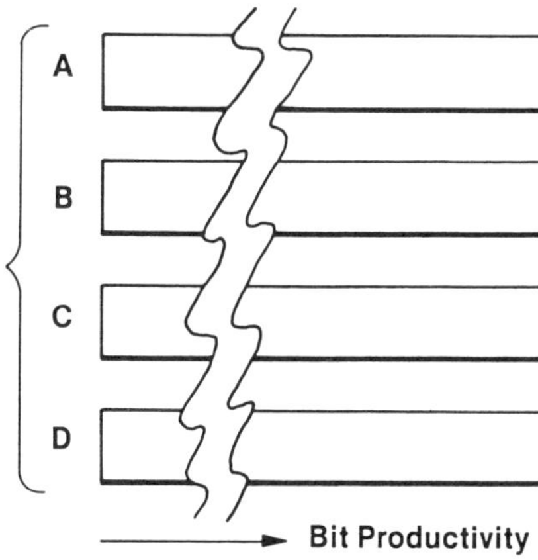

Figure 3. Chip productivity. The number of expected useable bits per wafer at wafer final test are shown as relative quantities. The scheme with the least amount of redundancy resulted in the highest productivity.

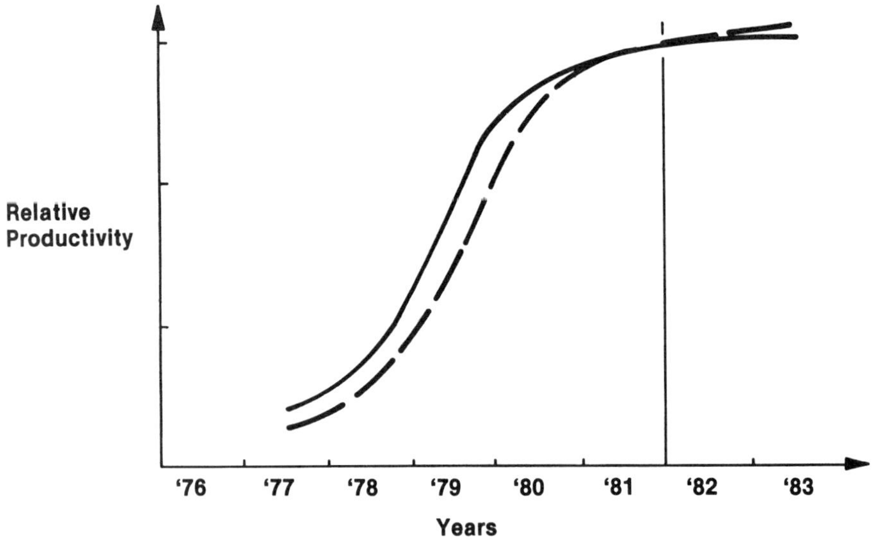

Figure 4. Productivity learning curves. Chips with less redundancy tend to be more productive when the yields are high.

RESULTS

Results of this nature depend primarily on the defects caused by the manufacturing process. The chip used as an example in this paper was to be introduced in a mature manufacturing line with low defect levels. An example of the result of defect learning is shown in Figure 4. The two curves represent the bit productivity as a function of time for cases A and D of Figures 2 and 3. The results indicate that the productivity of case A exceeds that of case D only at high productivity levels. This, however, is precisely the point at which the chip was to be introduced into the manufacturing line. The chip without redundancy, using 7/8 partially good chips was, therefore, designed. It was introduced into the manufacturing line four years after this yield-productivity study was actually made. The relative yields projected with the yield model and the relative manufacturing yields are shown in Figure 5.

CONCLUSION

It has been shown here that yield models can be used effectively for optimizing, redundancy, and productivity in the manufacture of random access memory chips. The model used for this example has been useful in planning the productivity strategy of 36K, 64K, 256K, 288K, and 1M DRAMs in IBM integrated circuit manufacturing plants. Improved versions of this model are also used to map out the productivity of future products.

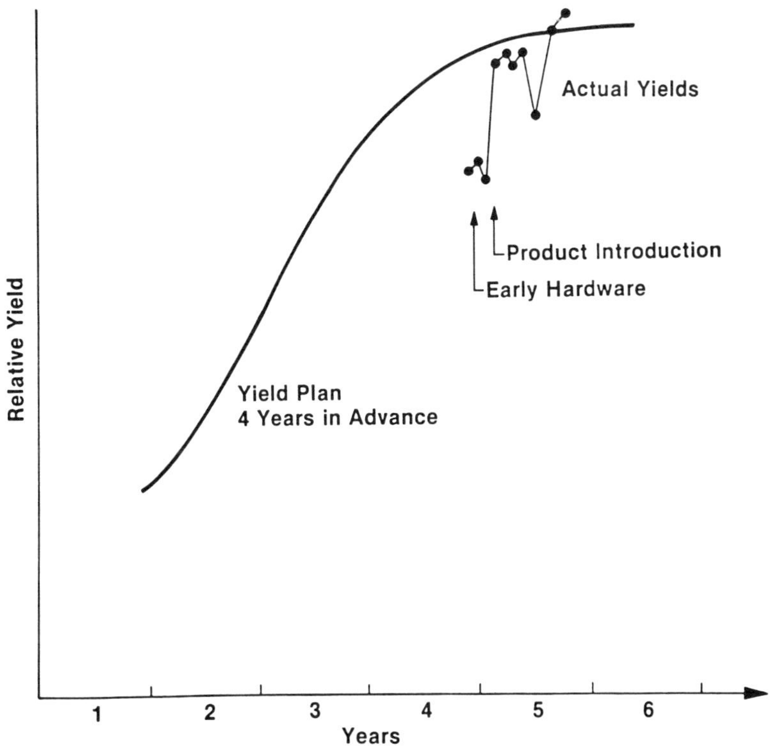

Figure 5. Planned and actual yield. The yields of this product were close to target. As a result cost objectives were met.

REFERENCE

Stapper C H ,McLaren A N and Dreckmann M 1980 Yield Model for Productivity Optimization of VLSI Memory Chips with Redundancy and Partially Good Product, *IBM J. Res. Dev.* **24** 398-409

2.3 WSI: A Technology for Reliability

Ray McKirdy and Mike Lea

INTRODUCTION

For around two decades there has been intermittent research into the Ultra Large Scale Integration (ULSI) and Wafer Scale Integration (WSI) of computer systems, culminating recently with renewed vigour in the form of the UK Alvey and ESPRIT 1 and 2 projects.

There are however, problems associated with realising WSI systems, the most severe of which are: distributing power and signals with integrity across the large wafer area (Warren _et al_ 1985, 1986), physically packaging the devices whilst maintaining intimate thermal contact and good TCE matching to avoid stress (McKirdy and Lea 1986, McKirdy and Sumerling 1986) and providing sufficient fault tolerance to overcome process induced defects distributed accross the wafer (Hedge and Lea 1986).

Despite these problems, research still tends towards realising the goal of WSI and ULSI, due to the significant benefits potentially afforded by these technologies, namely: reduced system manufacturing costs, cost-effective processsor-memory bandwidth increases, reduced system power dissipation overheads, a more compact, lightweight system and increased reliability (McKirdy _et al_ 1986). It _is_ with the comparative reliability increases this paper is primarily concerned.

The comparative reliabilities of VLSI, ULSI and WSI technologies are evaluated, with the Brunel VLSI (SCAPE) (Lea 1985a, Jalowieki and Lea 1986), ULSI (WASP2000) and WSI (WASP8000) (Lea 1985b and 1986, Jones and Lea 1985) image patch processing systems as the research vehicles on which to carry out the comparison, using HRD4 (British Telecom) and MIL SPEC (US Military 1982) reliability models and data. Having evaluated the increased reliability from ULSI and WSI, a cost model is used to predict system maintainance costs for the VLSI, ULSI and WSI implementations.

A brief overview of the three systems is given in terms of physical composition, and architecture where relevant to fault tolerance and reconfiguration issues.

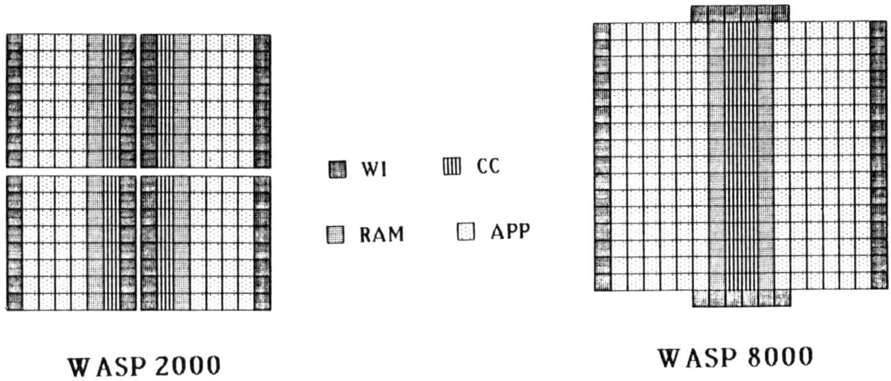

Figure 2. ULSI (WASP2000) and WSI (WASP8000) devices.

The WASP8000 device integrates an 8192-pixel image processing system on a single wafer, it replaces over 100 discrete components needed to implement the same system using SCAPE-based VLSI technology. The ULSI system is composed of four chips of approximately 4 sq.in. footprint. The impact of this at the PCB level is that there is no longer the need for 3 to 4 VME boards, associated edge-connectors and cabinet, but rather, one compact component which, with its low system pin out, can easily be interfaced to the host system on a single PCB or even plug-in direct to the host. The increase in package and interconnect density leads to a significant reduction in system size and weight.

It can be seen (Table 1) that a 12-fold increase in packing density is achieved with WASP compared with its VLSI (SCAPE) equivalent, and a 23% increase over the ULSI (WASP20000) implementation.

Table 1. Physical breakdown of WASP2000 (ULSI), WASP8000 (WSI) and SCAPE (VLSI)-based patch processors.

	SCAPE	WASP-2K	WASP-8K
number of packages	128	4	1
number of PCBs	4	1	0/1
area (Si) (sq in)	5.48	9.92	12.55
area (foot print) (sq in)	79.4	16	15.5
package density (area (Si)/area (fp)) (%)	6.9	62	80.97
system pin-count	4392	448	148

VLSI, ULSI AND WSI PATCH PROCESSING SYSTEMS

The structural organisation of a patch processing module, independent of which technology is used, is shown in Figure 1. For the VLSI system, a chain of SCAPE chips provides the ASP string for in-place processing of an image patch. The Associative String Processor (ASP) controller and patch buffer support the control interface and pixel data interface. The ASP Controller in the VLSI system is implemented with merchant bit-slice processing chips and the patch buffer is similarly composed from merchant RAM chips.

At the structural level, the WASP2000 and WASP8000 architectures feature a hierarchicaly fault tolerant tree-structured interconnection of Associative Parallel Processing (APP), Communication and Control (CC), RAM and Wafer Interface (WI) modules. The APP provides the ASP string, the CC the ASP controller and the RAM modules are the on-wafer patch buffers of the generic patch processor module in Figure 1.

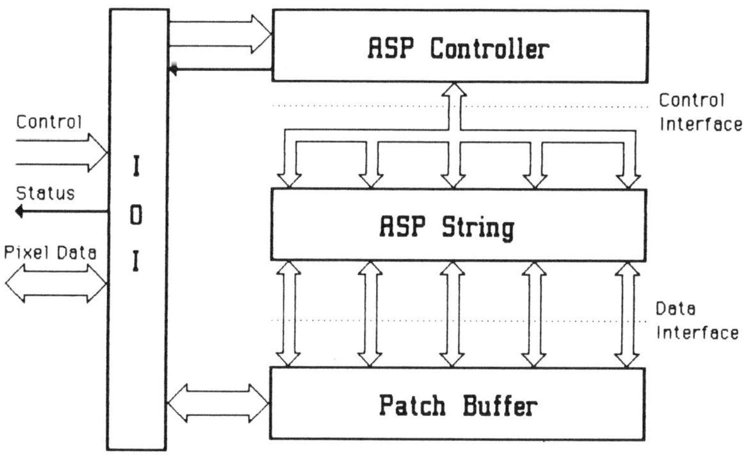

Figure 1. Patch Processing Module.

The equivalent system in ULSI and WSI technologies is shown in Figure 2. As can be seen, both the ULSI and WSI devices are composed of the same basic modules (viz. APP, CC, RAM, WI), stepped out differently on the wafer surface.

Having given on outline of the structural composition of the systems, we now turn to the physical breakdown of each implementation, as this is the major source of the predicted areas of reliability improvement for the ULSI and WSI systems.

From Table 2, it can be seen the system failure rate is 35 times reduced for WSI and 11 times for ULSI, over the VLSI system.

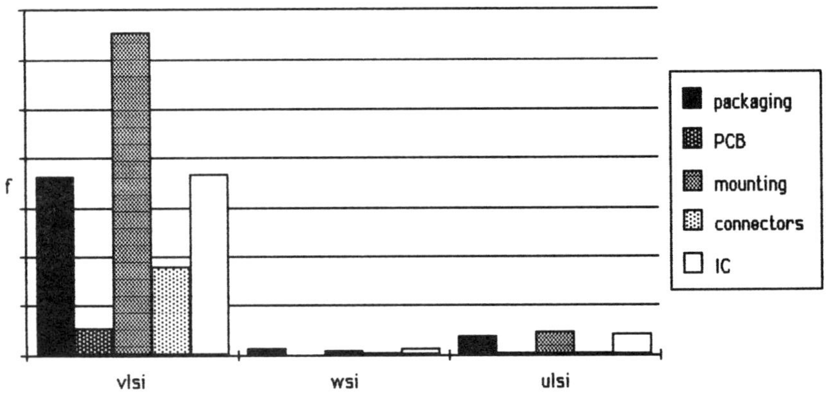

Figure 3. Relative component failure rates.

Looking more closely at the sources of system failure, packaging failure rate has been improved some 28 times for WSI and 9 times for ULSI. This is primarily due to the reduction in package count itself and the overall pin-count of the packages. Figure 4 shows that an individual VLSI pack (68 pins) is more reliable than an individual ULSI or WSI package. However, when we consider the system as a whole, there are 128 packages in the VLSI patch processor and only the one for WSI, and 4 for ULSI. The system reliability due to packages is therefore increased as shown in Figure 4.

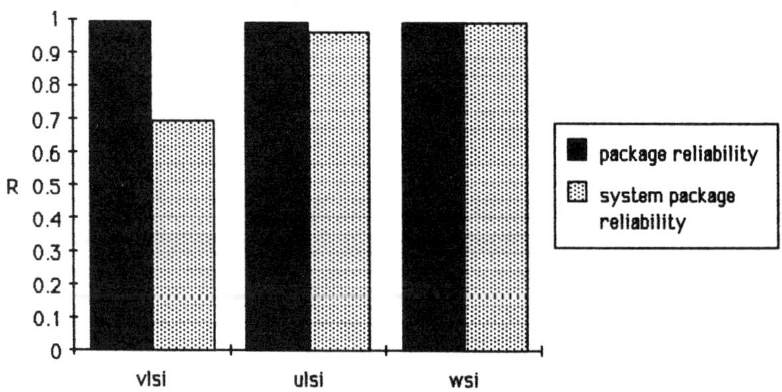

Figure 4. Package reliabilities.

Similarly, PCB and mounting (PTH) reliabilities are increased for WSI and ULSI as shown in Figure 5. The complexity of the boards needed to implement the connections in the VLSI system is greater than that for the ULSI and WSI cases, where interconnect is run on the silicon surface in Al tracks. The number of system PTH solder joints has also been significantly reduced for both ULSI and WSI systems.

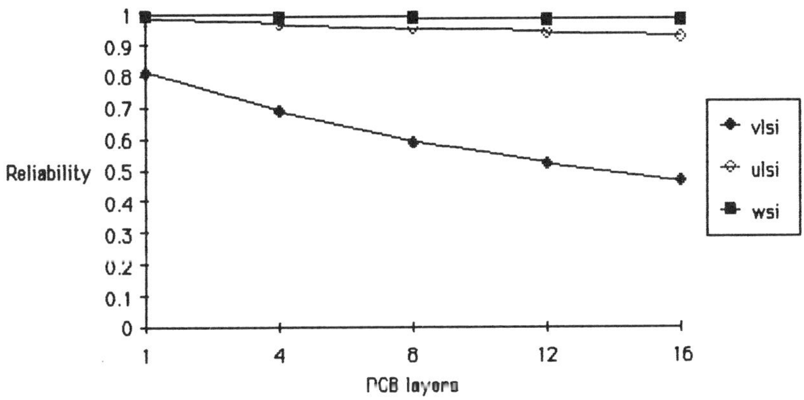

Figure 5. PCB reliabilities.

Connector failure rate contribution is increased for VLSI due to the increase in the number of PCBs (Figure 6.), and hence the number of edge connectors off to the backplane of the VLSI system.

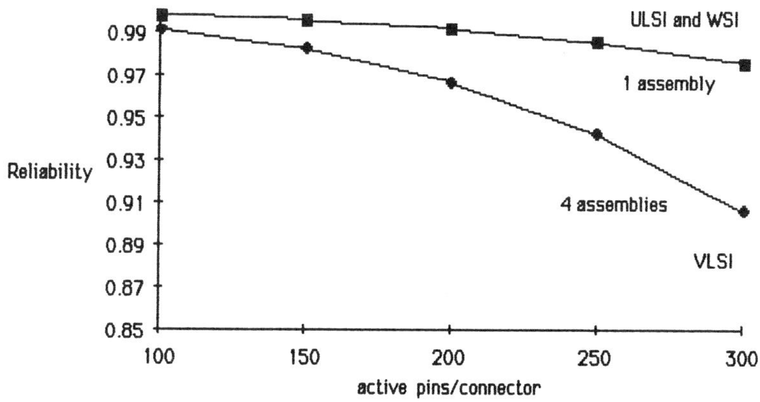

Figure 6. Connector reliabilities.

MOS IC failure rates are predicted to be increased for the VLSI system. The BFR is based on the number of gates or bits (for memories) monolithically integrated on the chip. However, the BT model predicts a levelling off of the failure rate in the ULSI and WSI gate count region, with the overall effect that the WSI system gate count on the one slice of silicon is more reliable than that of the VLSI case, spead over 128 discrete components. This is purely a mechanical/physical prediction and does not take any kind of ´soft´ reliability (viz. redundancy and reconfiguration) into account for the ULSI and WSI systems.

Redundancy and reconfiguration

As outlined earlier, the WASP architecture is composed of a tree-structured interconnect of APP, CC, RAM and WI modules (Figure 7).

All four blocks incorporate greater or lesser degrees of redundancy, with testing and configuration logic such that small-scale defects can be tolerated. Indeed, the APP ´leaf´ modules are composed of fine-grain Associative Processing Elements (APEs) which, due to their associative nature, are easily amenable to fault tolerance by including spare APEs and operating a sub-block bypass scheme (Hedge and Lea 19876) to avoid faulty elements. However, in the event of large-scale defects (viz. clusters) whole APP modules may be similarly bypassed. For the CC and RAM, the entire locality (branch) is bypassed via the CC module.

In addition to the inrease in mechanical reliability, this ´soft´ electronic configuration used on WASP to provide tolerance to process induced faults, could be used to confer the capability for automatic post-delivery reconfiguration. This facility would allow WASP to perform in-service self-repair, should a failure occur. This may be crucial in applications where systems reliability and minimum down-time is of paramount importance.

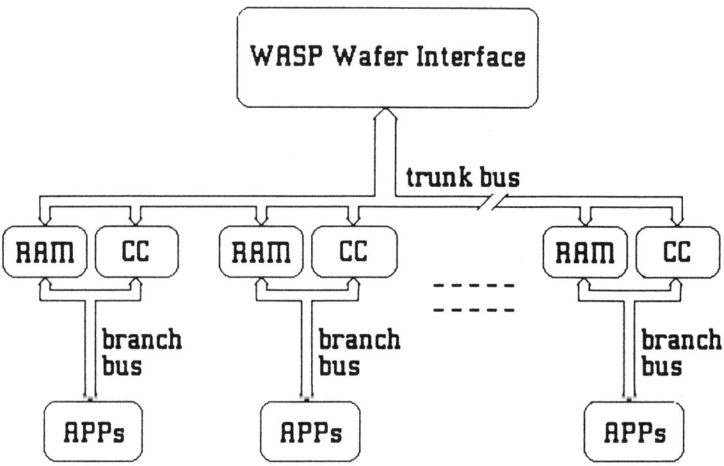

Figure 7. WASP tree-structure.

The reduction in bulky PCB and system interconnect mediums for ULSI and WSI also contributes to the compactness of these systems over the VLSI implementation, due to a natural partitioning of the processing boundaries across the monolithic silicon area, resulting in a very low system pin-out, and significantly improving reliability for the ULSI and WSI patch processing modules.

COMPARATIVE EVALUATION

In the previous section we have seen that a dramatic reduction in component, pin and PCB count has been realised by ULSI and WSI technologies. A reliability model of all the three systems has been created, based on the British Telecom HRD4 and MIL HDBK 217D reliability handbook guidelines. Basic failure rates (BFR) are based on data collected from components operating (and failing) in the field.

A constant failure rate model is assumed for all the components in each system, which is to say, the evaluation is carried out over the useful life of the system, ignoring the infant mortality and wearout periods (Arsenault and Roberts 1980). The useful life of the system has been taken as approximately five years.

The overall system reliability is calculated from a series reliability model of the individual components making up each system. The following distinct failure sources are considered: MOS IC, package, PCB and connector failures.

A breakdown of the failure rates for each of these components is shown in Table 2, and graphically illustrated in Figure 3.

Table 2. Normalised failure rate breakdown.

	SCAPE	WASP2000	WASP8000
packaging	1140	125	41
PCB	181	9	1
mounting	2050	147	29
connector	56	14	14
chip	1156	137	47
system	4503	432	132
system reliability	0.23	0.87	0.96

The WASP reconfiguration facility also allows the device to continue functioning, possibly with a reduced efficiency, even in the event of multiple failures. The failure of modules beyond the capacity for self-repair (viz. less than 8192 APEs usable on the wafer) would not lead to total loss of image, but rather in portions of the image being distorted. This distortion could provide sufficient warning for wafer replacement before a total failure occurs, while a useful level of functionality is maintained in the interim period.

COST CONSIDERATIONS

Having seen the increased reliability from the ULSI and WSI implementations, the cost implications are evaluated in terms of maintainability overheads (viz. system ownership costs).

Table 3. Normalised system costs for VLSI, ULSI and WSI patch processors.

cost	SCAPE-PPM (VLSI)	WASP-PPM (ULSI)	WASP-PPM (WSI)
non recurrent design	23	28	26
system manufacture	65	31	30
system ownership	12	2	1
total system	100	61	57

The system ownership costs are related to the reliability of each system (McKirdy and Lea 1987). The more reliable the system, the longer its MTBF will be, resulting in minimised down-time and maintainance costs. The reduction in the ULSI and WSI costs are a direct result of the substantial decrease in the number of mechanical failure sources (viz. solder joints, packages and connectors).

For the ULSI and WSI cases, the system ownership costs represent a much less significant fraction of overall system cost. The above costs are based on the relative likelihood of the system going down over the 5 year useful life period, and the subsequent repair and down-time costs incurred.

The systems have been assumed to be in a benign environment (viz. ground, fixed) which infers lower failure rates, but perhaps more importantly, that the system is accessible to maintainance. However, it may be that the environment is air or seabourne, in which case maintainance is infeasible, then the Mean Time To Failure (MTTF) is important, and the ULSI and WSI systems have a distinct advantage, especially if 'in-field' reconfiguration is implemented.

CONCLUSIONS

This paper has investigated the comparative reliability of the same image patch processing system, implemented in VLSI, ULSI and WSI technologies.

The WSI and ULSI systems have been shown to be significantly more reliable than the VLSI, over the useful life of the systems. With WSI and ULSI failure rates some 35 and 11 times less than VLSI.

The increase in system reliability comes from the reduction in mechanical failure sources such as solder joints, PCB complexities, component and connector counts.

In addition to the mechanical reliability increase, the fault tolerance necessary to survive process defects in ULSI and WSI devices, confers the capability for 'in the field' reconfiguration in the event of a module failure.

As a result of the inherent reliability of ULSI and WSI systems, the down-time and maintainance costs are reduced when compared to the VLSI implementation.

All these factors strongly suggest that ULSI and WSI are indeed technologies for systems reliability.

ACKNOWLEDGEMENTS

The authors gratefuly acknowledge the enthusiastic help of the Brunel WASP team and the support of the SERC.

REFERENCES

Arsenault J E and Roberts J A 1980 Reliability and Maintainability of Electronic Systems, Pitman Publishing Ltd.

British Telecom 1987 Handbook of Reliability Data (HRD4), London Information (Rowse Muir) Ltd.

Hedge S J and Lea R M 1986 Intra-Module fault-tolerant strategies for the WASP device, IFIP Workshop on Wafer Scale Integration, Grenoble University.

Jalowiecki I P and Lea R M 1986 SCAPE- A High Performance Image and Signal Array Processor Chip, Proc. 3rd Silicon Design Conference 345-354.

Jones S R and Lea R M 1985 Interconnection Strategies for the WASP (WSI Associative String Processor) Device, Wafer Scale Integration (eds. Jesshope and Moore), Adam Hilger Publ. 82-91.

Lea R M 1985a SCAPE: a VLSI chip architecture for image processing, Microarchitecture of VLSI Computers (eds. Antognetti, Anceau and Vullemin), Martinus Nighoff Publ. 285-294.

Lea R M 1985b WASP: a WSI Associative String Processor for Structured Data Processing, Wafer Scale Integration (eds. Jesshope and Moore), Adam Hilger Publ. 140-147.

Lea R M 1986 A WSI Image Processing Module, Wafer Scale Integration (eds. Saucier and Trilhe), North Holland Publ. 43-58.

McKirdy R D, Abdelrazik M B E A, Bolouri H S, Hedge S J, Jones S R, Warren K D and Lea R M 1986 Comparative appraisal of the implementation of an image processing module using WSI and VLSI associative string processors, Proc. of an International Workshop on Systolic Arrays, paper L5, Oxford University.

McKirdy R D and Lea R M 1986 Physical Design Issues for WSI, Wafer Scale Integration (eds. Saucier and Trilhe), North Holland Publ. 311-319.

McKirdy R D and Lea R M 1987 A Comparative Appraisal of WSI, ULSI and VLSI Associative String Processors, EDA Conference (ed. Ambler), Electronic Design Automation Publ. 291-301.

McKirdy R D and Sumerling G W 1986 WSI: the limits of packaging ?, IEEE Workshop on VLSI Packaging, Paris.

US Military 1982 MIL HDBK 217D, Reliability Prediction of Electronic Equipment.

Warren K D, Abdelrazik M B E A, Bolouri H S and Lea R M 1986 Power, Clock and Signal Distribution in the WASP Image Processing Device, IFIP Workshop on Wafer Scale Integration, Grenoble University.

Warren K D, Abdelrazik M B E A, McKirdy R D and Lea R M 1985 A Power Distribution Strategy for WSI, Wafer Scale Integration (eds. Jesshope and Moore), Adam Hilger Publ. 54 61.

2.4 Use of the Focused Ion Beam for Inspection, Characterization and Repair of Defects in VLSI and Wafer Scale Integration

John McDonald, How Tzu Lin, Sri Gupta and Ravi Selvaraj

INTRODUCTION

The FIB employs a liquid metal ion source to produce an extremely bright ion beam. The LMIS gun consists of a needle which is held by a ribbon heater containing a sample of the liquid metal or alloy which is melted by the heater resulting in wetting of the needle (see Figure 1). By applying an extractor voltage of approximately 10 KV the electric field in the vicinity of the tip is high enough to cause ions to emit from the liquid surface.

The specific focused ion beam examined here is the VG IBL-100S shown in Figure 2. High voltage (up to 110 KV) and heater power are applied through the cables at the ion beam column. The heater must deliver as much as 28 watts of power to the source for certain alloys and compounds to melt.

As shown in the cutaway drawing in Figure 3 the ions proceed down the column through a Wein E X B mass filter which selects the appropriate ion species. Such a separation is important if the source charge is an alloy or compound, but also helps remove other species sputtered from the apertures and other column components. This filter is not sufficiently selective, however, to separate closely spaced isotopes. Therefore an additional aperture is used at the base of the column to screen out these unwanted ions. If this is not done then multiple ghost images will result from the isotopes whenever a pattern is written.

Ions continue down through the condenser lens, the beam blanking plates, octupole stigmators, and objective lens finally passing between the 900 V electrostatic deflection plates and emerging from the column into the main writing chamber to impinge on the wafer surface. The full writing field ranges between 0.05 mm X 0.05 mm and 1.0 mm X1.0 mm depending on the size of features desired. The minimum beam deflection step size is 0.033 microns.

Operation of the focused ion beam as an ion microscope requires that the secondary electrons ejected from the wafer by the impact of the high voltage accelerated ions be

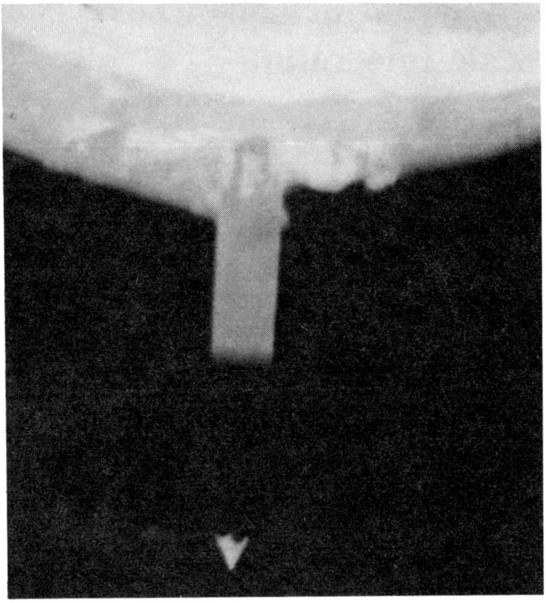

Figure 1 Liquid Metal Ion Source (LMIS). Molten droplet at top of gun needle consists of metal alloy or chemical compound containing the desired ion source species.

Figure 2 VG IBL-100s Focused Ion Beam System

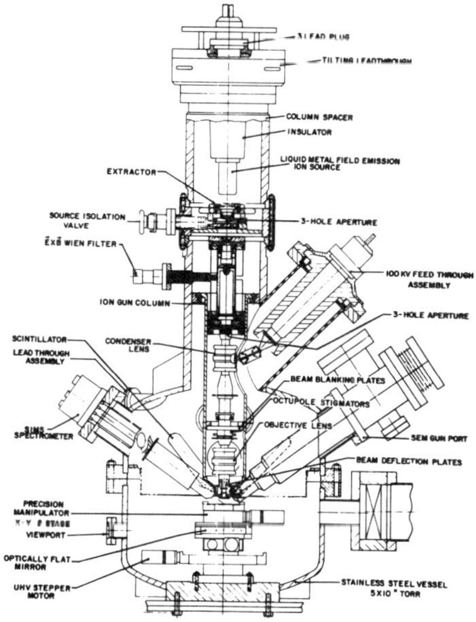

Figure 3 Cutaway Drawing of VG IBL-100s

detected. A scintillator (channeltron) detector collects events while the ion beam is scanning. The resulting signal can be converted and stored in digital form in frame freeze or computer memory, or displayed in real time producing a video image. This video image is useful for visually acquiring a registration mark for subsequent refinement under computer control.

Another option on the FIB is the Secondary Ion Mass Spectrometry (SIMS) attachment. Using a quadrupole mass separator, this instrument provides a means for determining the identity of ion species sputtered from the surface by the impinging beam. By carefully choosing the impinging ion species, the atomic makeup of the surface can be analyzed. In effect an image can be constructed in which the identities of the recoiling atoms can be tagged. In order for the SIMS analyzer to detect these atoms they must themselves be ionized.

A spare attachment port is provided for inclusion of a focused scanning electron microscope. This attachment can be used to provide a noneroding, noncontaminating means of imaging or aligning the wafer. During the blanking time of the ion beam the electrons generated by the electron microscope can provide a precision means for neutralizing the accumulating charge resulting from the primary beam.

Control of the wafer positioning table involves use of precision stepping motors which can locate the table within 1.0 micron of the desired position. An HP 5501A laser interferometer and optically flat mirrors mounted on the table are used to measure stage displacements in real time.

APPLICATIONS TO INTERCONNECTIONS IN VLSI AND WSI

The authors have examined the electromagnetic and fabrication requirements of interconnections in WSI (Donlan et al 1985 and Jesshope and Moore 1986), many of which must apply for VLSI or ULSI. Ultra clean fabrication facilities are needed for 16 to 64 megabit DRAM's and 64 bit microprocessors. The projected die size for these circuits is 2 to 4 sq. cm. involving half to quarter micron feature sizes. Automated Class 1 clean room modules will be mandated, resulting in 0.1 defects per sq. cm. or better. Completely sealed in-situ UHV processing environments (Steckl 1986) have also been considered.

Interconnections are generally considered to be the largest source of fabrication faults in integrated circuits, especially for silicon based technology. One of the highest yielding metallization patterning techniques is the lift-off process (Bergendahl et al 1985). Referring to Figure 4, first a layer of insulator (spin-on polyimide or SOG) is applied to the wafer. The action of spin-on followed by a curing operation tends to planarize the surface.

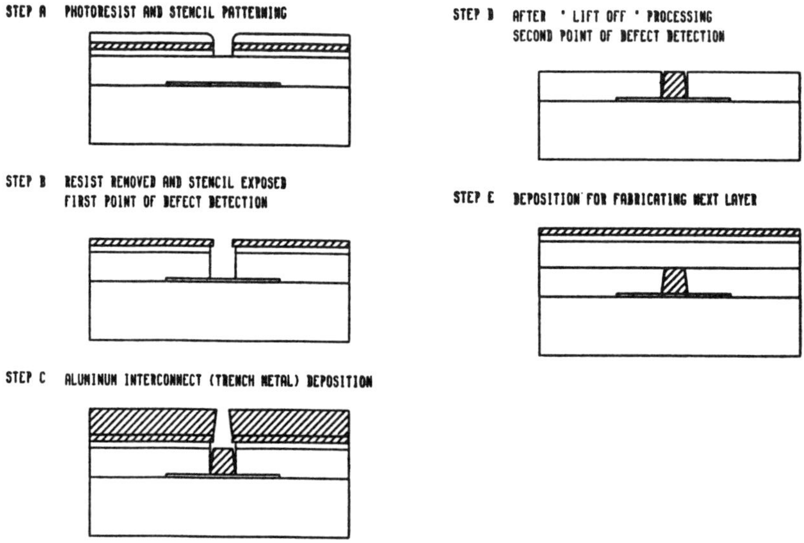

Figure 4 Thick Film Lift-Off Process

Next the surface of the wafer is coated with a thin layer of release (R) material. If the release material is not strong enough to withstand the other etching steps required, then a stencil or shield layer is added. After coating the stencil with resist, exposing the interconnect pattern and developing the resist, the stencil is etched open. Using a reactive ion etcher the unshielded sections of the wafer are anisotropically etched providing a trench with extremely straight vertical walls. A brief isotropic etch is used to undercut the shield. Finally the metal desired for the interconnections is directionally deposited into the trench and on top of the release layer. The release layer is attacked and the excess metal on top of it is "lifted off" under mild agitation. The entire process is then repeated for each layer of interconnection and also for plug or stud vias between these interconnections.

We will now examine the specific defects which tend to arise using this processing sequence. In thick film lift-off of aluminum, the growth of large columnar grains can cause cracking and open circuits. Any rough surface features at the bottom of the lift-off trench can nucleate large grains and precautions are needed to avoid this (Selvaraj et al 1987).

Assuming that the aforementioned precautionary steps are taken then the collection of faults which can occur can be readily identified:

D1) Stencil under etched causing missing trench section resulting in a open circuited line. This can be due to gel slugs in the resist or other particulate contamination.

D2) Large stencil pin holes or over etch which results in intralayer short circuits.

D3) Improper thickness control or failure to control reactive ion etch during isotropic or anisotropic operation resulting in rip-out of metal layer.

D4) Trash metal resulting from the lift-off step and which is not cleansed properly from the wafer. The presence of the crevice on the sides of the trench metal seems to provide a residual surface topographic feature which forms a snag point for this excess metal.

D5) Improper via etch. This results in an interlayer open circuit or a resistive via.

The most common of the defects we have observed using lift-off is the open circuit traceable to defect type D1 (see Figure 5). Presumably by reducing the defect level in the electron resist and using a dry etch (a wet etch was used here) the number of these defects could be minimized provides an irreducible minimum path for contamination to enter the process. Only resistless processing or a nonliquid method of applying the resist could avoid this problem.

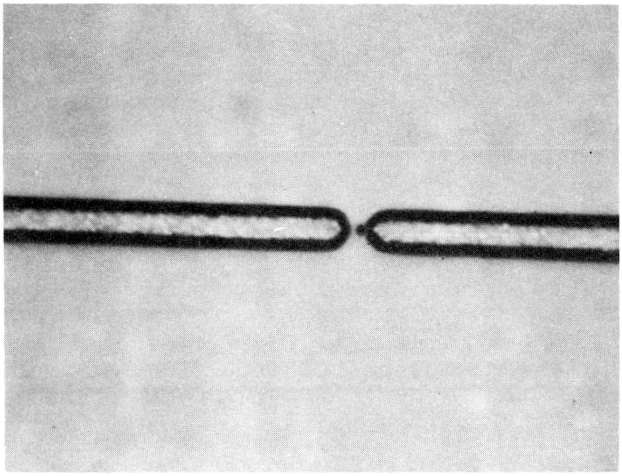

Figure 5 Open Stencil Defect (D1)

Defects of the type D3 involving improper thickness tolerance can be corrected by using the FIB. However, the beam time required to repair such faults would be excessive. These defects are best eliminated by optimizing the tightness of control over such processes as reactive ion etch operating parameters and spin-on and cure conditions.

A FOCUSED ION BEAM STRATEGY

In this section we describe repair strategies for coping with the most common defects such as those of type D1 and D2. We note that the stencil gap defect, D1, can readily be detected from secondary electron images obtained using the FIB as a digitized ion microscope. The contrast between metal and insulator is quite good. Hence by scanning the stencil layer after it has been patterned, but before this pattern has been transferred to the insulator in the form of a trench by reactive ion etching, it is possible to catch this type of fault before it becomes a permanent open circuit in the interconnection metallization. Figure 6(a) depicts a stencil gap illustrating how the lift-off trench is incomplete and the metal line possesses an open circuit. Figure 6(b) shows a strategy by using ion milling to open the stencil where it was unetched. The SEM photograph (Figure 7) shows the result of using the FIB to effect the repair. The LMIS used in this experiment is gallium. The accelerating voltage is 75 KV with beam diameter of 0.5 micron and target current of 0.25 nA. The total milling time to open the stencil is several minutes. The computer generated repair pattern is slightly larger to ensure proper coverage of the fault in the stencil. The few holes shown in the SEM photograph are the result of longer beam dwell time during SIMS analysis.

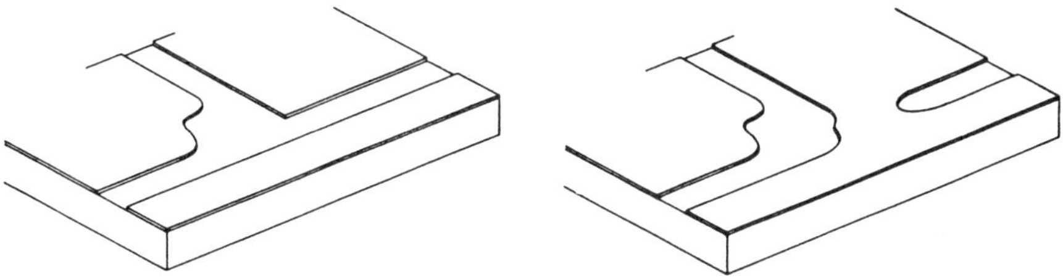

Figure 6 Stencil Repair by Ion Milling

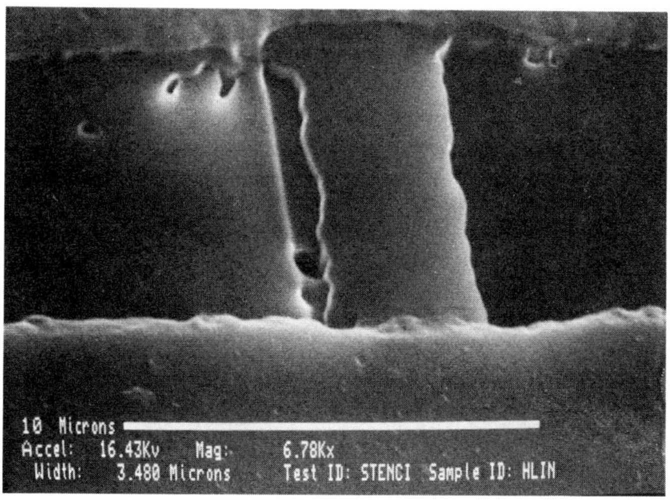

Figure 7 SEM Photograph of Stencil Repair by Ion Milling

Defects of type D2 are similar to defects of type D4 in that metalization already deposited into the lift-off trench must be cut. For ULSI and WSI the thickness of the metallization may be several microns. Consequently, the scanning of the ion beam for ion milling must be optimized with a view toward the tendency for the sputtered metal to redeposit on the vertical walls of the lift-off trench (see Figure 8). Figure 9 shows a metal line which is 5 microns thick that has been cut using the FIB. In order to verify that the line has been cut to the bottom of the metal layer, the secondary electron detector current can be monitored.

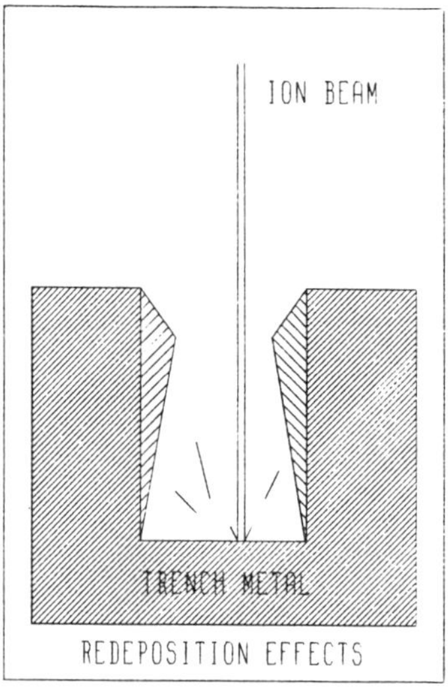

Figure 8 Effect of Ion Redeposition

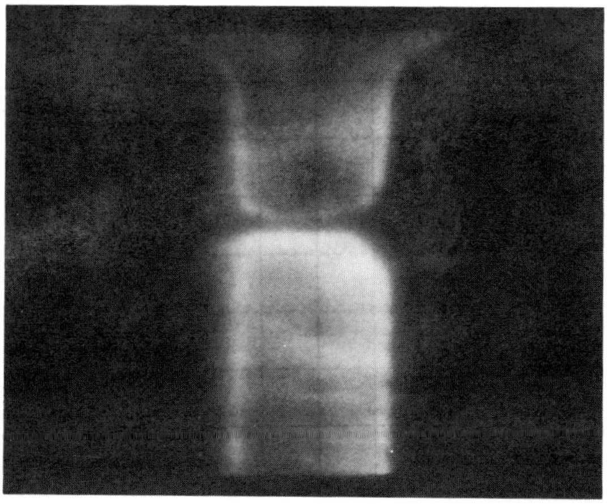

Figure 9 Cut Metal Line

Figure 10(a) shows the signal generated during the start of the ion milling operation to cut the metal line. Due to the high efficiency of secondary electron generation near the edge of the metal line the contrast has peaks near these edges. Figure 10(b) shows sweep signal near the end of the cutting operation. This signal appears quite adequate to give an end point indication in this case. Charging can be a severe problem in milling thick lines and some form of charge neutralization is required.

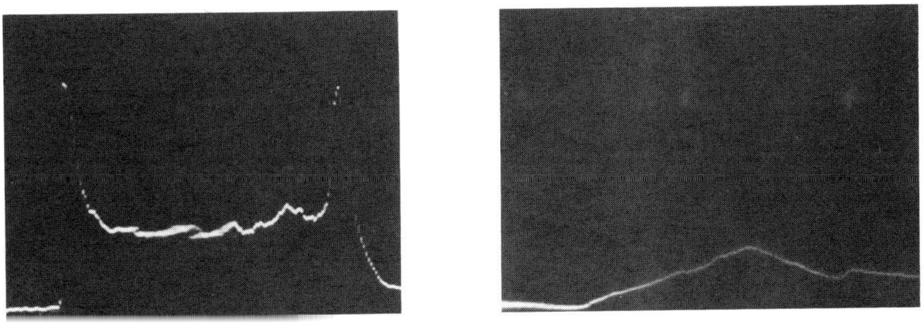

(a) (b)

Figure 10 Secondary Electron Signal at
(a) The Start of Ion Milling
(b) The End of Ion Milling

VIA VERIFICATION AND SIMS ANALYSIS

The cleanliness of vias is extremely important in assuring low resistivity through this path. When using aluminum for interconnect metalization a thin layer of oxide quickly forms when its surface is exposed to the atmosphere. In addition, when opening the via in the lift-off process preparatory to deposition of stud vias the reactive ion etch may leave some contamination at the bottom surface of the lift-off trench. These residual contaminants coat the surface onto which the metal studs must be deposited. Removal of this residual interface contamination can be accomplished by using an argon plasma ion beam sputtering step prior to metal deposition. However, certification that all vias are open and clean would be a useful step before metalization, since a failure in a via would be difficult to repair. Heating the wafer in an attempt to liquefy the material near the interface can result in a reduction of the density of the contaminant by dissolution. However, with some metals this would require excessive heating, a step which should be avoided to reduce thermal stress in the multilayer interconnection structure.

Figure 11 illustrates the use of the ion beam as a SIMS probe for via verification. One problem which arises concerns the shadowing of the back sputtered material by the via wall itself. This wall intercepts ions traveling along a line-of-sight trajectory to the SIMS detector from the point of impact for the incoming ion beam. Hence an attracting lens is needed to collect ions from other trajectories to produce a satisfactory SIMS signal.

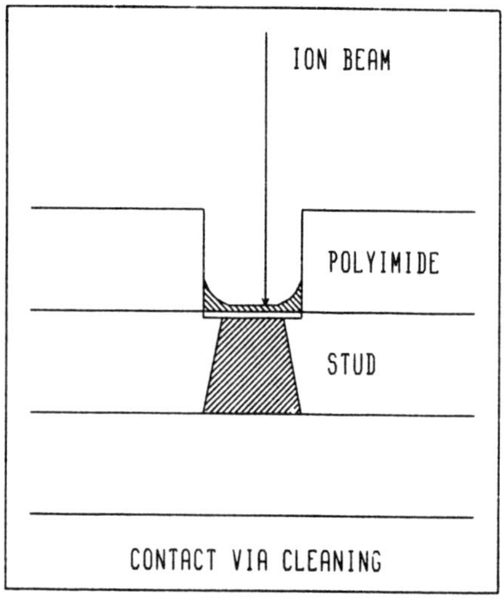

Figure 11 Contact Via Cleaning

Figure 12 shows a SIMS line scan and mass spectrum plot. A gallium ion source is used to produce these results. The accelerating voltage is 75 KV with 2 micron beam diameter and 1.55 nA target current. The SIMS line scan plot shows the result of an ion beam scan across a 20 micron wide aluminum wire. The SIMS mass spectrum plot, some of the peaks are Al(27), Si(28), Ga(69) and Ga(71).

SPEED OF MILLING

The speed with which ion milling occurs is the determining factor for establishing whether the focused ion beam is a suitable tool for practical defect imaging, characterization and repair. This in turn depends on the ion beam current which can be generated while still maintaining suitable focus and preventing charge up of the surface. Most of the experiments conducted for this paper were at low beam currents (typically only a few tenths of a nanoamp). At these beam currents the rate of milling of aluminum by an 85 KV gallium beam is in the range of cubic microns per several minutes. Successful experiments at ten times this beam current have been made leading to the removal of hundreds of cubic microns in a matter of minutes.

However, in order for this strategy to be successful several orders of magnitude of higher milling rate must be obtained. At the current time two methods appear to hold promise for improving this milling rate. The first involves introducing minute quantities of a gas near the point of impact for the incoming ion beam. The second involves increasing greatly the incoming ion current without loosing resolution. We note that prior to the current limiting aperture the existing gun produces several milliamperes of ion current. Techniques for increasing this current by a factor several hundred have already been found (Melngailis <u>et al</u> 1986).

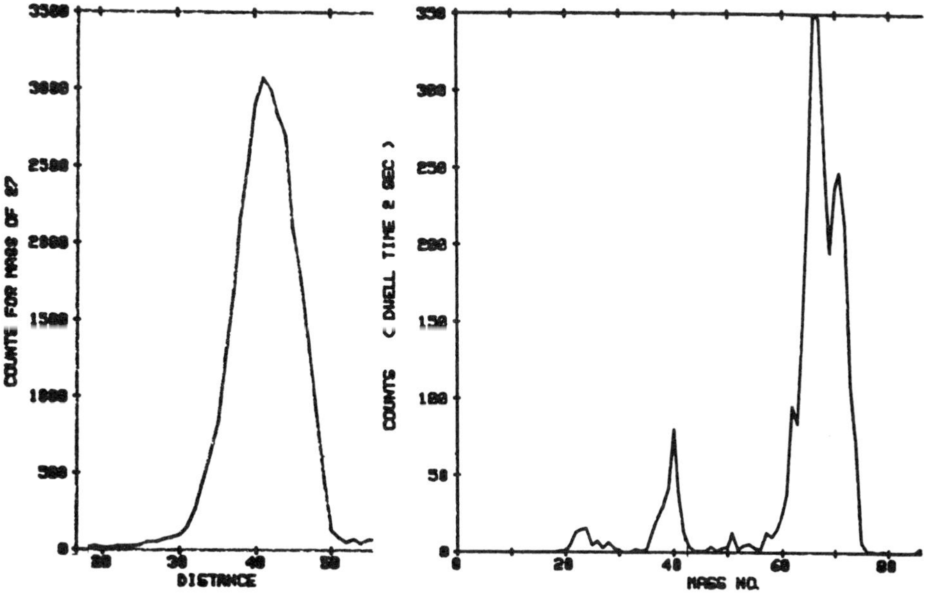

(a) (b)

Figure 12 SIMS Analysis of Aluminum Wire
(a) Spatial Scan (b) Mass Scan

CONCLUSIONS

The focused ion beam is a relatively new system which can be used for inspection, characterization, and repair of many classes of defects in VLSI, ULSI and WSI. We have examined some of the applications of the focused ion beam to repair of interconnections in this paper. Strategies for using the focused ion beam have been presented which appear appropriate for repair residual defects in the high yield lift-off fabrication procedure. The primary advantage of this scheme compared with others which have been presented is the maintenance of extremely low resistance throughout the interconnections and their vias. However, the ultimate success of this approach will require the development of high current ion beam sources capable of much higher milling rate.

ACKNOWLEDGEMENTS

The authors wish to express their gratitude for the sponsorship of the corporate sponsors of the Center for Integrated Electronics, especially GE/CRD, Raytheon, and IBM/Endicott. Special thanks are due to the Semiconductor Research Corporation for their support of the focused ion beam project at Rensselaer.

REFERENCES

Bergendahl A S, Donlan Capt. B J, McDonald J F, Steinvorth R H and Taylor G F 1985 A thick film lift-off technique for high frequency interconnection in wafer scale integration, Proc. Second VLSI-Multilayer Interconnection Conference (June 25-26) 154-162

Donlan Capt. B J, McDonald J F, Taylor G F, Steinvorth R H, and Bergendahl A S 1985 Computer-aided design and fabrication for wafer scale integration, VLSI Design (April) 34-42

Jesshope C and Moore W 1986 Wafer scale integration, Adam Hilger, Bristol & Boston 31-45

Melngailis J et al 1986 The focused ion beam as an integrated circuit restructuring tool, J. Vac. Sci. Technol. B4(#1) (Jan.-Feb.) 176-180

Selvaraj R, Yang S N, McDonald J F, and Lu T M 1987 Ion cluster beam metallized interconnections for wafer scale integration, Proc. Fourth International IEEE VLSI Multilevel Interconnection Conference (June 15-16) 440-448

Steckl A J 1986 Particle-beam fabrication and in situ processing of integrated circuits, Proc. of the IEEE 74 No.12 (December) 1753-1774

3 CATASTROPHIC YIELD LOSS MODELS

As was suggested in Chapter 1, one of the key issues in yield modelling for VLSI circuits is the relationship between the yield and the characteristics of the defects. In this chapter the relationship is discussed in more detail.

In the first section, Maly *et al.* (§3.1) describe a defect characterization methodology that is illustrated by the results of a manufacturing experiment. Reported data clearly indicate that the defect size distribution and topology of the IC layout must be included as essential information in the yield prediction and the authors show how such information may be obtained.

Koren (§3.2) reaches a very similar conclusion. He uses a simple yield model to show the relationship between the topology of the IC and the yield. By analysing the yield in terms of the layout scaling factor, Koren shows that the circuit critical area (and therefore the yield) is strongly related to the defect size distribution.

The next section by Gandemer *et al.* (§3.3) addresses a similar problem again, but uses both theoretical models and experimental data. They demonstrate that not only the circuit's sensitivity to defects of various sizes but also the effects of strongly non-uniform defect densities are important in yield prediction.

In the fourth paper of this chapter, Rivier (§3.4) deals with the same basic problem but uses a different method of investigation. Rivier suggests another version of the Monte Carlo approach to yield estimation. He illustrates the concept of critical area and shows that the simple approach of Monte Carlo simulation can be used appropriately to relate yield to both the topological detail of the IC layout and the defect size distribution.

Hence, the four papers of this chapter seem to support the common conclusion that the functional yield prediction, performed for the VLSI circuits, must take into account both variations in the densities of the defects of various sizes and the detail of the IC layout.

3.1 Characterization of Type, Size and Density of Spot Defects in the Metalization Layer

Wojciech Maly, Michael Thomas, Jeffrey Chinn and David Campbell

INTRODUCTION

Yield prediction and yield estimation have been and will remain very important issues for IC design , as well as, process development. Thus, such problems as the modeling of spot defect related yield losses have been investigated since the early nineteen sixties and a significant number of various yield models have been proposed as an immediate outcome of these investigations.

The majority of these models do not take into account realistic relationships between defect size and defect location, with respect to the detail of the IC layout and therefore, are inaccurate.

The necessity for a better description of the spot defect size in the yield analysis has been already noted and models dealing with the distributed nature of defect size have been published by Maly and Deszczka (1983), Stapper (1983) and Ferris-Prabhu (1985). The accuracy of the new proposed models depend, however, on the accuracy of the information that describes the defect size distribution and defect density. In this paper, we present a method that can be used to determine the density of spot defects and defect size distribution. The proposed method is suitable to investigate spot defects in the metal layer, but it can be extended to other layers.

The proposed method uses a test structure, which was introduced by Maly (1985) and then described by Maly et al. (1987). We call this test structure a double-bridge test structure and in this paper we discuss its design, features, and applications.

DOUBLE-BRIDGE TEST STRUCTURE

Double-Bridge Test Device

Let us consider a device whose structure is defined by three masks shown in Fig. 1. It is composed of a polysilicon meander and a set of parallel metal strips that cover straight segments of the polysilicon meander. The metal strips are connected to the polysilicon meander through the set of contacts etched in the dielectric deposited on top of the polysilicon meander. In all other regions, the metal is isolated from the polysilicon.

The resulting device is shown in Fig. 2 a. The most important feature of this device is that straight segments of polysilicon between contacts are shorted by the metal bridges (see Fig. 2 b). Consequently, the equivalent circuit of the device created in this manner can be represented by a chain of resistors shown in Fig. 2 c. Resistors labeled R_1 and R_2 represent folded and straight segments of the polysilicon meander, respectively. Resistors R_3

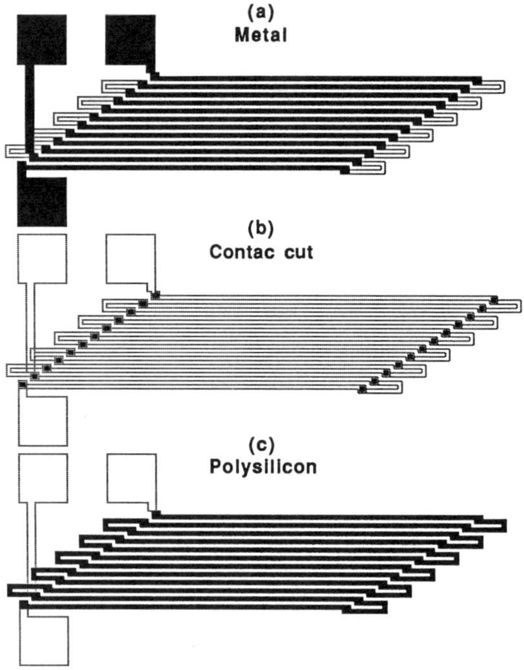

Figure 1: Metal (a), contact cut (b), and poly layers of the double-bridge test structure (indicated in black).

represent metal strips.

The double-bridge test structure has three terminals labeled 1, 2, and 3. Note that the resistance that can be measured between terminals 1 and 2, R_{12} , is practically equal to the resistance of the folded segments of the polysilicon path represented in black in Fig. 2 d.

Straight segments do not contribute to this resistance because they are shorted by the metal strips. The resistance that can be measured between terminals 2 and 3, R_{23} , represents one folded element of the polysilicon path. Hence, by measuring resistances R_{12} and R_{23} and computing their ratio one should obtain an integer equal to the number of parallel metal strips in the structure minus 1.

Spots of extra metal (shorts)

Let us assume that on the surface of the metal layer, due to a disturbance in the manufacturing process, an additional spot of metal has been deposited (See Fig. 3 (a)).

Such a spot causes a short between some of the metal strips and some of the folded elements of the polysilicon path become shorted (See Fig. 3 c). Consequently, the resistance R_{12} measured between terminals 1 and 2 is smaller than the equivalent resistance measured on the test structure without a spot defect. Therefore, in the case when the value of R_{12} measured in the test structure is smaller than expected, we assume that the test structure contains a spot of extra metal.

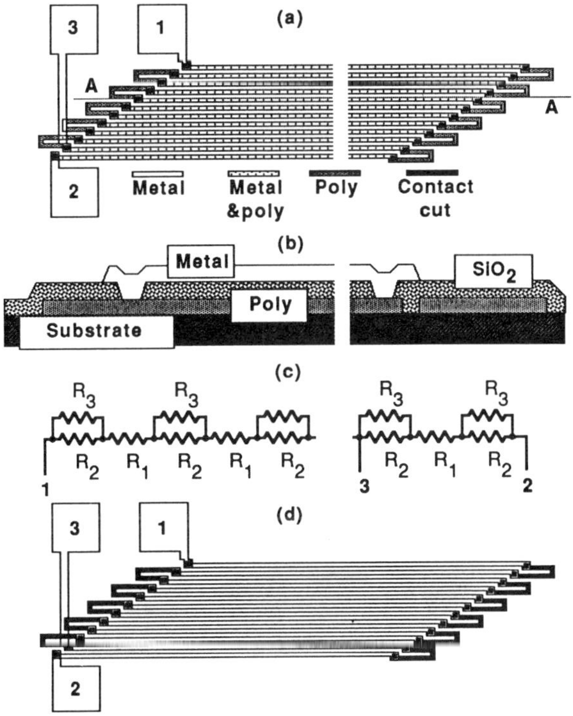

Figure 2: Layout of the double-bridge test structure (a), its cross-section (b) the equivalent circuit (c) and segments of the polysilicon meander that contribute to the resistance R_{12} (in black) (d).

Note that by measuring R_{12} and R_{23}, and computing the ratio R_{12}/R_{23}, one can evaluate the diameter of the metal spot. The difference between the obtained value and number of metal strips in the test structure indicate the number of parallel metal strips that are shorted by the metal spot (See Fig. 3 c). Since the pitch of the metal pattern is known, the diameter of the defect is known as well.

Regions with missing metal (breaks)

The double-bridge test structure can also be used to measure the size of a region with missing metal. This capability is illustrated in Fig. 4.

Note that due to the spot of missing metal some of the straight segments of the polysilicon path are no longer shorted. Thus, in the test structure with broken metal lines, the total resistance seen between terminals 1 and 2 (see Fig. 4 b) must be equal to the resistance, R_{12}, measured on the test structure without a spot defect plus n_m times R_2 where n_m is the number of broken metal connections in the structure (see Fig. 4 b).

Hence, the double-bridge test structure can be used to determine with simple electrical measurements type and the size of the defect in the metal layer. (Note that the idea presented above can also be applied for other than metal layer but upper layer, i.e. bridging layer, must have resistivity much smaller than the lower layer.)

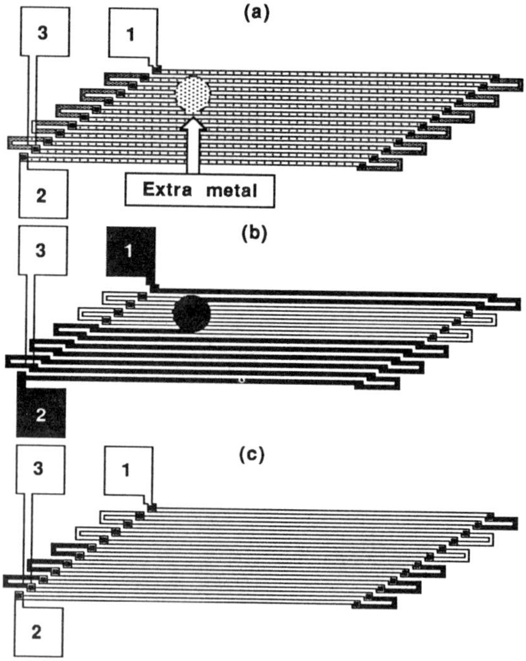

Figure 3: The double-bridge test structure with the spot of extra metal (a), current path between pads 1 and 2 (b), and segments of the polysilicon path that contribute to the resistance seen between pads 1 and 2 (c).

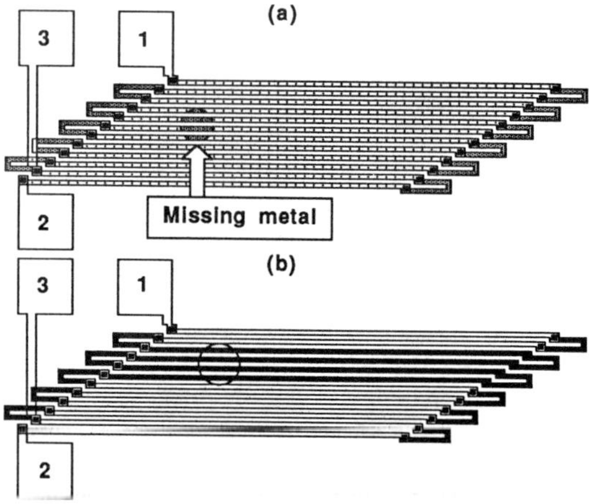

Figure 4: The double-bridge test structure with the region of missing metal (a), and segments of the polysilicon path that contribute to the resistance between pads 1 and 2 (b).

TEST DIE

To test the practical implementation of the ideas presented in the previous section a test die with the double-bridge test structures was designed and then fabricated. In this section we describe the details of this design.

Floor plan of the test die

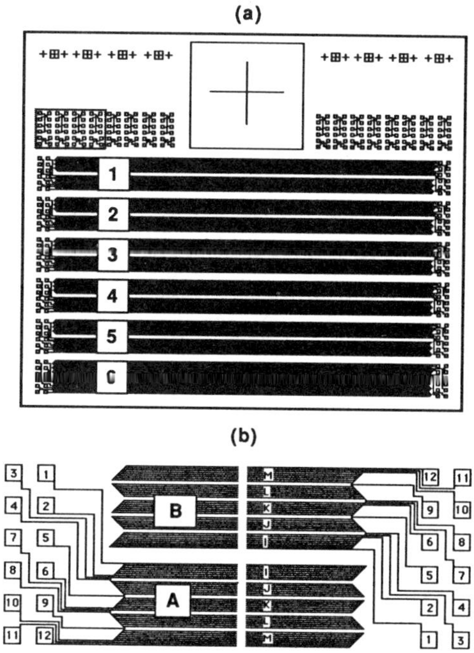

Figure 5: The test die (a) with six double strips of double-bridge test structures (b) and four sets of split-cross-bridge and contact resistor test structures.

Fig. 5 shows layout of the test die. It contains:

1. Six identical strips with double-bridge test structures;
2. Four groups of structures that are used for the evaluation of :
   a. polysilicon line registration error;
   b. metal line registration error;
   c. polysilicon sheet resistance;
   d. metal sheet resistance;
   e. contact resistance;

In the lower portion of this die six strips with the double-bridge test devices are placed. Each of the strips in the test die is composed of two elements, A and B, each containing five devices labeled I,J,K,L, and M, respectively.

Each of the two elements is connected to 12 pads. The pads at the left hand side of the die are connected to the element A of each strip and the pads located on the right hand side of the die are connected to the B element of each strip. The upper part of the die contains four blocks of the split-cross-bridge and contact resistor test structures. One split-cross-bridge and contact resistor structure are connected to pads arranged the same way as those of the double-bridge test structure.

Layout of a single strip of the double-bridge test structure

A single element of each strip in the test die, shown in Fig. 6, contains five double-bridge test devices.

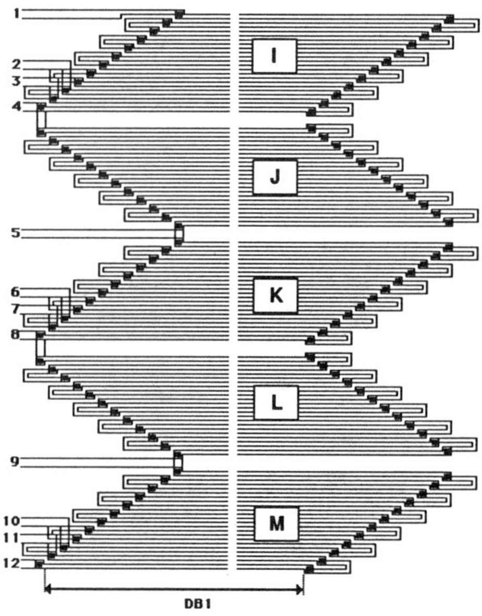

Figure 6: Layout of one element of the double-bridge test structures. Numbers prescribed to the structure terminals are pad numbers.

There are 12 terminals connected to such an element. Between terminals :

A. 1 and 4, 4 and 5, 5 and 8, 8 and 9, 9 and 12 one can measure the resistance of each device;

B. 2 and 3, 6 and 7, 10 and 11, one can measure the resistance of two contacts and one folded polysilicon element;

C. 3 and 4, 7 and 8, 11 and 12, one can measure the resistance of four contacts, two metal strips and one folded polysilicon element;

Fig. 7 shows the detail of the junction between the knee of the polysilicon path and two double-bridge straight elements.

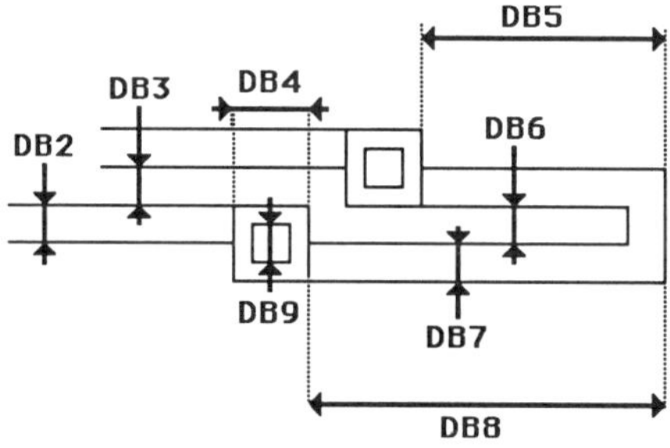

Figure 7: Layout of the single knee of the double-bridge test structure.

The split-cross-bridge resistors applied in the test structure is described by Buehler (Buehler 1983).

DESIGN RULES FOR THE DOUBLE-BRIDGE TEST STRUCTURE

There are three important issues that have to be considered in the design of the double-bridge test structure. The first issue is related to the size measurement resolution that can be achieved with the double-bridge test structure. The second is related to the disturbances in the size measurements due to spot defects in other than the metal layers. The third issue deals with the area required for the defect size evaluation. In this section, we show how these three issues are related one to another. An optimal design is presented that assures the required resolution and, at the same time, minimizes vulnerability to the disturbances in the process.

Resolution in the defect size measurement

As was shown previously, the double-bridge test structure can be used to classify the size of defects with extra or missing metal. To describe this classification procedure in more detail we assume that :

R_{14} - represents the value of the resistance measured between the end terminals of a single double-bridge device (indices correspond to the labels in Fig. 6);

R_{34} - represents the value of the resistance measured between the terminals 3 and 4 and describes the resistance of one folded element of the polysilicon path;

DB2 - is the width of metal and poly straight strips (see Fig. 7);

DB3 - is the space between metal strips (see Fig. 7);

SQM - is the number of squares in the metal and poly straight strips;

SQP - is the number of squares in the folded element of the poly path;

n - is the number of metal strips in the device (therefore n - 1 is the number of folded polysilicon elements in this structure);

The classification procedure is as follows:

Step 1. Find the location of the next test device;

Step 2. Measure R_{14} and R_{34} ;

Step 3. Compute $n_r = R_{14} / R_{34}$;

Step 4. Determine whether $n_r = n - 1$? If yes go to Step 1. If no continue.

Step 5. If $n_r > n-1$ then a defect, with a missing metal, has been detected. The diameter of this defect, d_m, must be in the range defined by the following formula:

$$DB2 + (n_m-1)\,(DB2+ DB3) < d_m < DB3 + (n_m + 1)(DB2 + DB3)$$

where n_m is a number of broken metal strips that can be evaluated from the formula:

$$n_m = (SQP/SQM)\,(R_{14} / R_{34} - n + 1); \qquad (*)$$

If $n_r < n-1$ then a defect which has shorted metal lines has been detected. The diameter of this defect, d_e, must be in the range defined by the following formula:

$$DB3 + (n_e-1)\,(DB2+ DB3) < d_e < DB3 + (n_e + 1)\,(DB2 + DB3)$$

where n_e is a number of shorts between metal strips that can be evaluated from the formula:

$$n_e = (n - 1 - R_{14} / R_{34}); \qquad (**)$$

Step 6. Go to Step 1.

The above procedure is a direct extension of the considerations presented in the previous section. It is very important to note, the relation between the number of shorted or broken lines and the size of the defect. This relation is schematically outlined in Fig. 8.

It can be seen that the diameter of a spot defect can be determined in terms of a range whose size is equal to the pitch of the test structure (DB2+DB3). Thus, the first limitation in the resolution of the double-bridge test structure is determined by the limits in the resolution with which the test structure can be fabricated. Therefore, the smallest possible pitch should be used in the design of the double-bridge test structure.

The other limitation in the double-bridge test structure resolution is of a more complicated nature. According to the considerations presented in the previous section, n_e and n_m should be integers. But, the actual values of n_e and n_m are computed using values of the

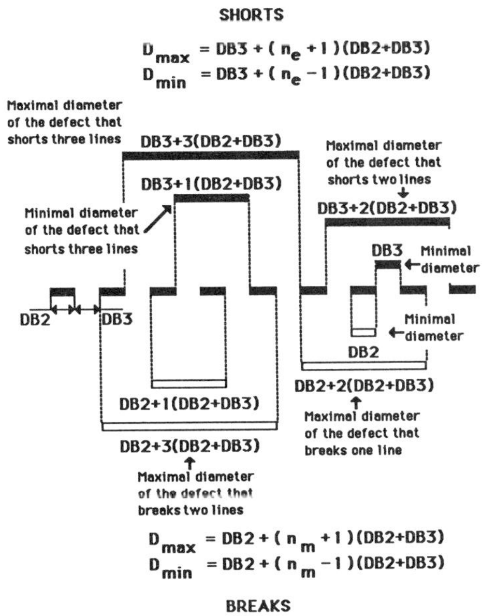

Figure 8: The relationships between defect size and the number of shorted or broken lines.

ratio R_{14}/R_{34} which are in general, real numbers. This situation occurs because the sheet resistance, as well as, the width of the polysilicon path are not ideally uniform. In addition, these nonuniformities will have a random nature. Thus, as one would expect, values of n_e and n_m will be random variables and must be rounded to the closest integer. (An illustration of the above feature is presented in Fig. 9.) Of course, such an approximation is justified only in the case when the rounded number has a value that is close enough to an integer.

Note that to be realistic, we have to assume that some of the values computed from formulas (*) and (**) (see Fig. 9 (c)) cannot be classified as integers. These values will represent defects that cannot be measured with the double-bridge structure. A larger number of such defects degrade the resolution of the test structure.

Fortunately, the above problem can be solved by an appropriate choice of the number of strips in a single device. Namely, the number of strips in the single device must be chosen so that the change of the resistance seen between the end terminals of the test device that is caused by missing or extra spot of metal, is much larger than all anticipated random changes of this resistance due to inherent random instabilities in the process. In other words, if we assume that the spread in the resistance measured between the end terminals of the single device, without a spot defect, is ΔR_{41} then R_{14} - (n-1) $*R_{34}$ must be substantially larger than ΔR_{14} . Assuming for instance, that ΔR_{14} is 5% of R_{14} one has to chose a number of strips in the test structure to be smaller than R_{14} / ΔR_{14} i.e. less than 20. A choice of this kind assures that any short or break causes larger changes in the resistance R_{14} than the process induced noise.

Hence, the pitch of the double-bridge test structure and number of strips in the structure are main factors that determine its resolution.

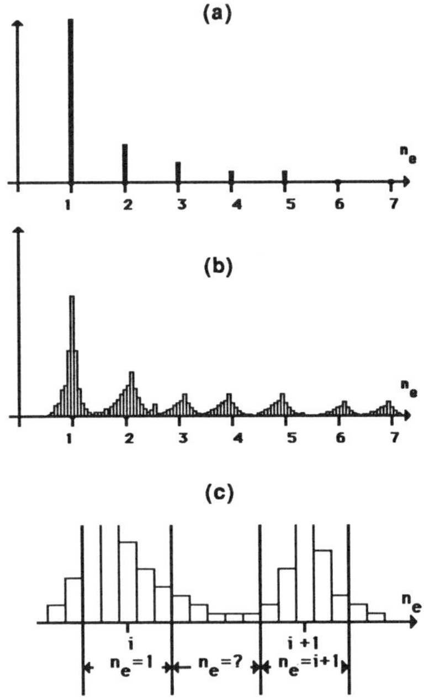

Figure 9: Histograms that illustrate the resolution problem. The ideal (a) and actual (b) histograms of the number of shorted segments in the double-bridge test structure. The test structure resolution is determined in terms of the number of samples that can be classified as integers to the number of samples that cannot be classified as integers (c).

Sensitivity to the spot defects in the polysilicon layer

In the previous section the role of spot defects was discussed with respect to the metal layer only. In the actual test structure, spot defects will occur in all layers. In this subsection, we consider the sensitivity of the structure to shorts and breaks in the polysilicon layer.

First, consider a spot defect that shorts two segments of a polysilicon meander. (See Fig. 10 a). Such a short changes the electrical equivalent of a segment of the test structure between nodes labeled A and B in Fig. 10 a in the way that is depicted in Fig. 10 b. Note now that the resistance $R_{AB} = R_1||(2R_{21}||R_{22} + R_{defect})$. The equivalent resistance without a spot defect is equal to R1. Thus, the resistance measured between the end terminals of the double-bridge test structure can be disturbed by a spot defect in the polysilicon layer only in the case when R_{21} or R_{22} is comparable with R_1. This only occurs when the spot defect is located close to the contacts. If we can build the test structure such that $R_2 >> R_1$ i.e., where the double-bridge segments of the structure are much longer than the folded elements, then for the majority of locations of the spot defects in the polysilicon layer $2R_{21}|| R_{22} >> R_1$. This means that the majority of polysilicon spot defects will not change the value of R_{AB} in such a structure .

Breaks in the polysilicon path will disturb the resistance measurements only in the case

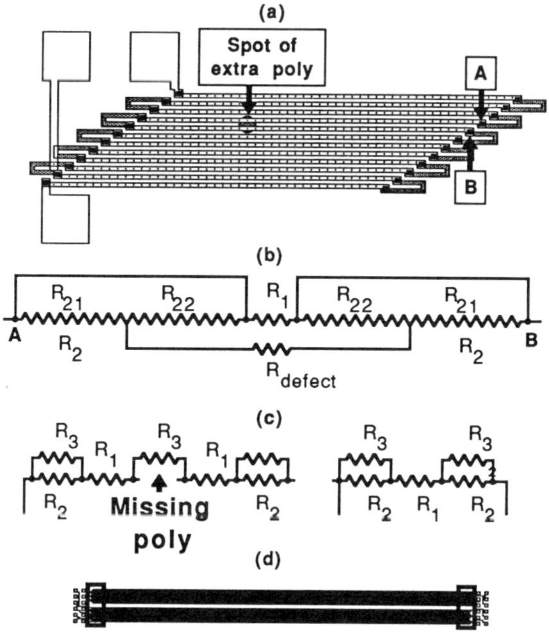

Figure 10: Extra and missing regions in the poly layer. A spot of an extra poly (a) and the equivalent circuit for the short caused by such a defect (b). The equivalent circuit for the test structure with the break in the polysilicon path (c). The area of the double-bridge test device that is prone to spot defects in the polysilicon layer.

Breaks in the polysilicon path will disturb the resistance measurements only in the case when they are located in the neighborhood of the folded segments of the structure (see Fig. 10 (c)). Based on this analysis, it is desirable to design the test structure such that folded polysilicon elements are much shorter than double-bridge elements.

Hence, in the test structure such as shown in Fig. 10 one can minimize the area that is prone to the spot defects in the polysilicon layer. This area, indicated in Fig. 10 (d) by the two boxes, is comparable to the double area covered by the folded segments of the polysilicon paths.

Test structure area considerations

There are two area related factors that have to be considered in order to determine appropriate sizes of the double-bridge test devices.

The first is the area utilization factor i.e. the ratio of the area that is sensitive to the spot defects in the metal layer to the total area. As far as this factor is concerned it is desirable to maximize the area covered with metal strips and minimize all other areas. Since, the area of pads and the width of a single device is determined by the resolution requirements described earlier, the only remaining parameter that can be manipulated is the length of the device. From previous considerations, it is evident that the test device should be as long as possible. It should be noted, however, that the longer the device the larger the nonuniformity in the geometry and resistivity of the polysilicon and metal paths. In addition, an increase in the area of a single device is equivalent to an increase in the probability that two defects may occur for a single device. The increase in length of the structure translates to an increase in the resistance that is measured between the end terminals of the structure for a break in the metal lines. Hence, the length of the single double-bridge device should be chosen such that :

1. An estimate of the probability of the occurrence of two defects in the same test device is small.

2. Random differences between resistances of folded elements of polysilicon path at two ends of the structure are small.

3. The largest resistance that may be seen between the end terminals of the test device (resistance measured when all metal lines are broken) can be measured with available electronic equipment.

DESCRIPTION OF THE FABRICATED TEST DIE

The test die described above was fabricated on a large number of wafers. Each wafer contained 52 die. Die length and width were 11430 μm and 9753 μm , respectively. The total die area was 1.067 cm^2 and the area sensitive to the spot defects was 0.48 cm^2 (45% area utilization) . Further details are as follows.

Layout

Each strip in the test die is composed of two elements, A and B, and each element contains five devices. **Each device contains 20 metal-poly bridges and 19 folded polysilicon elements.** The layout of each strip with double-bridge test devices is identical but elements A and B in these strips are different. They have the same **pitch**, equal to **4 μm,** but in element **A** the metal line **width** and the **spacing** are equal to **2 μm** and in element **B width** and **spacing** are **2.5 μm** and **1.5 μm**, respectively. The **lengths** of the **metal** strips for both elements is **1 cm** so that resulting number of squares is **5000** and **4000** for elements **A** and **B**, respectively. The numbers of squares in the **folded** elements of the **polysilicon** paths are **100** and **80** for elements **A** and **B**, respectively. Other dimensions are specified in Table 1.

Table 1: Dimensions of the double-bridge test device specified for two strips A and B of a single testing strip. All dimensions in μm. (For the definitions of the distances see Fig. 6 and 7).

	A	B
DB1	10 000	10 000
DB2	2	2.5
DB3	2	1.5
DB4	4	4
DB5	96	76
DB6	2	2
DB7	2	2
DB8	102	82
DB9	2	2

Process

The test wafers were fabricated on 4 inch wafers using standard processing steps. Values measured during the process were as follows:

1. Poly sheet resistance 250 Ω / square

2. Metal sheet resistance 50 mΩ / square

MEASUREMENTS AND COMPUTATIONS

Wafers were fabricated and test data measured. This section will present the practical details involved with probing the test die, organizing the data, and making computations to detect defects that exist on the wafers. Statistical analysis results will be given on both the raw measured data and on the computed values.

Measurements

A wafer has 52 test die, a test die has 12 elements (6 "A" and 6 "B"), and each element contains has 5 devices (labeled I,J,K,L,M). This means each test wafer has 3120 individual defect sensors that must be measured and analyzed.

Each element on a test die has 12 probe pads attached, and 11 resistance measurements are made between different pairs of those pads. The 11 values are in turn used to determine the defect types and sizes that exist on the 5 devices within that element.

The wafers were tested on a system controlled by an HP9836A computer. A standard technique of forcing 1 volt and reading current was used to determine resistance between the appropriate pairs of pins.

During testing, care is taken to record the location of the flat and use a consistent order of testing the 52 die on the wafer. This allows accurate wafer maps to be generated.

Computations

The following definitions are used in this section:

R1 = Resistance of overall device;

R2 = Resistance of one poly end-wrap;

R3 = Resistance of opposite poly end-wrap plus 2 metal lines

Fig.6 shows that R1 includes R1-4, R4-5, R5-8, R8-9, and R9-12 as measured on devices I, J, K, L and M, respectively. R2 includes R2-3, R6-7, and R10-11 (measured on devices I, K, and M), and R3 includes R3-4, R7-8 and R11-12 (also on devices I, K, M).

Basic characteristics of the measured data

To illustrate the nonuniformity of the polysilicon paths this section presents the basic characteristics of R2 and R3 measured electrical values. Since different line widths are used for the A and B elements, the data for the two elements are presented separately. Before computing the results given, the data was screened using limits of 15.0 through 70.0 kΩ to throw out wild values of R2 and R3.

Table 2 shows the range of data and basic statistics for the electrical measurements taken.

Table 2: Characteristics of R2 and R3. (Values given in kΩ.)

Var. Name	Element	Average Value	Standard Deviation	S.D./Avg.	Minimum Value	Maximum Value
R2	A	38.725	2.6698	6.9%	29.900	64.980
R3	A	36.142	2.7726	7.7%	16.660	54.440
R2	B	30.801	1.9637	6.4%	15.510	51.200
R3	B	29.449	2.0498	7.0%	21.130	66.180

Computations of defect size distributions

The average of the R2 and R3 measured values is used to represent the resistance of a single polysilicon folded element. This minimizes the effect of poly non-uniformity over the horizontal "long direction" of the device.

It is assumed that the vertical differences are insignificant since the distance between devices is small. Because of this, the number of pads and the test time can be reduced by measuring R2 and R3 on every other device (I, K, and M) of the five devices on an element. Also, if bad data for R2 or R3 is found on one of the five devices, then a nearby value can be used in its place.

Keeping the above practical considerations in mind, the following specific equations are used to compute the defect types and sizes found on the test wafers.

```
Ravg = (R2 + R3) / 2.       (R2 or R3 may be substitute value)

 A   = (R1 / Ravg) - 19     (R1 is never a substitute)

 If   -.5 < A < .5          then no defect is indicated.

 If      A < -.5            then a defect causing a short is
                            indicated, and INT(A) is the number of
                            line shorts (i.e., the defect size).

 If      A >  25.           then a defect causing broken metal
                            lines is indicated, and INT(A/50)
                            is the number of broken lines.
```

(Where function INT(x) is the value of x rounded to the nearest integer.)

Densities and size distributions of spot defects in metal layers

This subsection gives statistical results on the computed defect data from the test wafers. The results given are only for defects causing shorts because number of breaks found was very small.

Fig. 11 shows the distribution of defects by size category for all A element devices. Fig. 12 shows a similar distribution for the B element devices. A defect falls in category 1 by having a calculated defect value between -1.5 and -0.5, category 2 between -2.5 and -1.5, and so on.

Note that the smaller space between metal lines for the B elements results in a significantly larger number of defects in the smaller size categories.

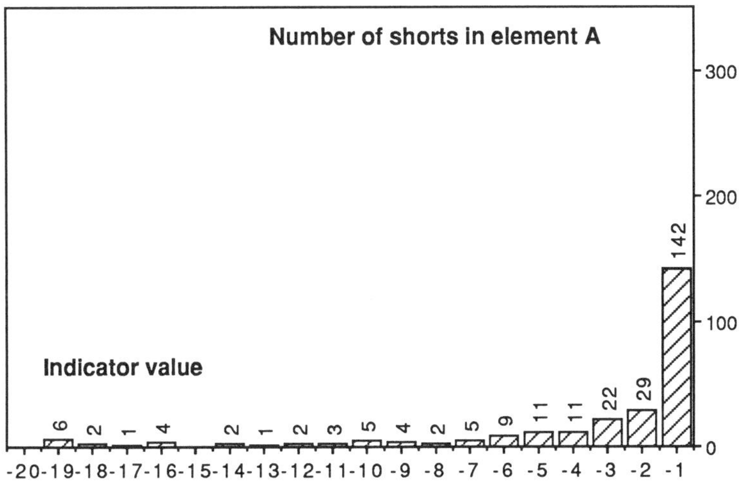

Figure 11: Distribution of the number of shorts. Element A.

One of the previous sections of this report describes the "resolution" of the test structure related to how tightly the calculated defect values group around integer values. Figs. 13 and 14 show histograms of the defect values with the scale expanded to make the individual distributions visible around each integer value.

Fig.15 shows an SEM photo of a defect causing a short between 7 metal strips. This defect was found by observing the "6" on the wafer map.

As it was already indicated the number of breaks on the analyzed wafers was very small. But they have been detected correctly. Figure 16 shows break of a two lines detected on a device with the indicator equal to +2.

Some of the data produced noninteger values of the indicator that have been reported as "bad values". Fig. 17 shows an SEM photo of the die that was found to produce such a bad value.

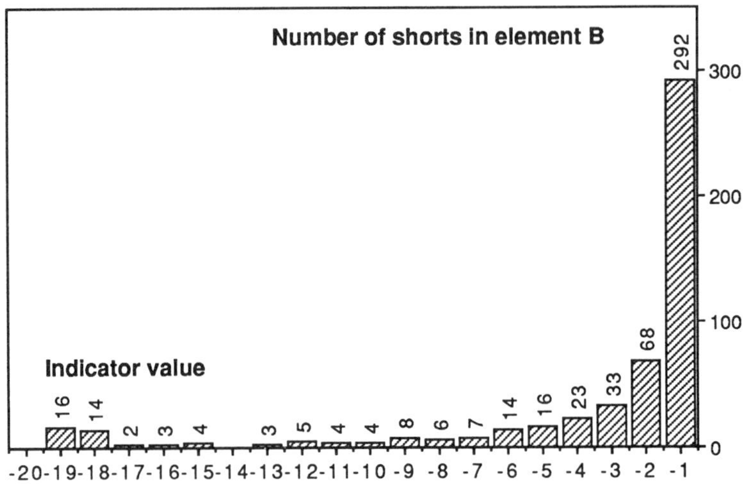

Figure 12: Distribution of the number of shorts. Element B.

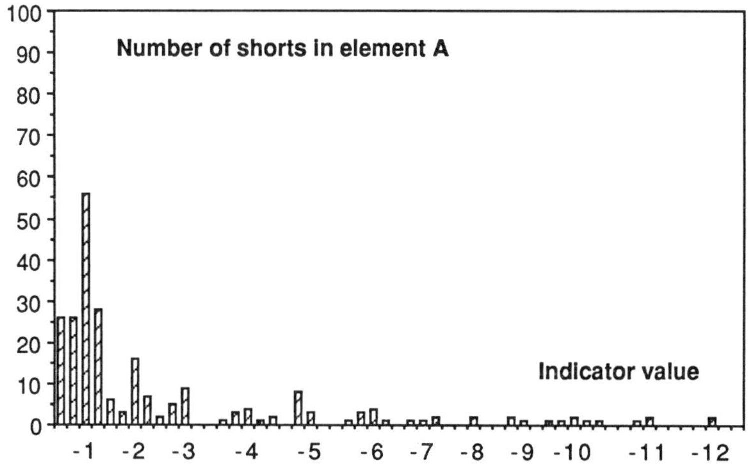

Figure 13: Histogram of the number of shorts. Element A.

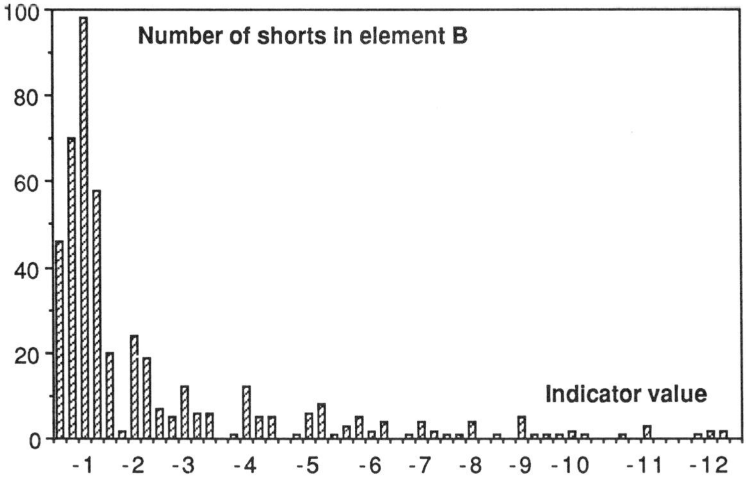

Figure 14: Histogram of the number of shorts. Element B.

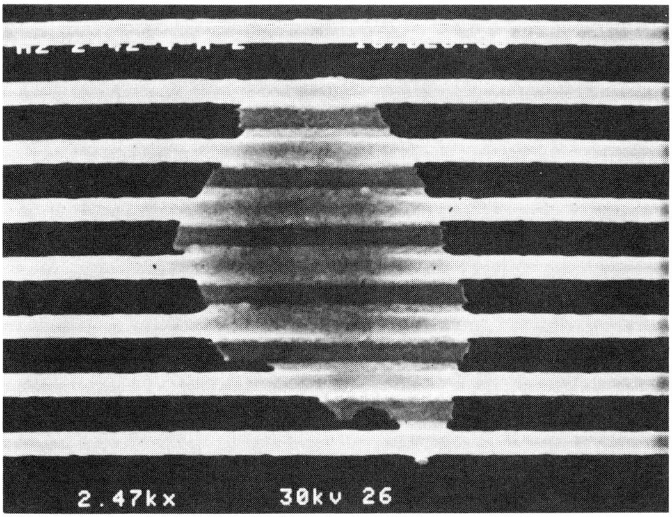

Figure 15: SEM photo of a size 6 short.

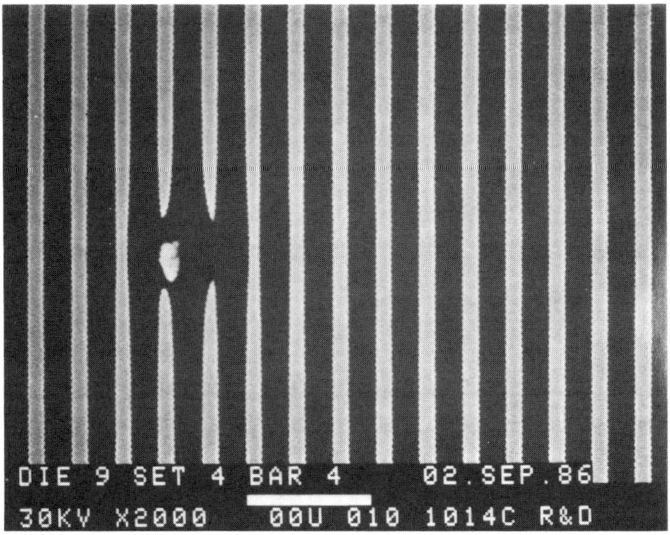

Figure 16: Photo of a size 2 break.

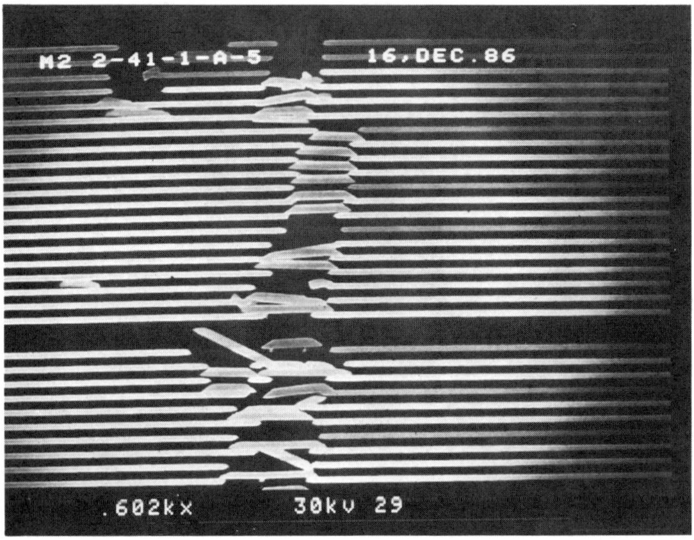

Figure 17: SEM photo of non-spot defects.

SUMMARY OF THE OBTAINED DATA AND CONCLUSIONS

In this paper it was demonstrated that the double bridge test structure can be used to electrically determine the type and size of the defects introduced during interconnect processing. The capability of the structure to do this has been verified by the optical inspection and identification of the defects.

Analysis of the obtained data suggests, however, that the resolution of the defect "1" type must be improved because there are a small fraction of defects producing indicator values around -0.5. Shorter structures and a smaller number of metal strips in a single device should therefore be used in the future design of the double-bridge test structure to insure a higher measurement resolution.

Overall evaluation of the double-bridge test structure seems to be very positive. Especially important is the fact that this structure is at least 20 to 40 times more efficient in area utilization than comb structures that are commonly used to classify defect sizes.

Initial analysis of the measurement data also suggests that:

- The dominant metallization loss mechanism observed is due to bridging between adjacent lines. Losses due to line breaks are small in comparison to bridging. This situation can change if metal line width is dramatically reduced while holding pitch constant. The largest number of defects were observed as single bridging events between 2 metal lines.

- Small changes in the line to space ratio for a fixed pitch has a large impact on the relative yield of the two defect structures examined. (As expected, the yield was observed to drop due to a greater number of bridging events when the space between metal lines was reduced.)

- Results from initial studies using the test structure yielded a defect frequency vs. size distribution in which the number of observed defects is inversely proportional to the square of the defect size.

- In the electrical analysis, a small fraction of the total failure distribution ($<10\%$) could not be defined due to gross defects on the wafers.

This leads to the conclusions that:

- Yield prediction cannot be performed without precise analysis of the layout-defect size relationship.

- There is no universal and unique function describing defect size distributions as is suggested in the literature. It appears that the distribution is totally controlled by the fabrication environment.

- Efficient defect size identification is a crucial step in yield the estimation process.

Therefore by employing the double-bridge test structure and methodology of its application described in this paper it should be possible to obtain an accurate defect size distribution for a specific fabrication area. These results can then be used for accurate yield prediction for VLSI circuits using appropriate models.

ACKNOWLEDGMENTS

This research was sponsored in part by grants from Fairchild Semiconductor Corporation and from Engineering Design Research Center, Carnegie Mellon University.

References

Buehler M G et al. 1983 Product Assurance Technology for Production Reliable, Custom LSI/VLSI Electronics, JPL Publication 83-70, Jet Propulsion Laboratory, California Institute of Technology, Pasadena

Ferris-Prabhu A V 1985 Modeling of Critical Area in Yield Forecasts, IEEE Journal of Solid-State Circuits SC-20(4) 874-878

Ferris-Prabhu A V 1985 Defect Size Variations and Their Effect on the Critical Area of VLSI Devices, IEEE Journal of Solid-State Circuits SC-20(4) 878-880

Maly W and Deszczka J 1983 Yield Estimation Model for VLSI Artwork Evaluation Electronics Letters 19(6) 226-227

Maly W 1985 Topics in Point Defect Related Yield Losses, Research Report CMUCAD-85-41, SRC-CMU Center for Computer-Aided Design, Department of Electrical and Computer Engineering, Carnegie-Mellon University

Maly W 1985 Modeling of Lithography Related Yield Losses for CAD of VLSI Circuits, IEEE Trans. on Computer-Aided Design of Integrated Circuits and Systems CAD-4(4) 166-177

Maly W et al. 1987 Double-Bridge Test Structure for the Evaluation of Type, Size and Density of Spot Defects, Research Report CMUCAD-87-2, SRC-CMU Center for Computer-Aided Design, Carnegie-Mellon University

Stapper C H 1983 Integrated Circuit Yield Statistics, Proc. of the IEEE 71(4) 453-470

3.2 The Effect of Scaling on the Yield of VLSI Circuits

Israel Koren

INTRODUCTION

As IC technology advances, the minimum feature size of VLSI circuitry continues to decrease. The smaller feature size of transistors has the potential of speeding up the circuits and increasing the yield of the manufactured chips. The latter is due to the smaller silicon area of the chip resulting in a lower average number of defects per chip. Manufacturers of existing VLSI chips also attempt to take advantage of the possible reduction in feature size by scaling (shrinking) the existing designs of their VLSI chips.

The effect of scaling the physical dimensions of VLSI circuits on their electrical characteristics (and consequently on their speed of operation) has already been studied (e.g., Saraswat and Mohammadi 1982). The effect of scaling on the yield has been studied until now only for special cases like interconnection buses (Stapper 1984, Koren *et al* 1987). The subject of this paper is the effect of scaling on the yield of more general VLSI circuits.

The analysis of the effect of scaling on yield is complicated by the variety of types of manufacturing defects that can occur in VLSI circuits. These include dielectric pinhole defects, photolithographic defects, junction leakage defects etc. The number of these defects clearly depends on the physical dimensions of the chip. However, the dependence of one type of defects on the physical dimensions does not necessarily equal that of another type. Thus, the various types of manufacturing defects have to be considered and the effect of scaling on each of them has to be analyzed separately. Only then can the total effect of scaling on yield be determined.

In this paper we consider several types of manufacturing defects and analyze the possible changes in their effect on the yield, when scaling of a VLSI circuit is performed.

YIELD ANALYSIS

Defects may occur in the different steps of the manufacturing process. Some of these defects may result in missing patterns or open circuits while other defects may result in extra patterns or short circuits. Since these defects happen in separate manufacturing steps, they may have different statistical characteristics and can usually be considered statistically independent (Stapper *et al* 1983). This leads to the following general expression for yield,

$$Y = \prod_{i=1}^{m} Y_i$$

where Y_i is the yield associated with step i and m is the number of manufacturing steps. It has been generally agreed upon that the distribution followed by most defects is a Poisson distribution whose parameter λ is not a constant but a random variable. This allows the modeling of defect clustering, a phenomenon that has been observed in practice. A convenient choice for the distribution function of λ is the Gamma function resulting in a Negative Binomial distribution (for the defects) for which the yield Y_i is given by,

$$Y_i = \left(1 + \frac{\overline{\lambda}_i}{\alpha_i}\right)^{-\alpha_i}$$

where $\overline{\lambda}_i$ is the average number of circuit faults occurring in step i, and α_i is the corresponding clustering parameter.

Note that $\overline{\lambda}_i$ is defined as the average number of circuit faults rather than manufacturing defects. The reason for this is that not all manufacturing defects result in faulty circuits which affect the yield of the chip. For example, a open-circuit type photolithographic defect occurring in a conductor can be harmless if its size is smaller than the width of the conductor (Ferris-Prabhu 1985, Stapper 1984). A dielectric pinhole defect will result in a short-circuit only if it occurs in the overlapping area between two conductors that cross each other and were produced at different photolithographic steps.

The average number of manufacturing defects is usually defined as $d_i A$ where d_i denotes the average number of defects per unit of area and A is the chip area. To calculate the average number of circuit faults we define a probability θ_i that a defect in step i will result in a faulty circuit and thus, $\overline{\lambda}_i = \theta_i d_i A$. The product $A\theta_i$ is also called the critical area for defects occurring in step i (Stapper 1984, Ferris-Prabhu 1985) and we denote it by $A_c^{(i)}$. The yield is therefore given by,

$$Y_i = \left(1 + \frac{d_i A_c^{(i)}}{\alpha_i}\right)^{-\alpha_i}$$

The probability that a defect will cause a circuit failure may be constant for one type of defects and may depend on the relative size of the defects (compared to the physical dimensions of VLSI patterns) for a different type of defects. Hence, to study the effect of scaling on the yield we analyze in what follows each of these two cases separately. We consider first dielectric pinhole defects for which the above probability is constant and then photolithographic defects for which this probability is a random variable.

Dielectric Pinhole Defects

The size of a dielectric pinhole defect is usually very small (compared to the size of VLSI patterns) and the probability that it will cause a circuit failure is simply the ratio between the area of the overlapping region to the total area A. If the physical dimensions of all patterns within the chip are scaled by the same factor a, i.e, *ideal scaling*, then the above probability remains unchanged.

Consequently, the effect of scaling on the critical area $A_c^{(i)} = A * \theta_i$ for pinhole defects is determined only by the change in the area A. Note that this simple dependence holds for any other manufacturing defects for which the probability that they will result in circuit faults is constant and does not change with scaling. For ideal scaling with $a < 1$, the area of the chip reduces to $a^2 A$ resulting in a similar reduction in the critical area $A_c^{(i)}$ which in turn may result in a higher yield Y_i.

The final effect of scaling on the yield Y_i depends also on any possible change in the clustering parameter α_i. The dependence of α_i on the area has been studied by Stapper (1986) and experimental data have shown that the clustering parameter can either decrease or increase when the chip area decreases. Any increase in the clustering parameter will increase the yield of the chip and vice versa. Combining this with the expected reduction in the critical area, we may conclude that if α_i increases with scaling ($a < 1$) or remains unchanged, the yield Y_i due to dielectric pinhole defects will increase. However, if α_i decreases with scaling, then the final effect of scaling on Y_i can be in either direction.

There is a different type of dielectric pinhole defects whose average number is proportional to the length of one conductor crossing over a second conductor. In this case,

$$Y_j = \left(1 + \frac{\theta_j d_j l}{\alpha_j}\right)^{-\alpha_j}$$

where d_j is the average number of defects per unit of length (instead of unit of area) and l is the length of the overlapping region. Again, the probability that such a defect will cause a circuit failure is constant and the final effect of scaling on the yield due to this type of defects is similar to the effect of scaling on the yield due to the previous type of "area" pinhole defects. The only difference is that l is reduced by the scaling factor a rather than by a^2.

Photolithographic Defects

The situation is different when photolithographic defects are considered. Here, the defects have a randomly distributed size which is comparable to that of VLSI patterns. Therefore, the probability that such a defect will cause a circuit failure depends on the exact shape of the pattern and its dimensions relatively to the size of the defect. Hence, when scaling a given pattern (with $a < 1$), less photolithographic defects can occur in the smaller area on one hand, but smaller defects may now cause a circuit failure on the other hand. The area A decreases to $a^2 A$ but the probability θ_i may increase. The final change in their product (i.e, the critical area) depends both on the specific given pattern and on the distribution of the defect size.

For convenience, we adopt the assumption made by Ferris-Prabhu (1985) and Stapper (1984) that a defect is circle shaped and we denote its diameter by x. Experimental data on defects in many wafers lead to the conclusion that the diameter x of a defect has a density function $f(x)$ which increases as x^q up to the mode x_o of the distribution (i.e., the value of x for which the density function is maximal) and then decreases as $1/x^p$ up to a maximum value of x_M, i.e.,

$$f(x) = \begin{cases} cx^q/x_o^{q+1} & \text{if } 0 \leq x \leq x_o \\ cx_o^{p-1}/x^p & \text{if } x_o \leq x \leq x_M \\ 0 & \text{if } x > x_M \end{cases}$$

where q and p are not necessarily integers but can be real numbers and the constant c for $p \neq 1$ is given by (Ferris-Prabhu 1985),

$$c = (q+1)(p-1)/[(q+p) - (q+1)(x_o/x_M)^{p-1}]$$

Data gathered through many experiments showed that $q \approx 1$ and $p \approx 3$ with "typical" values for the latter between 2.85 and 3.1 (Stapper 1984).

With defects of size comparable to that of many VLSI patterns, the probability that a defect will result in a circuit failure is a random variable rather than a constant. We define therefore, the critical area for defects of diameter x, as the area in which the center of a defect (of diameter x) must fall in order to cause a circuit failure. We denote this critical area by $A(x)$ and compute its expected value A_c using,

$$A_c = \int_o^\infty A(x)\, f(x)\, dx$$

Note that we omitted the subscript i to simplify the derived expressions.

Expressions for the critical area $A(x)$ have been derived by Stapper (1984) and Ferris-Prabhu (1985) for a number of long conducting lines on a chip. For these long lines the width w of a conductor and the spacing s between two adjacent conductors are very small compared to the length L of these conductors; a fact that simplifies the derivation of an expression for $A(x)$. In what follows we analyze two basic geometrical patterns that can be found in ordinary VLSI designs and derive expressions for their critical area. Most VLSI layouts contain these basic patterns and analysis of these allows us to estimate the effect of scaling on their critical area.

Consider first a conductor of length L and width w as shown in Figure 1. Assume that an open-circuit defect will cause the conductor to fail if at any point along the conductor, a width of δ (or less) of material is left. To simplify the expressions to be derived for $A(x)$ we set this minimum allowable conductor width δ to be zero. This simplification does not change the functional dependence of $A(x)$ on the physical dimensions of the conductor.

To derive an expression for the critical area $A(x)$ for open circuit defects (in the conductor shown in Figure 1) note that this area consists of a rectangular part and four circular-shaped parts. The rectangular part was shown (Stapper 1984 and Ferris-Prabhu 1985) to have an area of size $(x-w)L$. The area of each of the four circular

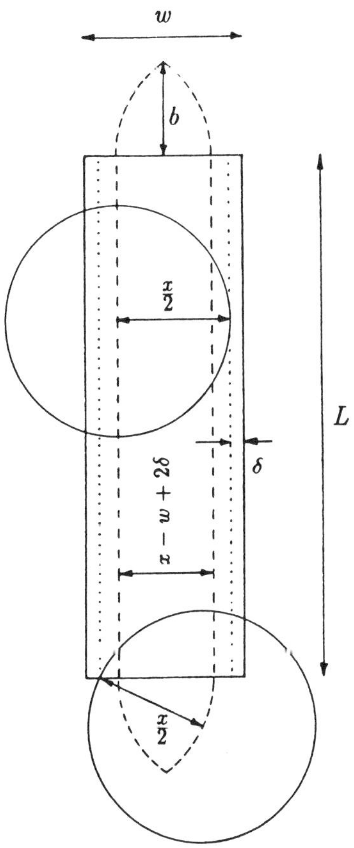

Figure 1: The critical area of a conductor of length L and width w.

parts is approximately equal to that of a triangle having a height of $(x - w)/2$ and a base of $b = \frac{1}{2}\sqrt{x^2 - w^2}$ (see Figure 1). The area of the triangle is,

$$\frac{1}{8}\,(x-w)\sqrt{x^2-w^2}$$

Finally, we obtain,

$$A(x) = \begin{cases} 0 & \text{if } x < w \\ (x-w)L + \frac{1}{4}(x-w)\sqrt{x^2-w^2} & \text{if } x \geq w \end{cases}$$

Notice that the critical area is a quadratic function of the defect size. Also note that for $L \gg w$, the quadratic term in $A(x)$ becomes negligible. Thus, for very long conductors we may use the linear term only. An analogous expression for $A(x)$ for short-circuit

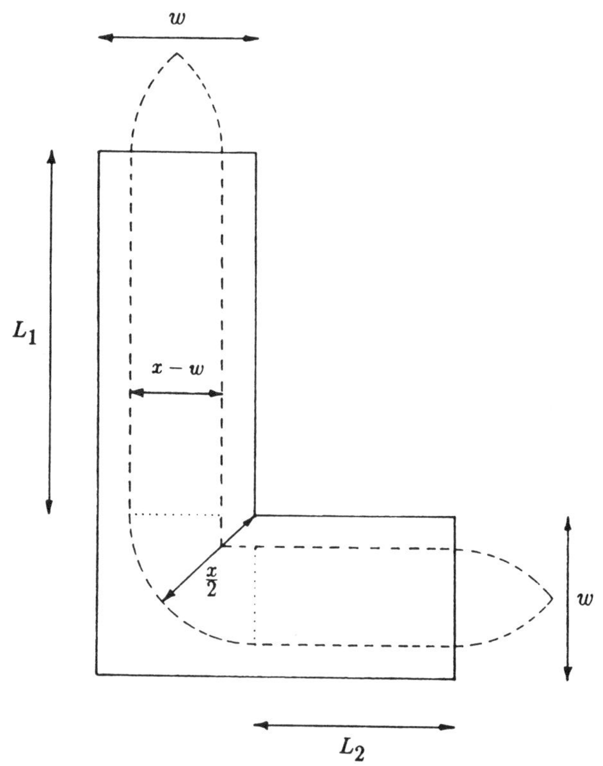

Figure 2: The critical area of an L shaped conductor.

defects in a rectangular area of width s (between two adjacent conductors) can be obtained by replacing w by s in the above equation.

Another very common pattern in VLSI layouts is an L shaped pattern, depicted in Figure 2. As is evident from this figure the shapes and sizes of most parts of the critical area are identical to those of the previous simpler pattern in Figure 1, except that the critical area at the corner of the L has the shape of a quarter of a circle of diameter x. The resulting approximation for the critical area is,

$$A(x) = \begin{cases} 0 & \text{if } x < w \\ (x-w)(L_1+L_2) + \frac{1}{2}(x-w)\sqrt{x^2-w^2} + \frac{1}{4}\pi x^2 - (\frac{x}{2}-w)^2 & \text{if } x \geq w \end{cases}$$

The expression for the critical area in this case is very similar to the one for the conductor in Figure 1 and again, if $(L_1 + L_2) \gg w$, the linear term in $A(x)$ is the dominant one.

Common VLSI layouts consist of several patterns of the shapes shown in Figures 1 and 2, in different sizes and orientations. Consequently, the exact expression for the critical area of one layout will be different from that of another layout. However, we may still analyze the possible effects of scaling on the critical area of an arbitrary layout by studying the effect of scaling on certain terms which may appear in the expression for the critical area.

Generalizing the results presented by Stapper (1984) we can conclude that the expression for the critical area of a layout will be a linear combination of the expressions for the critical areas of its components (i.e., the basic geometric patterns). Hence, we may expect to find in the critical area expression for any general VLSI layout, a linear term which will usually be the most dominant one. Analysis of this term can indicate the nature of changes in the critical area that will occur when scaling is performed.

Even the analysis of the linear term only is not trivial since the exact form of the corresponding expression for A_c depends on the relative size of the width w of the conductors (or the spacing s between two adjacent conductors if short-circuit defects are considered) and the mode x_o of the defect size distribution.

Let L denote some "equivalent length" of the general pattern (e.g., $L_1 + L_2$ for the pattern in Figure 2). The precise value of L is immaterial; as will be shown next we only have to know that it reduces to aL when ideal scaling is performed. The average value of the critical area due to a linear term of the form $(x - w)L$ is given by

$$A_c = \int_w^\infty (x - w)L \; f(x) \; dx$$

Because of the special form of the defect-size density function $f(x)$ we have theoretically, to analyze two separate cases, namely, $x_o < w$ and $x_o > w$. In practice however, it has been observed (Maly 1985) that x_o is close to the maximal resolution of the lithography process and therefore, only the case $x_o < w$ is of interest. For this case we obtain

$$A_c = \int_w^{x_M} (x - w)L \; \frac{c x_o^{p-1}}{x^p} \; dx$$

and for $p > 2$,

$$A_c = \frac{cL}{(p-1)(p-2)} \frac{x_o^{p-1}}{w^{p-2}} - \frac{cL}{(p-1)(p-2)} \frac{x_o^{p-1}}{x_M^{p-2}} \left[(p-1) - (p-2)\frac{w}{x_M}\right]$$

When scaling is performed, the resulting average critical area A_c' is calculated from the above expression by substituting L and w by aL and aw, respectively, yielding

$$A_c' = \frac{cL}{(p-1)(p-2)} \frac{x_o^{p-1}}{w^{p-2}} \frac{1}{a^{p-3}} - \frac{cLa}{(p-1)(p-2)} \frac{x_o^{p-1}}{x_M^{p-2}} \left[(p-1) - (p-2)\frac{w}{x_M}a\right]$$

We may now compute the ratio A_c'/A_c,

$$\frac{A_c'}{A_c} = \frac{\frac{1}{a^{p-3}} - (\frac{w}{x_M})^{p-2} a \left[(p-1) - (p-2)\frac{w}{x_M}a\right]}{1 - (\frac{w}{x_M})^{p-2}\left[(p-1) - (p-2)\frac{w}{x_M}\right]}$$

For $x_M \gg w$ and $p > 3$ the ratio A_c'/A_c increases approximately as $1/a^{p-3}$. The exact value of this ratio for three values of p, i.e., $p = 2.8, 3$ and 3.2, is depicted in Figure 3

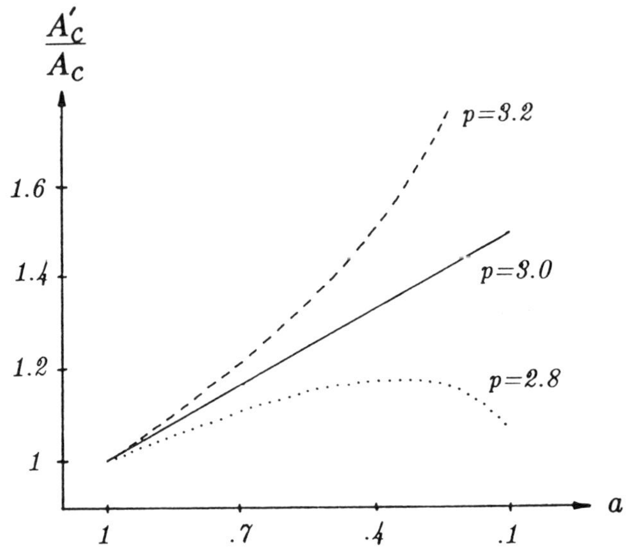

Figure 3: The critical area ratio in the case of $x_o < w$ for different values of p.

for $w/x_M = \frac{1}{5}$. Very similar curves were obtained for many other values of the ratio w/x_M. We may therefore conclude that scaling (with $a < 1$) results in an increase in the expected value of the critical area which in turn reduces the yield.

To finally determine the effect of scaling on the yield due to photolithographic defects, any possible changes in the clustering parameter α must be taken into account. For large area VLSI chips the clustering parameter will in most cases decrease when the area of the chip is being reduced. This will tend to lower the yield of the scaled chip. The expected increase in the average critical area A_c combined with the smaller clustering parameter, will result in a lower yield (due to photolithographic defects) for the scaled chip.

CONCLUSIONS

The exact effect of scaling on the yield of an arbitrary VLSI chip consisting of a variety of different geometrical patterns can not be predicted. However, we may still draw some important conclusions out of the analysis done in the previous sections. If the number of photolithographic defects is larger than the number of pinhole defects, the final effect of scaling on yield will mainly depend on the ratio A_c'/A_c for photolithographic defects. Consequently, as long as the scaled feature size is still larger than the mode x_o we may expect a reduction in yield. Note that this analysis does not take into account other factors, mostly fabrication related ones, which may influence the final yield of the scaled VLSI design.

ACKNOWLEDGEMENT

This work was supported in part by NSF under contract DCR–85–09423.

REFERENCES

Ferris-Prabhu A.V. 1985, Defect Size Variations and Their Effect on the Critical Area of VLSI Devices, IEEE Journal of Solid-state Circuits, *sc-20,* 878-880.

Koren I. and Pradhan D.K. 1987, Modeling the Effect of Redundancy on Yield and Performance of VLSI Systems, IEEE Trans. on Computers, *C-36,* 344-355.

Koren I., Koren Z. and Pradhan D.K. 1987, Designing Interconnection Buses in VLSI and WSI for Maximum Yield and Minimum Delay, submitted for publication.

Maly W. 1985, Modeling of Lithography Related Yield Losses for CAD of VLSI Circuits, IEEE Trans. on Computer-Aided Design, *CAD-4,* 166-177.

Saraswat K.C. and Mohammadi F. 1982, Effect of Scaling of Interconnections on the Time Delay of VLSI Circuits, IEEE Journal of Solid-state Circuits, *SC-17,* 275-280.

Stapper C.H., Armstrong F.M. and Saji K. 1983, Integrated Circuit Yield Statistics, Proc. of the IEEE, *71, 4,* 453-470.

Stapper C.H. 1984, Modeling of Defects in Integrated Circuits Photolithographic Patterns, IBM J. Res., Dev., *28, 4,* 461-475.

Stapper C.H. 1986, On Yield, Fault Distributions, and Clustering of Particles, IBM J. Res., Dev., *30, 3,* 326-338.

3.3 A Method for Determining Critical Areas and Critical Levels for IC Yield Estimation

Sophie Gandemer, Bernard Tremintin and Jean-Jacques Charlot

INTRODUCTION

Usually yield models are derived from mixed Poisson statistics as described by Murphy (1964). The major differences between models come from whether and how the defect density distribution takes into account defect clustering on wafers. The currently used model parameters (susceptible area, mean defect density and number of critical level steps in a technology) are fitted to the yield data without precise attention to design rules or defect density distribution with defect size. This paper describes a way to take into account the sensitivity of the layout of any circuit to the defect size in a given technology. We develop our analysis from the following remarks:

- It is possible to define a yield at every masking step; the global yield is the product of step yields.
- At every masking step, the yield may be calculated with areas representing a partition of the total area; each area is defined by the patterns included in it: their width is comprised between two limits.
- Small patterns are more sensitive to defects.
- Defect density depends on defect size.

These remarks are developped in the following sections of the paper. In the second section, we present the yield model that is going to be used; details on subsurface calculations at each level and critical coefficient calculations are given. Third section describes the experimental procedure we take for subsurface extraction and for defect counting at every level. In the fourth section the defect counting and subsurface extraction are used for calculation on three different circuits.

THEORETICAL

General Model Equations

Mixed Poisson statistics (Murphy 1964) are based on the fact that the defects are distributed according to some distribution function $f(\lambda)$, where λ is the defect density per unit area. The expression for the yield is given by the integral over the distribution:

$$Y = \int_0^\infty e^{-\lambda A} f(\lambda)\, d\lambda \quad \text{where A is the die susceptible area.} \tag{1}$$

Stapper (1973) assumed the gamma approximation for the defect density distribution:

$$f(\lambda) = \frac{\lambda^{a-1} e^{-\lambda/b}}{\Gamma(a)\, b^{a}} \tag{2}$$

where $\bar{\lambda}$ = ab is the mean value of defect density and var(λ) = ab² is the defect density variance. Therefore, (1) can be written as:

$$Y = \left[1 + \frac{A\, \mathrm{var}(\lambda)}{\bar{\lambda}} \right]^{-\bar{\lambda}^2/\mathrm{var}(\lambda)} \tag{3}$$

A large number of distribution function can be fitted by equation (2) using a and b parameters. As we choose a=1 and b=λ, the distribution function becomes:

$$f(\lambda) = \exp(-\lambda/\bar{\lambda})/\bar{\lambda} \tag{4}$$

which fits with the data measured on several levels on our wafers. In these conditions, the yield equation becomes:

$$Y = \frac{1}{(1 + A\, \bar{\lambda})} \tag{5}$$

Equation (5) takes into account the clustering effect we have on our wafers and the yield degradation due to random defects. Other defects as mask defects, scratches, misalignments... have to be taken into account through an other yield equation that will give an overall yield equation:

$$Y = \frac{Y_0}{(1 + A\, \bar{\lambda})} \tag{6}$$

where Y_0 is the parametric yield and will not be considered here. For each process masking step, equation (5) will be applied to specific areas (as defined later) of every mask.

Subsurface Calculations

Applying equation (5) to a specific level i will result in the following equation:

$$Y_i = \frac{1}{(1 + A\, \lambda_i)} \tag{7}$$

where Y_i is the yield at level i, λ_i is the defect density at that level, and A the susceptible chip area. The product A λ_i is the average number of defects at level i. In order to take into account the defect size distribution at any level, the mask surface is partitioned into areas which are sensitive to a minimal defect size: the subsurface a_{si} is defined as the area where all pattern geometries are included between two limits of size. These calculations have been made for pattern widths as well as pattern spacings. Every subsurface a_{si} will be sensitive to a given defect density λ_{si} which depends on the minimum pattern size. In these conditions, the average number of defects susceptible to be killer is given by:

$$A\ \lambda_i = \sum_{s=1}^{n} a_{si}\ \lambda_{si} \qquad (8) \qquad \text{with} \quad \sum_{s=1}^{n} a_{si} = A_T \qquad (9)$$

where A_T is the chip area and λ_{si} the real defect density determined by defect counting on wafers. The level i yield can be expressed as:

$$Y_i = \frac{1}{1 + \sum_{s=1}^{n} a_{si}\ \lambda_{si}} \tag{10}$$

Critical Levels Calculation

The defect size distribution usually found in the literature (Stapper 1973, Ferris-Prabhu 1985a, Ferris-Prabhu 1985b) appears on Figure 1. Figure 1a is the analytic curve representation while Figure 1b is the discretized one. The discretization ranges are defined to be the same as the subsurface ranges. Here the discretization is made with the variable called s.

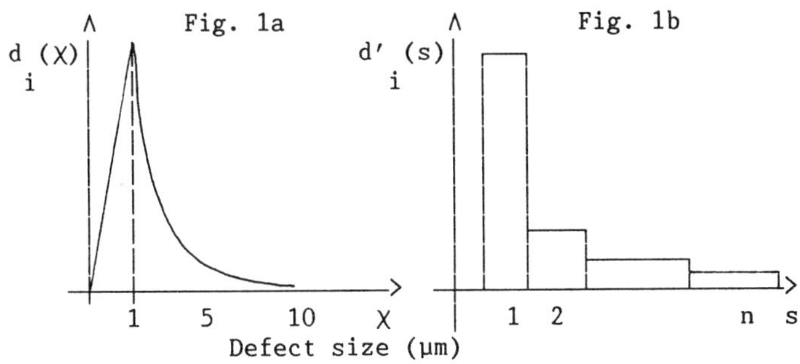

FIGURE 1 - DEFECT SIZE DISTRIBUTION
(a): Analytic representation (b): Discretized representation

On these figures, the mean defect density at one level is given by:

$$\lambda_i = \int_0^{\infty} d_i(X)\ dX = \sum_s d'_i(s) \tag{11}$$

On Figure 2 the discretization of subsurface is shown as an example.

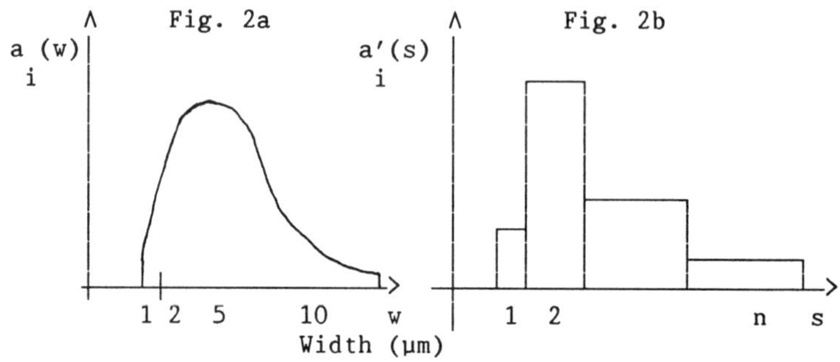

FIGURE 2 - MASK SURFACE DISTRIBUTION
(a): Analytic representation (b): Discretized representation

The die area is given by:

$$A_T = \int_0^{\infty} a_i(w)\, dw = \sum_s a'_i(s) \tag{12}$$

In order to calculate the λ_{si} defect density, we have to consider that a given subsurface will be sensitive to all defects the size of which starts from the minimum pattern dimension, which gives equation (13). Figure 3 is the λ_{si} representation.

$$\lambda si = \sum_s^n d'i(s) \tag{13}$$

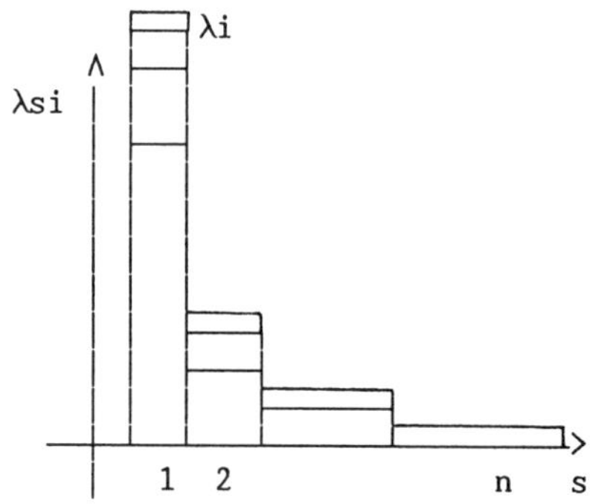

FIGURE 3 - DISTRIBUTION OF DEFECT DENSITY APPLICABLE TO EACH SUBSURFACE

The important thing to notice here is the fact that the mean defect density on a given level is only applied on a reduced part of the mask patterns. Critical coefficients are introduced in our equations as we want to be able to make calculations with the total die area A_T. Equations (7) and (10) give the yield per level. Using a critical coefficient α_i we express the level yield:

$$Y_i = \left(\frac{1}{1 + A_T \lambda_i} \right)^{\alpha_i} \tag{14}$$

where A_T is the total die area. Using equations (10) and (14) leads to the α_i equation:

$$\alpha_i = \ln(1 + \sum_{s=1}^{n} a_{si} \lambda_{si}) / \ln(1 + A_T \lambda_i) \tag{15}$$

In these conditions, it will be possible to calculate the die yield as the product of the yields per level:

$$Y = \prod_i Y_i = \prod_i \left(\frac{1}{1 + A_T \lambda_i} \right)^{\alpha_i} \tag{16}$$

Another simple way to calculate the critical coefficients is to use the same mean defect density distribution at each level:

$$Y_i = \left(\frac{1}{1 + A_T \lambda} \right)^{\alpha_i} = \frac{1}{1 + \sum_{s=1}^{n} a_{si} \lambda_s} \tag{17}$$

where λ is the mean defect density for a given process at each level and λ_s is the mean defect density value depending on defect size. Using equations (16) and (17) the die yield will be given by:

$$Y = \left(\frac{1}{1 + A_T \lambda} \right)^{\alpha} \tag{18}$$

$$\text{where } \alpha = \sum_i \alpha_i \tag{19}$$

We will show in section 3 that using equation (15) or (17) to calculate the α_i gives the same result since the α_i are slightly sensitive to the λ_{is} variations. So it will be possible to calculate the critical coefficients using equation (15) and use them in equation (16) or (18) for yield calculation.

EXPERIMENTAL PROCEDURE

Subsurface Calculation on Three Different Products.

Subsurfaces have been distributed within each level so that patterns inside the former have the same size limits. Distribution calculations have been made for pattern width as well as spacing between patterns. Total mask surface is then partitioned into areas which are sensitive to a minimal size defect. We have divided the surface into S1, S2, S3 and S4 subsurfaces so that:

S1 $0.0\mu m < \chi \leq 2.5\mu m$
S2 $2.5\mu m < \chi \leq 5.0\mu m$
S3 $5.0\mu m < \chi \leq 10.0\mu m$
S4 $10.0\mu m < \chi$

S2 is sensitive to defects greater than 2.5µm, S3 to 5µm and S4 to 10µm. We chose these limits to maintain a homogeneous defect number for each subsurface because of the $1/X^3$ distribution of defect density. We use a design rule checker (DRC) program to partition each level mask surface into more or less critical surfaces. This program uses geometric mask information recorded on a CALMA tape. We use UNDERSIZE and OVERSIZE functions and such logical program functions as OR, AND, XOR and NOT. A 1µm UNDERSIZE makes a 1µm side reduction on all masking level figures; thus if a tape is less than 2µm wide it disappears. Then a 1µm OVERSIZE causes a 1µm increase on all figure sides. All lines less than 2µm wide have disappeared. By subtracting from the starting figure we are left with a surface of lines less than 2µm wide (Fig.4).

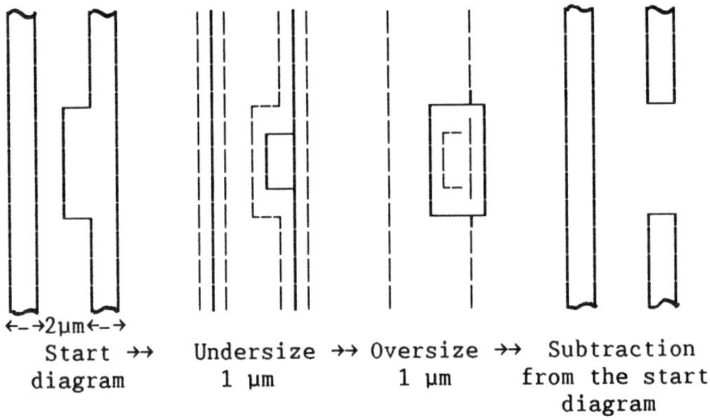

FIGURE 4 - METHOD FOR DETERMINING REGION WIDTH OF LESS THAN 2µm

We begin with an OVERSIZE to carry out the same operation on spacings so that the patterns will be in contact if the spacings are too narrow. Next, using an UNDERSIZE, the narrow spacings are transformed into full lines. It is thus possible to partition the total surface of each level in order to compare them. The DRC program has been used to determine surfaces on three different circuits. The circuits were fabricated using the following CMOS technologies:

1. 2kx8 Memory (2.5µm technology, double poly, single metal)
2. 16kx1 Memory (1.6µm technology, double poly, single metal)
3. Microprocessor 80C51 (1.6µm technology, single poly, single metal)

Defect Counting

We have counted defects after each photolithography plus etch step and taken data from the three products and technologies from two different manufacturing lines as mentioned before. Defect counting was done using dark field illumination on a microscope. Defects were counted on five dies (one in the center and four on a half wafer radius circle) for two wafers per lot, which gives 10 dies per level on each lot.

EXPERIMENTAL RESULTS AND DISCUSSIONS

Subsurface Calculation on Three Different Products.

The subsurface calculations have been made for the 2Kx8, 16Kx1 and 80C51 circuits. Table 1 gives the results obtained on the 2Kx8 memory. These results are given as a percentage of the total die area.

TABLE 1 - DIVISION OF 2Kx8 MEMORY MASKS INTO FOUR SUBSURFACES

	S4	S3	S2	S1
P- IMP	99.79 %	0.19 %	0.02 %	0 %
NIT	48.75 %	35.59 %	11.23 %	4.43 %
AGN	99.98 %	0.02 %	0 %	0 %
POLY1	42.62 %	28.74 %	23.09 %	5.54 %
N+ IMP	96.96 %	2.86 %	0.17 %	0 %
P+ IMP	96.96 %	2.86 %	0.17 %	0 %
B.APS	94.71 %	2.84 %	2.45 %	0 %
POLY2	78.31 %	14.35 %	5.31 %	2.03 %
CONT	95.57 %	1.76 %	0.76 %	1.91 %
METAL	23.81 %	56.90 %	6.51 %	12.77 %
PADS	100 %	0 %	0 %	0 %

We can immediately notice which levels are going to be the most critical ones: they have the weakest percentage area in subsurface S4 (Metal, Poly and Nitride masks).

Defect Counting Results Analysis

The available literature gives the general trend of defect-density/defect-size evolution. Stapper (1973) gives the following law:

$$\chi < \chi_0 \implies dd = k\ \chi^p$$
$$\chi > \chi_0 \implies dd = k\ 1/\chi^n$$

where χ is the defect-size, dd the defect density and k a multiplicative coefficient. The distribution has a maximum and discontinuous slope at defect size χ_0. For convenience, we choose p=1, n=3 (Ferris-Prabhu 1985a and 1985b, Fig.5).

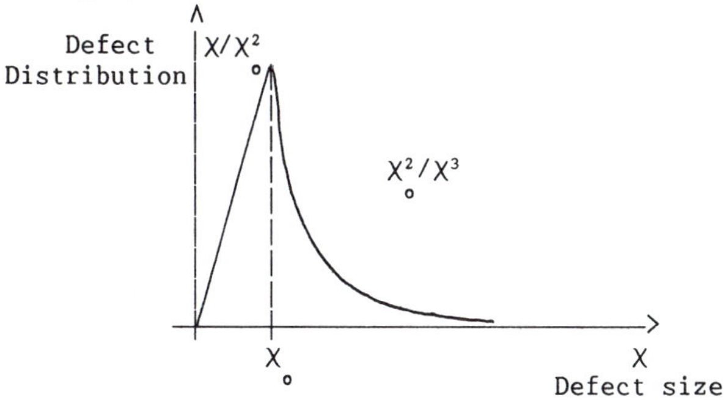

FIGURE 5 - DEFECT DISTRIBUTION VERSUS DEFECT SIZE

Table 2 presents the defect size classification for each level of the 2Kx8 memory process.

TABLE 2 - SIZE CLASSIFICATION OF LEVEL DEFECTS NOTED ON 2Kx8 MEMORY CHIPS

	1 <<2.5μm	2.5<<5μm	5 <<10μm	10<<20μm
P- IMPLANT	83	20	4	4
NITRIDE	40	4	1	3
AGN	38	4	0	1
POLY1	67	16	4	2
N+ IMPLANT	82	26	5	5
P+ IMPLANT	84	30	2	11
BURIED APS	79	8	3	2
POLY2	171	35	8	15
CONTACT	80	22	3	1
METAL	63	17	2	3
PADS	119	25	3	9
AVERAGE	82	19	3	5

Chi-Square analysis done on the data indicates that defect-number data versus defect-size are homogeneous for each level. The resultant distribution is the average for each level with a 10% margin of error. We could obtain the defect density per unit size/defect size curve, density being the average on all 11 levels. The integral of the curve determines the defect density per unit area. We found that n=3.07 for defects smaller than 10μm, using a log/log expression to determine the regression coefficient, which verifies the above mentioned trend. By using the previous subsurface division and the previously verified $1/X^3$ law to determine the greater than 2μm defect density, we obtain the average defect-density values reported in Table 3.

TABLE 3 - DEFECT DENSITIES APPLICABLE TO SUBSURFACES

Defects greater than	Defect Density
2μm	3.64 cm^{-2}
2.5μm	2.54 cm^{-2}
5μm	0.75 cm^{-2}
10μm	0.48 cm^{-2}

TABLE 4 - LEVEL YIELD AND CRITICAL COEFFICIENTS FOR 2Kx8 MEMORY

LEVEL	Y	α
P- IMP	0.907	0.176
NIT	0.832	0.332
AGN	0.907	0.176
POLY1	0.796	0.412
N+ IMP	0.905	0.180
P+ IMP	0.905	0.180
B.APS	0.897	0.196
POLY2	0.872	0.247
CONT	0.894	0.202
METAL	0.801	0.401
PADS	0.907	0.176
	ΠY=22.7%	Σα=2.68

Critical Coefficient Calculation.

We calculated level-i yields, critical coefficients α_i, total calculated yield Y, and sum of critical coefficients α for each product. In Table 4 we present the results obtained with the 2Kx8 memory. We have seen above that because we use step probability for killer defects too many such defects are counted. This is the reason that we present defect density results divided by 2, 3 and 5. Table 5 shows results obtained with the 2kx8 and 16kx1 Memories and Microprocessor for different densities of killer defects λ:

-λ = λa, λa/2, λa/3 or λa/5, where λa is the average density,
-the most critical level name is written,
-α is the sum of critical level coefficients,
-Y is the calculated yield (product of all level yields).

There are 11 technological levels for 2Kx8 and 16Kx1 memories, and 9 for microprocessor.

TABLE 5 - CRITICAL COEFFICIENTS VERSUS PRODUCT TYPE AND VERSUS DEFECT DENSITY

		2Kx8 MEMORY	16Kx1 MEMORY	MICROPROCESSOR
		METAL-POLY1	METAL	METAL
λa	α	2.68	3.1	2.36
	Y	22.7%	32.3%	19.9%
		METAL-POLY1	METAL	METAL
λa/2	α	2.45	2.94	2.14
	Y	46.2%	55.8%	42.6%
		METAL-POLY1	METAL	METAL
λa/3	α	2.38	2.85	2.05
	Y	59.4%	67.4%	55.9%
		METAL-POLY1	METAL	METAL
λa/5	α	2.29	2.86	1.98
	Y	73.0%	78.6%	70.1%

We note two interesting points:

1.- The metal level is the most critical for the three products (for the 2Kx8 memory the POLY1 is equivalent).
2.- The number of critical levels for a product is nearly constant (decreasing slightly) when the considered density is divided by 2, 3 or 5.

The first point confirms that the critical levels are defined first by the relative area of surface patterned on the chip and second by the pattern size. That is why contact level (which is considered critical for processing because of contact size) appears not very critical compared to metal level in our analysis: the relative area occupied by contacts is very small.

The second point shows that the critical coefficients determination are quasi independent of defect-density, as we will see next.

Coefficients Dependance on Defect Density and Defect Size.

It is interesting to predict model responses to modifications in distribution from a given situation where defect density $\lambda(\chi)$ is fixed. That is why we studied the model sensitivity to a variation in λ and χ. In both cases the calculation of critical coefficients is poorly dependant on the defect density value, on the defect size variation. For classic values, variations in critical coefficients are never more than 10%.

CONCLUSION

We have described a method for calculating critical coefficients to be included in a I.C. yield model for each masking step. The yield model is a function of the chip surface, of the defect density and of a mathematical law. The method reposes on the partitionning for each level of total mask area into subsurfaces sensitive to a minimal defect size using a Design Rule Checker Program. Defect density with respect to size was experimentally determined and the selected mathematical law corresponded to our data. Yield per level Y_i, was determined by applying the model to critical subsurfaces with their corresponding defect densities. The calculation of α_i was carried out using the equation $Y_i = 1/(1+A_T\lambda)^{\alpha_i}$, where Y_i is the i-level yield, A_T the total chip area and λ the average value of defect density per level. We found that the resultant α_i values were stable in a given environment whatever the technology, the average defect density value and the design rules. In conclusion, this method permits one to construct a predicting yield model for a given circuit in a given production line.

REFERENCES

FERRIS-PRABHU A.V. 1985a Role of Defect Size Distribution in Yield Modeling, IEEE Trans. Electron Devices 32 1727-1736

FERRIS-PRABHU A.V. 1985b Modeling the Critical Area in Yield Forecasts, IEEE Solid State Circuits 20 874-878

MURPHY B.T. 1964 Cost-Size Optima of Monolithic Integrated Circuits,Proc. IEEE 52 1537-1545

STAPPER C.H. 1973 Defect Density Distribution for LSI Yield Calculations, IEEE Trans. Electron Devices 20 655-657

STAPPER C.H. 1983 Modeling of Integrated Circuit Defect Sensitivities, IBM J. Res. Develop. 27 no 6 549-557

STAPPER C.H. 1984 Modeling of Defects in Integrated Circuit Photolithographic Patterns, IBM J. Res. Develop. 28 no 4 461-475

3.4 Random Yield Simulation Applied to Physical Circuit Design

Michel Rivier

Introduction

Yield simulation can be traced back almost to the origins of the IC industry (Murphy 64, Lawson 66, Dennard 66, Seeds 67). Refinements of early yield models based on Poisson statistics evolved into the formulation of the generalized negative binomial model which fits a large body of experimental data (Stapper 76, Stapper et al. 83, Stapper 85). With a sound model, the next step in yield prediction was to understand the relationship between defects and circuit layout. This was done by several authors who have looked at spatial distribution of defects (Stapper 83, 84, Ferris-Prabhu 85, Maly 85) and their impact on layout. A number of computer programs have been written based on this framework in IBM (Carter 84, Wine 86) as well as outside (Walker and Director 86). Today, manufacturing yields on bipolar and CMOS products of 1.25 μm design rules can be predicted from a scaling of test site data.

For circuit designers, the knowledge of the average number of fauts falling on a chip is important. Even more important is to indicate on the layout areas most sensitive to faults. With this information, the designer will be able to correct, possibly interactively, a design for yield enhancement. In this paper, we present a method to compute random defect yield of a given layout, to select one or several levels most susceptible to faults, and display their fault sensitivity, before a corrective action can take place. We expect that such an approach will bridge the gap between process and design and help designers to assess tradeoffs between performance and manufacturability.

Yield calculation

According to Stapper (Stapper et al. 83), the manufacturing yield can be decomposed into a non random term Yo, called gross yield, and a random term, Yr.

$$Y = Yo \, . \, Yr \qquad [1]$$

The gross yield takes into account all non random components of yield (mask, tooling, test,...) as well as statistical process variations inducing parametric rejects. It varies little with time with a controlled manufacturing process.

Random defects contribute to more than 80% of the yield loss in existing stable manufacturing lines. Since they are uncontrollable by nature, it is important to minimize the impact of random defects on layouts. A first class of random defects is due to random inhomogeneities in dielectrics, also known as dielectric pinholes. A second class of random defects originate from particulates present in clean room air and in fluids (Hayakawa et al 87) deposited on the wafer during processing. The size of particulates is of the order of the minimum design dimension and their distribution is modelled as a 1/x3 law (Stapper 76). They impact photolithography in creating extra or missing material in the processed layer.

If we assume a non redundant scheme, the random yield is defined as the probability that a chip will be fault free. Its value depends on the statistical law used for the defect distribution. For a negative binomial law, we can express the effects of individual failure modes on the random yield using the equation:

$$Yr = \Pi \, (1 + \lambda i / \alpha i) \, \text{-} \alpha i \qquad [2]$$

where λi is the average number of faults for the failure mode i and αi a clustering factor. The product is done on all failure modes characteristic to the technology, 26 types for a 14 mask, 2 metal level CMOS technology. The average number of faults can be further decomposed into:

$$\lambda i = pi \, . \, A \, . \, Di \qquad [3]$$

where A is the chip area, pi the average (averaged over the defect distribution and the layout) probability that a defect will create a fault of type i, and Di the average defect density. One alternate way of writing the same expression is given by:

$$\lambda i = Ai \, . \, Di \qquad [4]$$

Ai being the average critical area of the layout for the fault i. Ai is a representation of the global sensitivity of a design to the fault i.

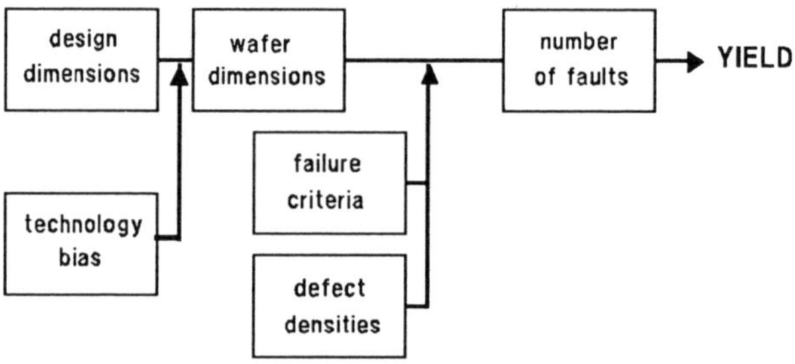

Figure 1: Random yield calculation technique

The flow diagram of the random yield program is given in Figure 1. A designer inputs his layout as a combination of shapes placed on different mask levels. The design dimensions are converted into nominal wafer dimensions using a technology bias file, specific to the technology considered. Two separate methods of random chip yield calculation can be used, the Monte-Carlo or the critical area approach, depending whether we use expressions [3] or [4] to obtain λ.
In the Monte-Carlo method, also known as dot-throwing approach, a level containing defect shapes randomly spread on the chip with the appropriate size distribution is created. Defects causing faults are screened and counted with shape checking programs using predefined failure criteria. One estimate of the probability pi is obtained and the relative number of faults is computed by [3]. Multiplying this value by the defect density Di gives the absolute number of faults from which the chip yield is computed.
In the critical area approach, the original layout shapes are manipulated in order to give directly Ai, the average critical area. This is done with subroutines of the shape checking program such as expansions, shrinks, unions and intersections. As defined in the next paragraph, critical areas are approximated by shapes having simple geometrical relationships with the starting layout. The average number of faults λi and the random chip yield are computed with equations [4] and [2].

We have made sure to be technology independent in keeping the technology file separate from the design file. The introduction of a new technology can be seen as an addition of a plug-in module containing data on design-wafer bias, failure types and corresponding defect densities.

Fault criteria and critical dimensions

Fault criteria are used to establish a link between physical defects and a functional chip failure. They are key elements in yield monitoring and predictions. They are process, layout and architecture dependent. In random logic applications, each circuit is unique and must behave properly for the chip to be functional. Any fault can be considered as catastrophic. For memory applications as well as for fault tolerant architectures, where redundancy is used, some faults are repairable, other are catastrophic. Redundant architectures can be extended from our study, using more complex fault criteria and statistics. For simplicity, we will consider in the remainder of this paper the random logic case, where each fault causes a functional failure.

As stated before, 20 to 30 failure criteria must be considered for existing CMOS or bipolar technologies. Fortunately, they can be usually split into the following generic failure modes (Stapper et al 83, Stapper 84, Walker 86), depending whether or not the defect has a size dependent probability of fail.

The first type of failure criterion is found for photodefects having a size distribution with a mean of the order of design dimensions. The probability of fail pi depends on the position of the defect with respect to the layout. This occurs in conducting levels fails and in contact fails.

- An open occurs when missing metal (or poly or diffusion) cuts totally or partially a conducting line. A short is created by a defect bridging two distinct conducting lines. The critical area (Lawson 66, Stapper et al, Ferris-Prabhu 85) is defined as the area in which the center of a defect must fall to create a fault. For a short circuit, the critical area for a defect of radius r is simply the union of the intersection of the layout shapes expanded by r. A representation of short and open faults and their respective critical areas is shown in Figure 2 a, b.

- As soon as extra or missing conducting material affects the gate or the active area, transistor characteristics are altered creating a fault. All defects touching these levels in the active area are considered as fails.

- contact fails occur when a given percentage of the contact or of the conducting layer above or below the contact is missing.

The second type of failure happens for defects much smaller than design dimensions such as dielectric pinholes. The probability of fail is one for defects located in the area where two conducting lines face each other and zero everywhere else. The critical area is equal to the area of intersection of the conducting layers. Dielectric pinholes and their critical area is represented in Figure 2c.

The third type of failure includes all other modes. Process specific fails, such as opens created at metal steps in a given technology can be considered as proportional to the step width and their number (Figure 2d). Electrical fails, such as diffusion leakage in CMOS technologies and pipes in bipolar technologies are defined as proportional to the perimeter of the diffusion or to the emitter area.

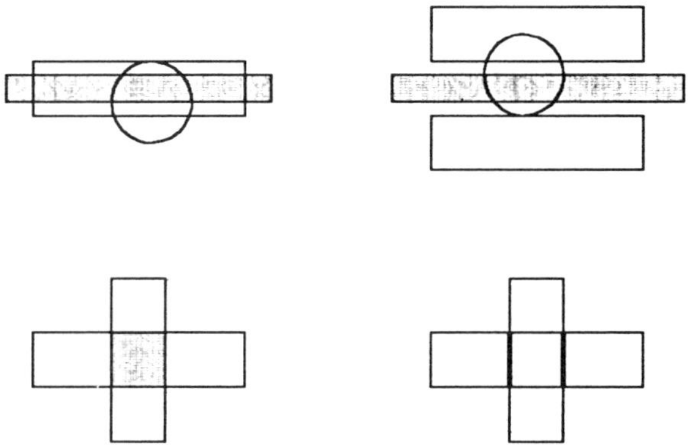

Figure 2 : Examples of faults and their respective critical area for open (a), short (b), dielectric pinhole (c) and metal step critical length (d).

Yield optimization

To increase yield, one must decrease the average number of faults per chip λ. Using [3], we see that we can accomplish it twoways:

- The first way is to decrease D, the average defect density. It can be done by decreasing the sources of contamination, through improved filtration of fluids, major process changes and improved photolithography. This is what is typically done when one changes from one technology to the next, with improved design rules.

- An alternate method of achieving the same result is to improve designs within a given technology and manufacturing environment such as to make layouts less sensitive to random defects. The optimization is based on the minimization of the product pi . A in expression [3] which is design dependent.

The method used in our work to optimize designs for yield is shown on Figure 3. We have seen that the result of the yield calculation is a list of number of faults impacting the layout. The designer can select one or several faults and display their relative physical distribution on the layout. This is what we call a map of fault sensitivities which is illustrated in the following paragraph. The designer can then locate areas most sensitive to faults and perform appropriate layout modifications. Knowing for instance that the fault type is open metal line, the designer will have the option to widen the metal lines where needed. A second pass on the modified layout will be done to compute the yield improvement and check that other fault types have not been impacted.

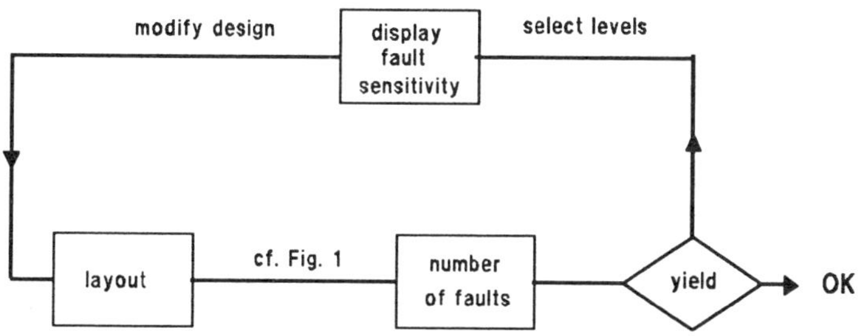

Figure 3 : Yield optimization flowchart

The reduction of total number of faults as provided by our program is important in the early stages of product life. It may mean early prototype availability and market capture. In the mature phase of the technology, when the yield (or defect density) is stabilized in time, optimized layouts will have a significantly higher yield than the unoptimized product, therefore offering a cost advantage over the competition.

More importantly the optimization will translate into improved manufacturability of the chip. The designer will try as much as possible to reduce the total number of faults as well as spikes in fault densities during the optimization cycle. A more uniform probability of fail over a layout will make a product less sensitive to drifts in manufactured process parameters.

Finally, the designer will gain freedom in his designs. A program will be available to him to optimize yield, where he could only count on his past experience and rely on rough checks such as chip area and number of shapes at minimum design rules. With this automatic approach, the designer will be able to test different designs, modify design rules and select quickly the best layout according to the tradeoffs which have to be made. We foresee an extensive usage of such an approach for development of large designs, where automatic tools are required or for the design of repetitive features such as memory cells or logic libraries, where the optimization of the unit cell is the key to the chip manufacturability.

Display of fault sensitivity

In this section, we will provide examples of the previously described program. We illustrate the sensitivity to shorts on the layout of a first metal level used for local wiring of transistors (Figure 5). We chose this example since it is the most complex in terms of shapes of a CMOS layout and its probability of fail cannot be obtained easily analytically. Yet, the failure criterion is simple enough so that the results can be easily checked for gross errors. The same approach has been used successfully to treat other failure criteria as well as other geometries.

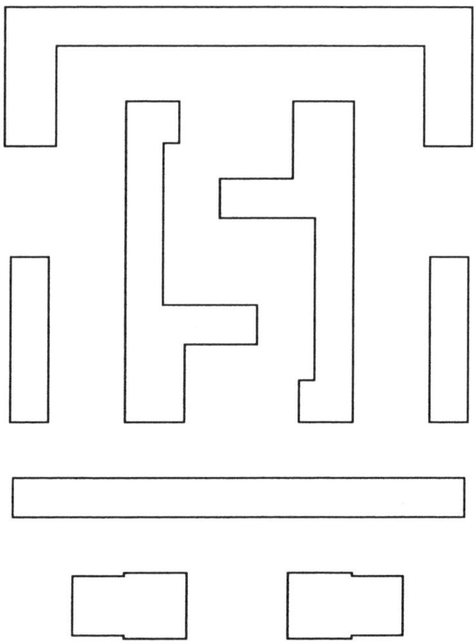

Figure 4 : Layout of a metal level

We first report results obtained by the Monte-Carlo approach. A square grid mesh is superimposed to the layout. The number of faults intersecting each unit grid area is counted, each fault being weighted by a factor proportional to the inverse of its area. Faults sensitivity is obtained by displaying classes of fault counts within each unit grid area. The precision of the method is dependent on the size of the population of defects as well as the grid area. Empirically, we have found that the mesh size must be smaller than the minimum radius of a defect creating a fault. If this criterion is met, a population of defects with densities in the range 10 to 100 defects per square microns must be generated to obtain good sensitivity maps for the CMOS designs we have looked at. A short sensitivity map for the example we have chosen is shown on Figure 5. Fault isosensitive areas are displayed in black, dark grey, light grey and white, the darkest colors corresponding to areas highly sensitive to faults. Minor density fluctuations are observed due to the population size. Ignoring the influence of other fault types on the same level or on other mask levels, the designer will increase the distance between the shapes where the highest density of shorts occur. On an actual design, a combination of faults per mask level (a minimum of open and short) or a combination of levels may be displayed simultaneously in order to modify a design and anticipate its impact on neighboring areas. This is where we expect our program to be useful for effective layout optimization.

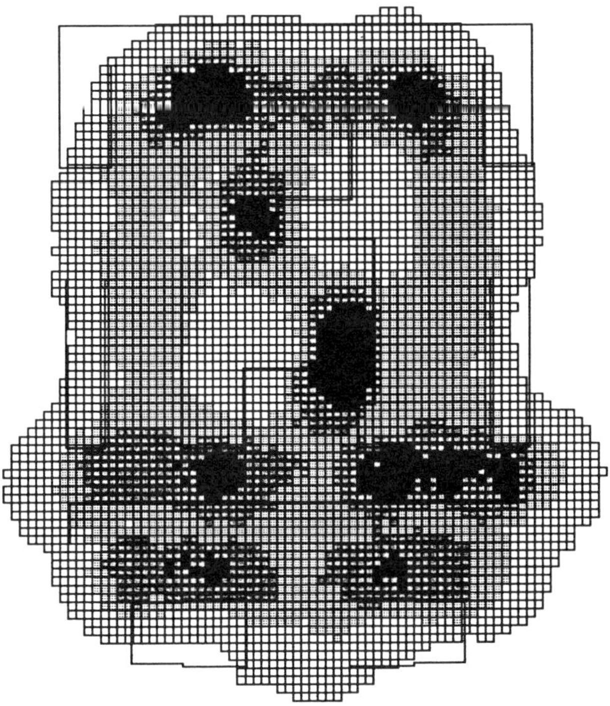

Figure 5 : Visualization of increasing probabilities of shorts on the layout of figure 4 defined by the Monte-Carlo method.

The second approach is to compute directly the critical area associated to the layout for the fault type considered. The method of computation of Ai for a defect of constant radius has been described in a previous paragraph. In order to display values of decreasing probabilities of short, we have visualized critical areas corresponding to increasing defect radiuses (Rivier and Guedj 87). Due to the simple relationship between critical area and starting layout, the CPU time will be much less than in the Monte-Carlo method. The display of short sensitivity obtained by the critical area method is shown on Figure 6. The qualitative information contained in this representation is identical to Figure 5, areas sensitive to shorts and gradients of sensitivity are identical. Furthermore, the average number of faults computed from the two methods fall within a few percents of each other. This is why we have used the critical area method for interactive displays or large areas of layout calculations.

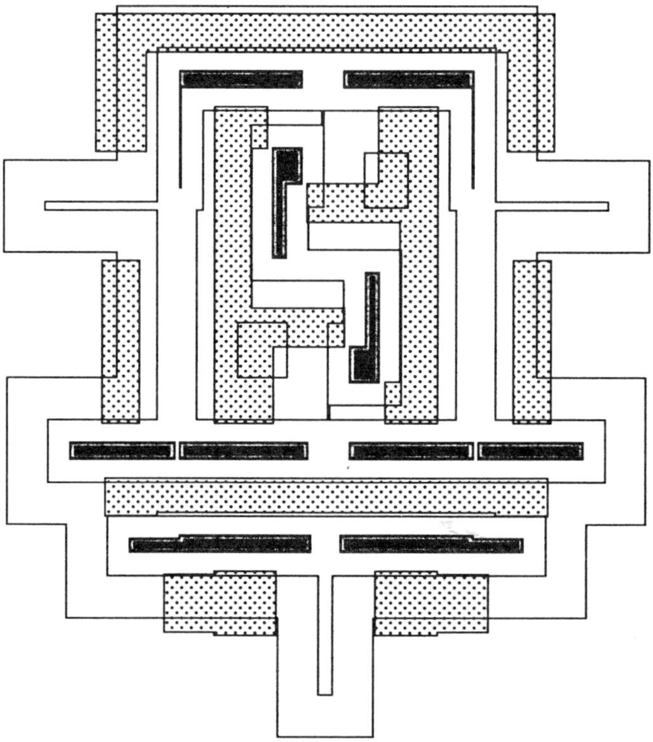

Figure 6 : Visualization of increasing probabilities of shorts on the layout of figure 4 defined by the critical area method.

Conclusions

We have described the several approaches which are used to compute the random yield of a layout. A Monte Carlo approach, CPU extensive is appropriate for small areas where a good precision is required. A critical area approach, more imprecise can be used for larger areas or whenever interactivity is required.

We have extended this calculation to provide to the designer a display of the faults sensitivity of a layout. Single fault, a combination of faults per masking level or a combination of levels may be displayed in order to modify and optimize a design for yield. This tool is aimed at an improvement of the manufacturability of future products through the usage of flexible design rules in automated design systems. Its display features will help the interaction between design, process development and manufacturing, a necessary ingredient for the future ULSI era.

References

Carter, J., private communication, 1984

Dennard, R.A, Reasearch Report **RC 1552**, IBM T.J. Watson research Center, Yorktown Heigths, N.Y, 1966

Ferris-Prabhu, A.V., IEEE Trans. Electron. Devices **ED-85**, 1985, pp. 1727-1736

Lawson, T.R., Solid State Technology **7**, 1966, pp. 22-25

Hayakawa, I., Fujii, S., Kim, K.Y., Proc. 32nd. Annual Technical Meetings, Institute of Environmental Sciences, Mt. Prospect, Ill., 1986, pp. 483-486

Maly, W., IEEE Trans. Computer-Aided Design **CAD-4**, 1985, pp. 166-177

Maly, W., Strojwas, A.J., S.W. Director, IEEE Trans. Computer-Aided Design **CAD-5**, 1986, pp. 114-130

Murphy, B.T., Proc. IEEE **52**, 1964, p. 1537

Rivier, M., Guedj, D., to be published, 1987

Seeds, R.B., IEEE Int. Conv. Rec. **6**, 1967, pp. 60-61

Stapper, C.H., IBM J. Res. Develop. **29**, 1985, pp. 87-97

Stapper, C.H., IBM J. Res. Develop. **28**, 1984, pp. 461-474

Stapper, C.H., IBM J. Res. Develop. **27**, 1983, pp. 549-557

Stapper, C.H., Armstrong, F.M., Saji, K., Proc. IEEE **71**, 1983, pp. 453-470

Stapper, C.H., IBM J. Res. Develop. **20**, 1976, pp. 228-234

Walker, H., S.W. Director, IEEE Trans. Computer-Aided Design **CAD-5**, 1986, pp. 541-556

Wine, D., private communication, 1986

4 PARAMETRIC YIELD LOSS

While the previous chapter was devoted to catastrophic yield prediction for design optimization purposes, this chapter focuses on the parametric yield loss issues. The three sections in the chapter address the manufacture of VLSI circuits more than their design.

In the first section, Walton (§4.1) describes parametric test structures and their applications to process characterization and diagnosis. He presents a brief overview of the placement of test structures covering such approaches as drop-in and scribe line structures. He then proceeds to describe several basic types of structure used to extract such parameters as resistivity, contact resistance and misalignment. He also includes examples of structures which are used to evaluate random yield.

MacConaill and Walton (§4.2) present an introductory overview of the Computer-Aided Manufacturing area. The bulk of the section is devoted to the motivation for CAM systems in IC manufacturing as well as a history of CAM system developments. The benefits of employing computerized manufacture are outlined in the context of tighter process control and yield enhancements. The section concludes with a discussion of totally automated manufacturing facilities.

In Section 4.3, Mozumder and Strojwas present a powerful statistical process control system which can be installed within existing CAM systems. The system contains a number of modules which can be applied to process optimization, redesign, diagnosis, feed forward control and statistical quality control. The latter module is described in detail both in terms of its theory as well as its implementation. Criteria for wafer/lot acceptance/rejection are developed and demonstrated in a computational example. The modelling software which is a key ingredient in this CAM system is described as well.

4.1 Parametric Test as a Tool for Increasing Engineering Yield

Anthony Walton

INTRODUCTION

The process control chip (PCC) has been used by the semiconductor industry since the inception of integrated circuits but their role, from this date, has changed from that of just characterising the process. They are now also used to predict yield, for reliability evaluations, as well as using them as a criteria for accepting or rejecting wafers which have been supplied by external vendors. This increasingly important role is reflected by the more sophisticated structures being employed and as a result, the design of the PCC is no longer treated as an after thought when a new process is being developed.

HISTORICAL DEVELOPMENT OF THE PCC

The process control chip in the production environment is typically inserted in four or five positions on the wafer replacing the product circuits as shown in figure 1.

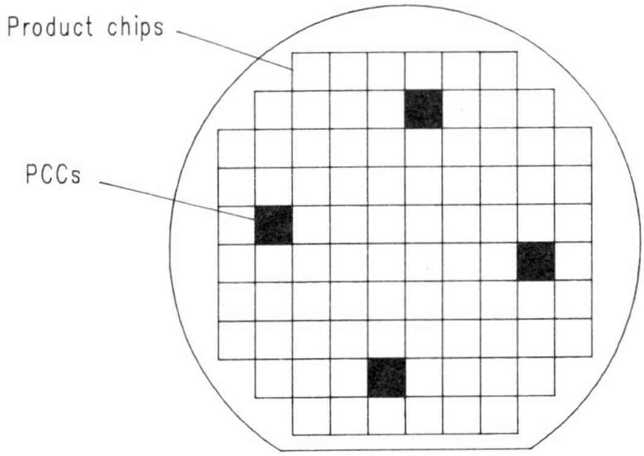

Figure 1. Process control chips as drop-ins on a product wafer.

In the early sixties, the minimum dimensions were very large by todays standards with the number of structures on the PCC typically in the region of 10 to 20 (Barone, Myers 1968, Penny, Lau 1972). These layouts were not specifically designed to be probed automatically since the manufacturer only used them to spot check that the process was under control. The size of the test chip was obviously constrained to that of the product circuit (2-3 mm^2) but this size steadily increased as the complexity of integrated circuits grew. Consequently more comprehensive PCCs were designed to evaluate the quality of processing and enough information could be extracted from these larger chips to conclude whether the wafer would contain good functional circuits. As a result, a standard could be established to enable meaningful comparisons to be made between batches which is not otherwise possible for those comprising of different products.

The extra area, which became available as chip size increased, introduced another factor which had to be considered. A larger number of test structures has a corresponding increase in probe pads which, if they are to be measured in a single pass, requires equipment with a very large switching matrix. Many manufacturers have faced this problem by only testing a subset of the PCC in a routine manner. Another approach has been to use a 2 x n probe pad array (Buehler 1979) as illustrated in figure 2 and step within the chip using an automatic prober.

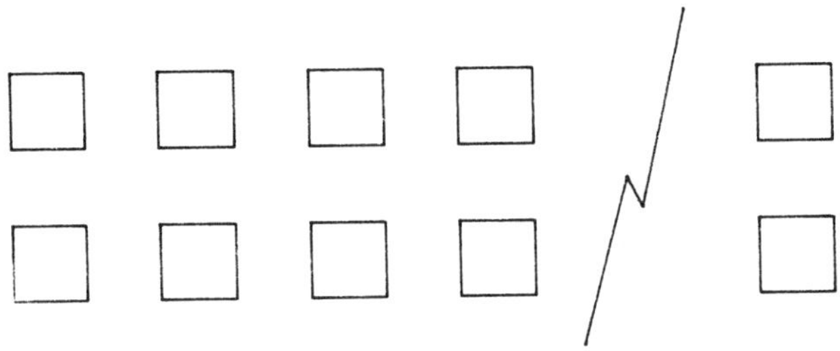

Figure 2. The 2 x n probe pad array.

The development of wafer stepper technology has largely led to the demise of the drop-in for small geometry processes and is due to the reduced throughput which results from the required mask changing. This problem has been addressed by placing the test structures in the scribe channel which removes the necessity for mask changing (Alcorn et al 1985). While this approach reduces the number of structures, it does enable parameters to be wafer mapped and this capability has become more important with the larger wafers now being employed.

Much of the development of test structures was pioneered at NBS and JPL and the ideas which have been developed are now routinely used to accept or reject wafers produced by silicon foundries. It is now widely recognised that the test chip has a very important role for ASICs where often only very small batches of wafers will be produced. In both these cases the number of designs on a wafer could be very large and the only manner in which to verify the process is the PCCs. The PCC can now certainly be considered to have come of age with a multi-million pound industry producing automatic parametric testers and semiconductor companies investing large amounts of money on equipment as a result of these measurements. This importance is further reflected by the development of a regular IEEE International Conference on Test Structures which has supplanted the special sessions at Conferences and reflects the increasing importance of the subject.

TEST STRUCTURES

In reference (Buehler 1983) it is suggested that test structures can be used for the following purposes:

(i) Device parameter extraction
(ii) Design rule checking
(iii) Process parameters extraction

(iv) Random fault analysis
(v) Reliability analysis
(vi) Circuit parameters extraction

Test structure can often be used to fulfill more than one of these roles and the following description of a selection of test structures is by no means comprehensive but gives a flavour of developments over the past 25 years.

Resistivity

The resistivity of conducting layers used for fabrication can be measured using a resistive strip with a known length and width. It is then a simple matter to derive the resistivity but the result is dependent upon the lithography and etching which defines the width. This degree of uncertainty can be eliminated by using a Greek cross (Buehler 1978a) as shown in figure 3.

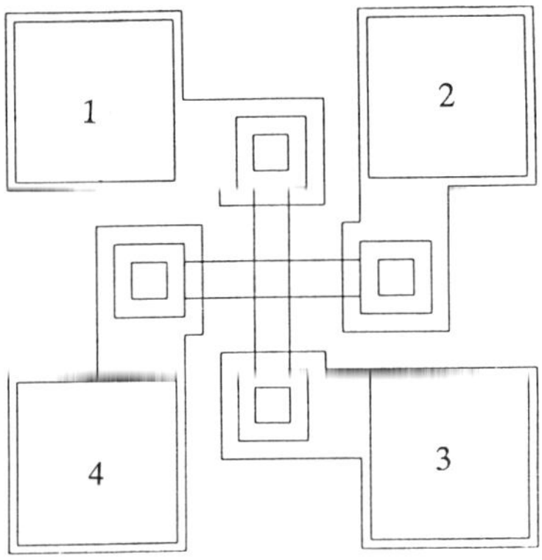

Figure 3 The Greek cross.

The theory behind this structure is due to Van der Pauw (1958) and the measured resistivity is that at the heart of the cross. This is simply obtained by forcing a current between terminals 1 and 2 and measuring the voltage between 3 and 4 (see figure 3). The resistivity is given by:

$$R_s = \frac{\pi}{\ln 2} \frac{V}{I} \quad \Omega/\blacksquare \tag{1}$$

The accuracy of the measurement can be improved by using the following procedure. First the direction of the current is reversed to obtain:

$$R_{0^o} = \frac{V_{12} - V_{21}}{I_{12} - I_{21}} \tag{2}$$

Current is then forced between pads 2 and 3

$$R_{90^o} = \frac{V_{14} - V_{41}}{I_{23} - I_{32}} \tag{3}$$

and the sheet resistivity calculated using:

$$R_S = 4.53 \left[\frac{R_{0^o} + R_{90^o}}{2} \right] \quad \Omega/\blacksquare \tag{4}$$

The length of the arms of the Greek cross must be equal or greater than their width to ensure that the structure is a good approximation to Van der Pauw's theory (Vernsel 1979). Provided this is the case the error is <0.01%.

Contact Resistance

Contact resistance is an important parameter which should be monitored to ensure the quality of the process. As dimensions have reduced its contribution to interconnect resistance has increased and research and development programs designed to reduce its value obviously rely on these types of structures. One of the simplest structures is that proposed by Berger (1969). This is shown in figure 4 and the contact resistance, R_c, is calculated as follows:

$$R_s = \frac{W\,(R_1 - R_2)}{L_1 - L_2} \tag{5}$$

$$R_c = \frac{(R_2L_1 - R_1L_2)}{2(L_1 - L_2)} \tag{6}$$

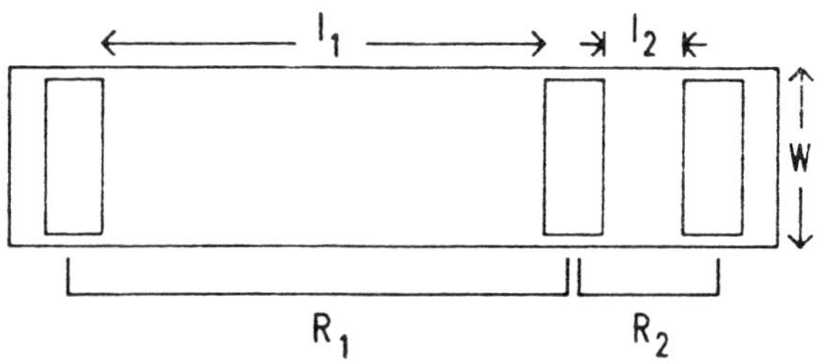

Figure 4 Structure proposed by Berger to measure contact resistance.

For equation (6) to be valid the width of the contact should be > 5μm (Berger 1969) but this is a rather large dimension for present day geometries. Another disadvantage of this structure is that the measurement of R_c depends upon the subtraction of two numbers which may have values which are of a very similar magnitude.

One design which overcomes these problems is the Kelvin structure shown in figure 5 (Proctor et al 1983).

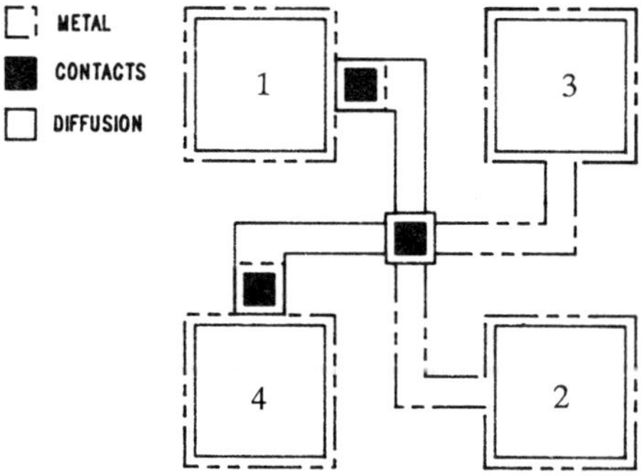

Figure 5 Kelvin structure for measuring contact resistance.

The resistance is obtained by forcing current between pads 1 and 2 and voltage between terminals 3 and 4. The direction of the current is reversed and the voltage once again noted. The terminals are then interchanged and the procedure repeated. The interfacial contact resistance is then given by the average of the four resistance values.

Misalignment

The registration between layers is of great interest both from the process control point of view and when developing new processes. Figure 6 shows a structure which can be used for electrically measuring misalignment between, for example, polysilicon and contacts (Russell et al 1977).

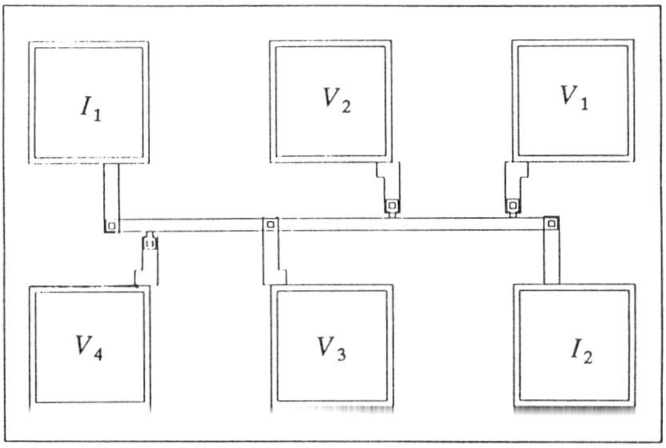

Figure 6 Structure for measuring misalignment between conducting layers and contacts

Current is forced between terminals I_1 and I_2. The voltage difference between the taps will be equal for no misalignment.

$$(V_4 - V_3) = (V_3 - V_2) = (V_2 - V_1) \tag{7}$$

If the contacts are misaligned then the voltage difference ΔV between the taps is

$$\Delta V = (V_1 - V_2) - (V_2 - V_3) \tag{8}$$

$$\Delta V = (V_1 - V_2) - (V_3 - V_4) \tag{9}$$

The sign of ΔV indicates the direction of misalignment and the magnitude is given by

$$e = \frac{\Delta V\ S}{V_4 - V_3} \tag{10}$$

To obtain the total misalignment, two of these structures at 90° to each other must be used. This design can be used to measure the registration error between contacts and any conducting layer or between two conducting layers if oversized contact windows are used.

One structure which can be used to measure misalignment between polysilicon and diffusion is shown in figure 7. Current is forced between pads I_1 and I_2 and the voltage (V_1 - V_2) and (V_3 - V_4) measured. The difference between the the resistance of these two lengths of diffusion is a measure of the misalignment.

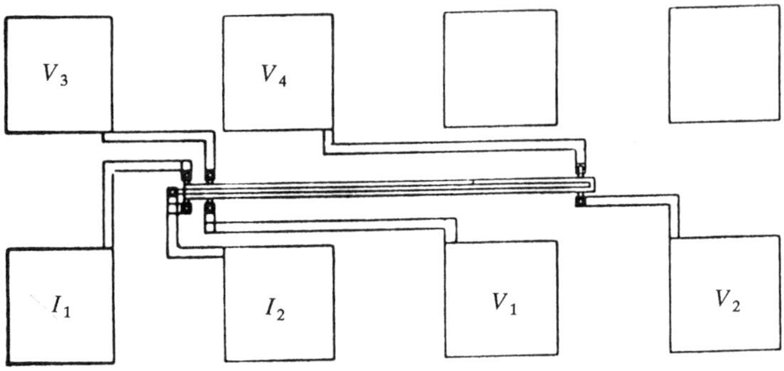

Figure 7 Structure for measuring misalignment between polysilicon and diffusion.

Linewidth

The structure shown in figure 8 can be used for measuring linewidth of conducting layers (Buehler 1978b).

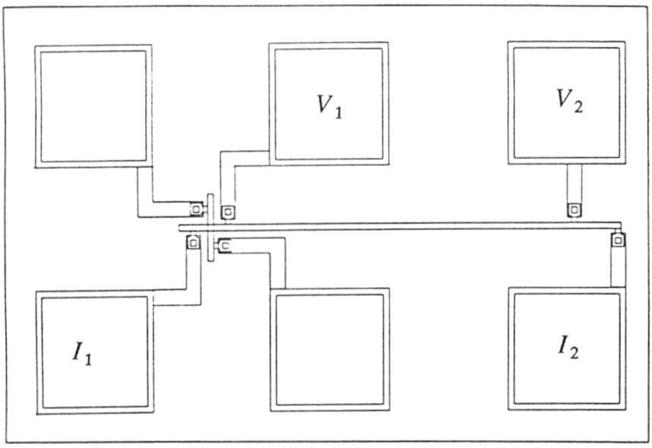

Figure 8 Structure for electrically measuring the linewidth of conducting layers

The sheet resistivity of the layer is measured using the Greek cross and then current is forced between terminals I_1 and I_2. The voltage (V_2 - V_1) is measured and the linewidth calculated using

$$W = \frac{L\ I\ R_s}{V} \qquad (11)$$

This design is sensitive to changes in linewidth of ± 0.1 μm provided that the distance between the voltage taps is very much greater than than their width.

Structures to evaluate yield

One of the problems of designing structures to evaluate yield is to ensure that they only fail with faults or defects they are designed to detect. For example, contact chains such as that shown in figure 9, can be used to examine the minimum sized contact for a given process (Mitchell Linholm 1981). However, the contact chain may fail due to interconnect problems

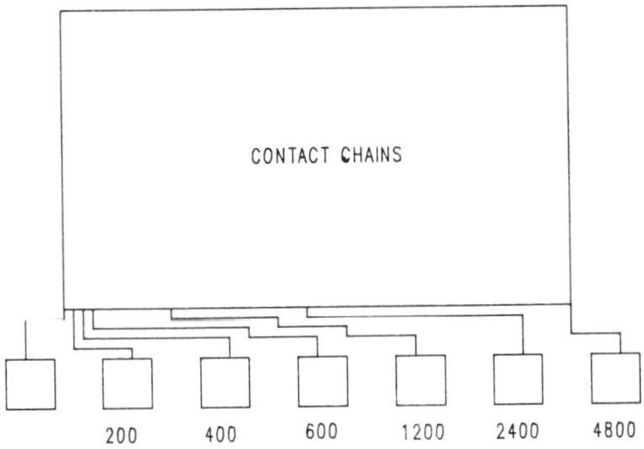

Figure 9 Schematic representation of a tapped contact chain.

or a random defect resulting from dirty photoresist or airborne particulates. As a consequence visual inspection is required to ascertain whether the fault is a result of over ambitious design rules or random defects.

The quality of step coverage and etching over uneven surfaces are important parameters to evaluate (Alcorn 1984) and figure 10 shows some possible serpentine and comb structures which are in widespread use.

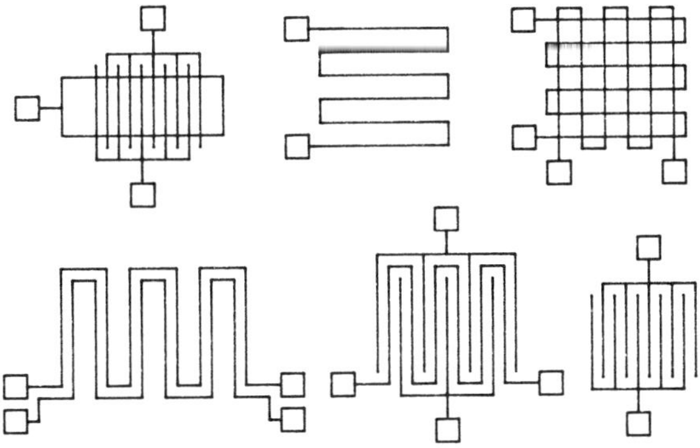

Figure 10 Some structures which can be used to examine the quality of processing.

These have become more important with the introduction of multilevel interconnect systems where metallisation is subjected to more severe requirements. This type of design may also be used for electromigration reliability studies and can be extended to examine etching such as bridging, fillets and leakage currents, all of which can be very simply tested.

Large area MOS capacitors can be used to examine oxide integrity (Buehler et al 1983). It should be noted that these devices may also be used for breakdown voltage studies and, in common with step coverage structures, the link between yield and reliability can be observed.

Any designs used to help evaluate both the nature of processing faults and their location must occupy a large area and consist of an array of addressable structures. The obvious candidate is to use memory which occupies a large area, and each cell is individually addressable. This enables both clustered and non-clustered faults to be identified providing that the addressing circuitry is functional.

One structure has had some success at both detecting faults and identifying their origin consisted of an array of 100 MOSFETs where the gate is connected to the drain (Linholm 1981). This layout is illustrated in figure 11 and the test results for 26,760 transistors located in the central portion of the chip are shown in table 1.

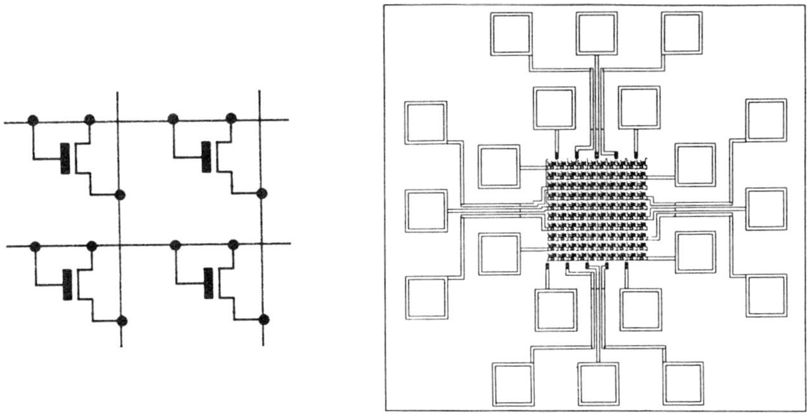

Figure 11 Array of 100 MOSFETS.

Number	Faults			Number of faults	
				Nonclustered	Clustered
1	Polycrystalline void/break			4	0
2	Epitaxial void			0	1
3	Metal void/break			0	0
4	Metal bridge			1	0
5	Gate short			4	0
	†V_T	I_L	V_B		
6	-	-	L	5	0
7	-	H	-	25	0
8	-	H	L	62	0
9	L	-	-	1	1
10	L	H	-	0	1
11	L	H	L	0	1
12	H	-	-	2	2
13	H	H	L	0	1

†V_T = threshold voltage; I_L = leakage current; V_B = breakdown voltage.
H = parameter too high, L = parameter too low.

Table 1 MOSFET array test result indicating causes of failure.

The disadvantages with this design is the area taken up by the probe pads and the long time to test it. Both the number of elements and the speed of testing can be improved upon by using decoders to address the individual transistors which also increases the area tested per chip (Walton et al). This approach has the disadvantage that the decoder must be operating before any meaningful measurement can be made.

TRANSISTOR MEASUREMENT

Obviously the engineering yield of the process is very largely determined by the performance of the transistors. SPICE parameters are widely used for circuit simulation and can be related back to the process parameters. This is an extremely important role of the PCC and there are a number programs for extracting these parameters. Examples are SIMPAR (Maes et al 1985), TECAP (Khalily et al 1985), MOSFIT (Silburt 1985), TOPEX (Bloom 1986) and PARAMEX (Gribben et al 1986).

DATA MEASUREMENT, DISPLAY AND STORAGE

The quantity of data which can be measured from PCCs is huge, especially if parameters are required to be wafer mapped. This poses three problems.

(i) Measurement time

(ii) Data display and examination

(iii) Storage of data for correlation with processing data.

The measurement time is largely determined by the measuring equipment and the switching time of the relays. Manufacturers of parametric testers are trying to speed up their operation but there are obviously physical limits which are fast being approached. The analysis of data poses severe problems simply because of its large volume. Wafer mapping can be used to examine uniformity in the form of shaded maps, contour and vector plots. Other methods of displaying the data are trend graphs and histograms which are in widespread use for process control. Data storage has until relatively recently been very expensive. This cost has reduced but not to the point where careful consideration about the data to be stored is not worthwhile. A consequence of storing large amounts of data is the increased time required to extract it from the database. As a result it is normal to store raw data for short periods and to later condense it for longer term storage.

FUTURE DEVELOPMENTS

Recent trends in fabrication technology have resulted in process control structures being located in the scribe channel with PCCs now mainly used for the development of smaller geometry processes. At these smaller dimensions linewidth and misalignment measurements require higher resolution measuring instruments. Digital structures have been proposed which overcome these problems (Henderson et al 1983). This direction has been taken further by putting the switching matrix on chip for yield type measurements (Walton et al 1985) and also SPICE parameter measurement of transistors (Ward et al 1987). This type of approach is now being used commercially for process control by Texas Instruments.

With the trend to ASICS and foundry production the requirement to determine wafer quality both in terms of its engineering yield, reliability and functional yield is now very evident. The determination of engineering yield is relatively simple with SPICE parameters and a few other measurements such as those previously discussed being sufficient. There is still a requirement for structures which can be used to predict yield and are also able to electrically identify the spatial distribution of faults and their causes. This is an area in which is now being widely addressed, both from the point of view of conventional ICs and wafer scale integration. It is to be hoped that this activity will lead to a better understanding of the inherent problems and will result in higher quality circuits with increased yield.

REFERENCES

C.N. Alcorn 1984 VLSI Multilevel Wiring Monitor, IEEE VLSI Multilevel Interconnect Conference, New Orleans, 252-258.

C.N. Alcorn, D. Dworak, N. Haddad, W Henley, P. Nixon 1985 Kerf Test Structure Designs for Process and Device Characterization, Solid State Technology, 229-235.

F.J. Barone, C.F. Myers 1968 Getting Beneath the Surface of Multilayer Integrated Circuits, Electronics, 41, 84-88.

H.H. Berger 1969 Contact Resistance on Diffused Resistors, IEEE International Solid State Circuits Conference, Penns. USA, pp 160-1614.

A. Bloom 1986 IP "C." 1986 Product Seminar, TMA, Palo Alto

M.G. Buehler 1978a An Experimental Study of Various Cross Sheet Resistor Test Structures, Journ. Electrochem. Soc., 145, 645-650.

M.G. Buehler 1978b Bridge and Van der Pauw Sheet Resistors for Characterizing the Line Width of Conducting Layers, Journ. Electrochem. Soc., 145, 650-654.

M.G. Buehler 1979 Comprehensive Test Patterns With Modular Test Structure: The 2 x N Probe Pad Array Approach", Solid State Technology, 74-89

M.G. Buehler 1983 Microelectronic Test Chips for VLSI Electronics in VLSI Electronics Microstructure Science Vol 6, ed N G Einspruch, G B Larrabee, Academic Press

M.G. Buehler, B.R. Blues, C.A. Pina, T.W. Griswold 1983 Pinhole Array Capacitor for Oxide Integrity Analysis, Solid State Technology, 26, 131-137.

A. Gribben, J.M. Robertson, A.J. Walton 1986 Accurate Physical Parameter Extraction for Small Geometry Devices, SEMICON, Birmingham, 186-202.

B.M.M. Henderson, A.M. Gundlach, A.J. Walton 1983 Integrated Circuit Test Structure which Uses a Vernier to Electrically Measure Mask Misalignment, Electronics Letters, 19, 868-869.

E. Khalily, P. Decher, A. Darrell 1985 TECAP 2 : An Interactive Device Characterisation and Model Development System, Hewlett-Packard Technical Data.

L.W. Linholm 1981 Semiconductor Measurement Technology: The Design, Testing and Analysis of a Comprehensive Test Pattern for Measuring CMOS/SOS Process Performance and Control, NBS Special Publication 400-66.

W. Maes, K. De Meyer, L.Dupas 1985 SIMPAR: A Parameter Extraction Program to be used for any User-Defined Analytical Expression in the Field of Process and Device Modelling", 15th European Solid State Research Conference, ESSDERC 85, Aachen, 153-154.

M.A. Mitchell, L.W. Linholm 1981 Test Pattern NBS-28 and NBS-28A: Random Fault Interconnect Step Coverage and Other Structures, NBS Special Publication 400-65

W.M. Penny, L. Lau (eds) 1972 MOS Integrated Circuits, Van Nostrand-Reinhold, New York.

S.J. Procter, L.W. Linholm, J.A. Mazer 1983 Direct Measurement of Interfacial Contact Resistance, and Interfacial Contact Layer Uniformity, IEEE Trans ED, ED-30, 1535-1542.

T.J. Russell, T.F. Leedy, R.L. Mattis 1977 A comparison of Electrical and Visual Alignment Test Structures for Evaluating Alignment in Integrated Circuit Manufacture, IEDM Technical Digest, 7A-7F

A.L. Silburt 1985 Automated Parameter Extraction and Modelling of the MOSFET Below Threshold, ESSDERC '85, Aachen, Germany, 160-161.

L.J. Van der Pauw 1958 A Method of Measuring Specific Resistivity and Hall Effects of Discs with Arbitrary Shape, Phillips Res. Rep., 13, 1-9.

W. Vernsel 1979 Analysis of the Greek Cross, A Van der Pauw Structure with Finite Contacts", Solid State Electronics, 22, 911-914.

A.J. Walton, J.M. Robertson, R. Holwill, B. Moore 1985 On Chip Switching for dc Parametric Testing, Electronics Lett., 21, 422-423.

D. Ward, A.J. Walton, J.M. Robertson 1987 The Measurement of Transistor Characteristics Using On-chip Switching for the Connection of Instrumentation, To be presented at ESSDERC 87, Bologna, Italy.

4.2 CAM—A Strategy to Improve the Quality and Efficiency of Processing

Alastair MacConaill and Anthony Walton

THE DEVELOPMENT OF CAM

For the last 30 years, there has been a continuing effort to apply the abilities of the computer to the problems involved in manufacturing. The first efforts were devoted to solving individual parts of the total problem - stock control, Bill-of-Materials (BOM) management, Work-in-Progress (WIP) monitoring, Machine Shop routing etc. These early systems were based on punched cards, magnetic tape files processed in batch mode overnight or at weekends and stacks of paper printout.

Although the pioneers who implemented these systems developed many concepts which have stood the test of time most of these early efforts failed to convince those people who they were meant to benefit. Their fundamental ideas are incorporated into today's systems but the procedures had fundamental shortcomings and table 1 outlines the main ones.

SLOW REACTION	-	Out of date information
ERRORS IN DATA	-	Multiple input
	-	Duplicating between systems
	-	Transcription errors
	-	Reproduction errors on punching
GRADUAL CORRUPTION	-	Overnight error lists not actioned
LIMITED AVAILABILITY	-	Massive listings with a few people
INFLEXIBILITY	-	Ad-hoc enquiries major programming exercises
RIGIDITY	-	Incapable of handling real life

Table 1 : First Generation Manufacturing Systems

As a result of these limitations, they were backed up by a host of private notebooks; these contained the *real data* and it was a constant source of complaint that the computer tabulations did not correspond to what was in the notebook!

In the mid-1960s, two pieces of computer hardware became available - the magnetic disc and the Visual Display Terminal (VDU). The ability of disc files to be *indexed* so that data could be retrieved selectively and in differing sequences was a great advance. Coupled together with the chance to distribute information via the VDU - which, because of its limited screen size, can only work effectively with selected data - pointed the way forward to more effective dissemination of the information. This was aided by the appearance of the first software tools to simplify the process of designing and programming enquiries. However, the high cost of these devices limited their application and in spite of or, perhaps, because of, easier and wider access to the data, the doubts as to accuracy (and the notebooks) remained. The position at the end of this period tended to be as shown in Table 2.

INDEPENDENT	-	Data duplication and conflict
OVERNIGHT	-	Transcription/Punching Errors
	-	Gradual Corruption
LIMITED ACCESS	-	Diluted responsibility for accuracy

Table 2 : Second Generation Manufacturing Systems

This was the position by the early 1970s. It was about this time that the word *database* first began to appear in common (computer world) currency. Two things happened around this time. Firstly, organisations and companies began to realise that their computerised files contained a large asset - if only it could be made as accurate and effective as many accounting systems had, by then, become. Secondly, a number of people, e.g. Codd (1970) began to publish results of their theoretical studies on the way data should be held and managed within the computer environment. The application of this work led to the availability of the first commercial *Database Management Systems*. Coincidentally with these developments, the technological revolution of the integrated circuit - and other technologies associated with information handling, led to the rapid decrease in cost of computer hardware.

The 1970's and the first half of this decade have seen further developments. The combining of separate files into a single integrated database, coupled with on-line entry of data through the VDU has led to the first generation of manufacturing systems which have a chance of being accurate, in terms of the information being reliable and up-to-date. On-line data entry, combined with immediate validation of the data entered, puts the responsibility where it belongs - on the person doing the job - and ensures that all errors are trapped at source, and, therefore, corrected *before* the transaction is passed through to update the database. Again, as the data is entered at the same time as the real transaction, enquirers have an up to the moment picture - a more accurate state can no longer be found by telephoning or going down to see the person in question. One result of these systems has been the disappearance of the classical Progress Chaser, the person who's job it was to inform management of the exact state of work on the shop floor.

The biggest hurdle to be overcome to achieve this third generation of computerised manufacturing systems has been to find suitable terminals for data entry on the shop floor. This is a problem that varies from industry to industry, robustness, flexibility and mobility being factos which have had to be considered for one application or another.

In parallel with the stages which are outlined above, two other developments have matured at a similar rate. These are the use of computing as a *design* aid and the installation of computers to control continuous or quasi-continuous processes in the chemical, steel, paper, oil, utility and similar industries.

Computing as a design aid dates back to before the advent of the electronic computer, in the use of mechanical devices to compute the ballistic trajectories of artillery shells with its most complex mechanical realisation being achieved in the gunnery control of warships. However, the mainstream in the immediate post 1945 years began with the batch processing of mathematical algorithms; the advent of the VDU with its graphical capabilities was the start of a revolution which is still working its way towards maturity. In the late 1970s, the words *Computer Aided Engineering* (CAE) began to be used to describe the various techniques available. This title rapidly became accepted and was sufficiently descriptive to lead those who had been dealing with the nuts and bolts of production systems to look for a similar mnemonic. In this way a series of mnemonics and abbreviations came into being as detailed in Table 3.

The genesis of CAE and CAM has been described. CAM today is a description of those systems which are used to manage the production process. Its true boundaries are hard to define. On the one hand, market forecasts, which some would see as being outside the manufacturing boundaries may be used as input to the basic manufacturing schedule; as they have an effect on what will be done on the shop floor, and it is difficult to exclude them from

CAE	-	Computer Aided Engineering
CAM	-	Computer Aided Manufacture
FMS	-	Flexible Manufacturing Systems
CIM	-	Computer Integrated Manufacture

Table 3 : Common Mnemonics within Today's Manufacturing Industry

consideration. At the other end, the results of design work lead to a specification for a product which lays down the processing variables permitted during manufacture and the degree of conformity required during testing.

At the heart of a CAM system is the need to know what we are manufacturing, how it should be processed and what, if any, deviations from the defined process have taken place. This is classical WIP (Work-In-Progress) tracking, based on a Process Model which defines the steps applicable to any given end product. Figure 1 shows this in diagrammatic form with extensions into the engineering and planning areas.

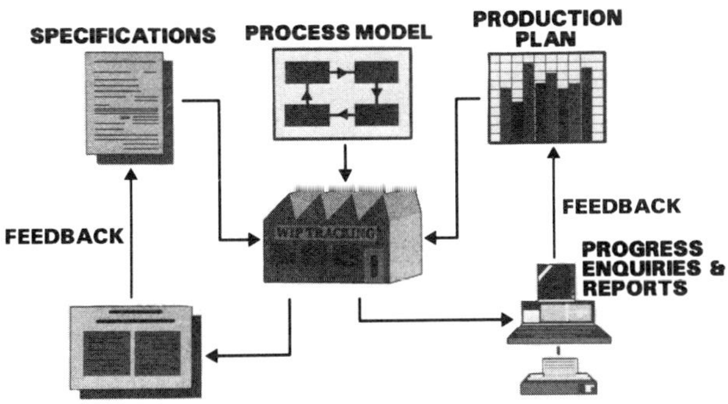

Figure 1. Outline of a basic CAM system.

Two other mnemonics have been indicated as being of interest. The first is FMS (Flexible Manufacturing Systems), which is commonly associated with the installation of robots for handling materials. This approach provides flexibility in the sense that reprogramming can allow a fixed robot to work on an entirely new product shape or, equally, a mobile robot (AGV on Automated Guided Vehicle) can easily be adapted to service a new layout of machinery.

Finally, CIM - Computer Integrated Manufacture. This, in concept, is the integration of design, production tools, production management and planning into a single consolidated system with data flowing naturally, as required, from one area to another without human intervention.

We have attempted attempted to outline the development of CAM within general manufacturing which can be directly related to the semiconductor industry. Large scale production of semiconductor devices, particularly integrated circuits, has evolved within the above timescale. Somewhat surprisingly, it was some time before CAM systems which had been developed for discrete part assembly, were adapted and refined for wafer processing (Singer 1984). However, the advantages gained by their implementation is just as great as in other areas of manufacturing and presents many interesting problems. There are no subassemblies in wafer fabrication and the nature and complexity of the processing steps involved make yield of good devices a fundamental issue within the industry. The remainder of this article illustrates some aspects of how a CAM system may be exploited to result in tighter control of the process and thus improve yield.

ERROR PREVENTION

One area where 100% yield can be achieved is in the prevention of errors in processing. This requires that the specified wafers arrive at the appropriate processing equipment and that the correct sequence of operations (or recipe) is performed. The WIP module, which is the basis of all CAM systems, is very important within this context. To ensure that processing is performed in the correct order requires a process model or menu. Before a batch starts the process engineer will have defined the recipe for each individual process and measurement step. These must detail every parameter; which for a furnace operation would include items such as, gas flow rates, temperatures (including ramp up and down times), the speed that the wafers are introduced and removed from the furnace etc. Other factors such as the maximum time between processing steps and the tolerance on measurements must also be specified. These recipes, when combined together, form the route which the wafers should follow.

This approach is fundamentally no different to that which has traditionally been used, except the controlling software checks that all the required conditions are set before a batch of wafers can start to be processed through any piece of equipment. This ensures that the correct route is chosen by not letting processing start, unless the CAM system agrees that the action the operator is proposing is the one specified.

Other factors which help to contribute towards increased yield are the monitoring of equipment performance and the cleanroom facilities. Every measurement which is specified has predefined tolerances and if these are exceeded the system will flag this problem to the process engineer and prevent any further processing being undertaken by the offending equipment. The services required in a cleanroom such as deionized water, humidity control, clean air, chemicals, gases etc can all be monitored and processing curtailed if critical problems occur.

The logging of equipment performance and processing time helps the maintenance engineer to both identify equipment problems before they become catastrophic and schedule planned maintenance in an optimum manner. This ensures the maximum uptime with the equipment performing within specifications which increases both the throughput and yield.

There are occasions when the equipment or an operator incorrectly processes a wafer and measuring or inspection equipment detects the fault. Provided that the error is non-catastrophic, and this is usually the case with lithography, the wafer can be re-worked. The CAM system can ensure that the appropriate route is taken and the wafers rejoin their original batch or are processed as part of a split lot.

To sum up, the CAM system can be used to prevent processing errors by

1. Ensuring the selection of the required batch at the appropriate time.
2. Making sure that the defined processing sequence is adhered to.
3. Checking that the maximum time specified between operations is not exceeded.
4. The correct recipe is displayed to the operator before processing of an individual step is embarked upon.
5. If any rework is required the appropriate route is taken.

ERROR DETECTION AND CORRECTION DURING PROCESSING

The detection of processing problems is of extreme importance if high yield is to be achieved. This requires comprehensive measurement to be performed throughout the process and is of fundamental importance if the fabrication line is to become totally automated. Systems, which are presently available, can flag errors, and call for assistance from the process engineer. This software cannot itself take action if a process step starts to drift from its mean and only calls for assistance once values are outside the warning or stop limits. In the future it should be possible to automatically feed information back from measurements systems to the equipment which then adjusts process parameters to reduce the tolerances. It should also be possible to feed this information forward to adjust equipment settings if required. For example, if the photoresist thickness is thinner than that required the exposure can be adjusted to compensate. This information should also be fed back to the spinner in order that adjustments can be performed if they are required. Figure 2 illustrates this concept ensuring that the line performance is continually tuned.

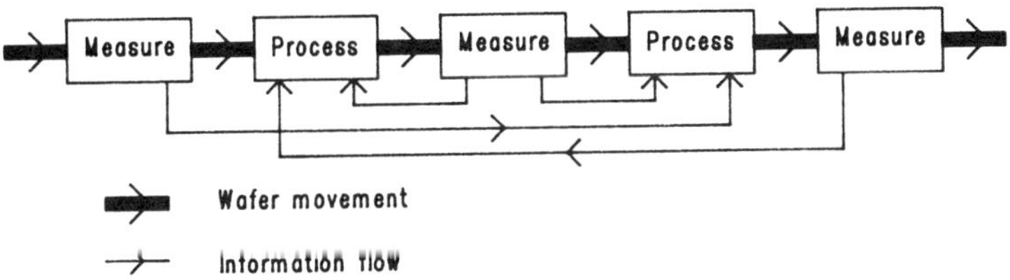

Figure 2. The feedback and feedforward of information.

Another example of adjusting parameters during processing is the use of process simulation to control the growth of oxide. It consists of monitoring the furnace temperature and using this to simulate the growth of oxide (Cameron et al). This information is then used to determine when the operation should be terminated. Figure 3(a) gives an example of oxide thickness as a function of time when the furnace temperature was dropped $50^{o}C$ for a short period as shown in figure 3(b). In spite of this the target oxide thickness of 250 Å was still achieved.

RETROSPECTIVE ERROR DETECTION

One big advantage of employing a CAM system from an engineering viewpoint is that all the measurements reside on one database rather than being distributed throughout the facility. This information can be used to correlate information to perform statistical quality control and to isolate the major causes of yield degradation. Obviously gross problems can be highlighted but hopefully these will have been isolated real time. The information contained within the database can be used in a much more subtle mode by looking for the factors which all the batches with a low yield have in common. This can identify problems such as equipment shedding particulates and operators who require more training. This approach is more in the fire fighting role but the database can also be employed to help increase yield. If all batches with a high yield are examined for common attributes then the same approach may be employed to improve the quality of processing.

One factor which is known to have a direct effect upon yield is the length of time taken to process a wafer from start to finish. This information can be extracted from the database and used with the scheduling facilities to ensure the maximum yield, commensurate with a given throughput, is achieved.

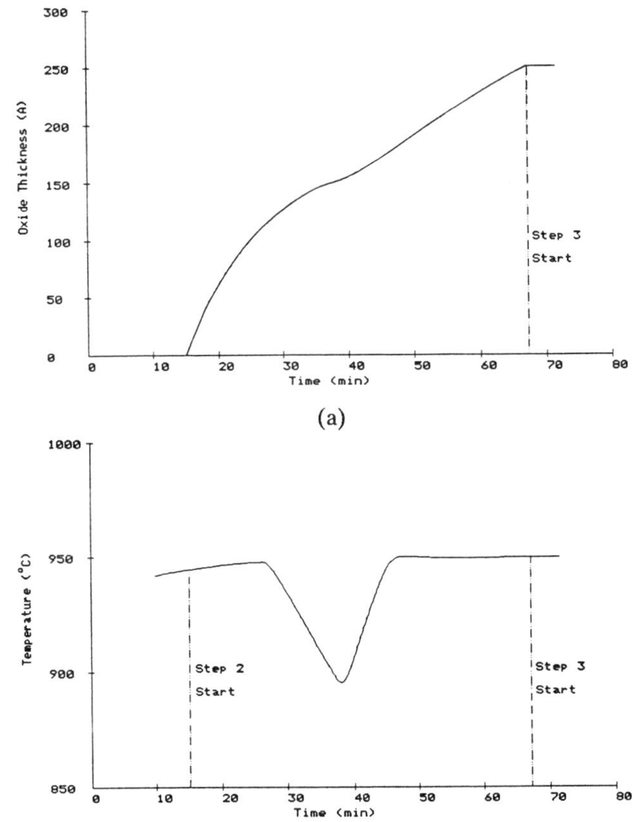

(b)

Figure 3. The variation of oxide thickness and temperature with time for an oxidation step.

THE TOTALLY AUTOMATED FACILITY

The totally automated facility has many attractions to the semiconductor industry. It removes people from the vicinity of the wafers which reduces both particulate contamination and the requirement for huge volumes of air to be filtered and processed to the specified temperature and humidity. This in itself could potentially save millions of pounds in a large factory but its realization requires many factors to be addressed before it becomes the normal mode of operation. Another advantage would include the elimination of processing errors due to the wrong wafers being selected or the equipment settings incorrectly set. The wafer handling systems, providing they are well designed are less likely to damage wafers than the human operator and should reduce the creation of particulates from the wafers themselves.

The majority of pieces of equipment are now available with cassette to cassette operation and the problems which must be solved are wafer transport, information transfer and process verification. There are a number of competing systems for wafer transport and storage systems and these include tracks, RGVs (Remotely Guided Vehicles) and SMIF (Standard Mechanical InterFace) boxes (Parikh, Kaempf 1984).

SEMI have defined standards for the communication between the CAM computer and equipment (SECS I and II) (SEMI 1986) and this is now available on a wide variety of equipment types. The CAM software can use this interface to download recipes and interrogate equipment status. The batches of wafers can then be routed through the next piece of equipment which is available.

The verification of process steps still remains a problem since this is not available for every process step. In the short term this will have to be addressed in the same manner as that used in conventional cleanrooms by ensuring that all equipment parameters are monitored and kept constant.

CONCLUSIONS

Totally automated facilities are already in production (Grenier 1986) but they are designed for high volume manufacturing where flexibility is not a primary requirement. There are many problems to be solved before the ultimate goal of total automation and flexibility is achieved. Today, in the semiconductor industry, we have CAM systems which are close to working. We are approaching accurate and reliable WIP data within a real production environment. We have begun to integrate the design data, albeit after manual conversion into specifications, within the overall system and the surface of the planning and scheduling problem is beginning to be scratched. The true effectiveness can be seen in the disappearance of the little notebooks and the decreasing time for which the CAM system can be off the air without stopping production, even in a plant where wafers are moved and handled by human operators.

The success to date of CAM systems, has been in significantly reducing cycle times and providing an engineering database which can be used to control and improve the process. The remaining problems which have yet to be fully addressed for totally flexible integrated circuit manufacturing include wafer transport and handling. Information storage must be addressed and new methods for the reduction of data are urgently required before we disappear beneath the sheer volume. The communication protocols between computers and equipment and equipment needs to be clarified since MAP (Donohue 1987) has clouded the issue. There is no question that with time, these issues will be resolved since it is widely recognised that the above approach is the only way to efficiently operate plants producing state-of-the-art silicon.

REFERENCES

E.Cameron, R J Holwill, J M Robertson 1986 Control of Oxide Growth by Real Time Simulation, Proc. 2nd International Conference on Simulation of Semiconductor Devices and Processes, Swansea, 405-411

E.F. Codd 1970 A Relational Model of Data for Large Shared Databases, Communications of the ACM, *13* 377-387

J. F Donohue 1987 Is MAP GM's revenge on us for not buying their cars?, Mini-Micro Systems, 43-45.

J. Grenier 1986 The Mitsubishi Saijo Factory: A New and Fully Automated IC Facility, Solid State Technology, 47-48.

M. Parikh, U Kaempf 1984 SMIF: A technology for wafer cassette transfer in VLSI manufacturing, Solid State Technology, 111-115.

SEMI 1986 Book of semiconductor standards, Vol 2 Equipment Division

P.H. Singer 1984 Computerizing the process line: A must for automation, Semiconductor International 68-73.

4.3 Statistical Control for VLSI Fabrication Processes

Purnendu Mozumder and Andrzej Strojwas

INTRODUCTION

IC fabrication facilities are subjected to a large number of constraints which can be broadly classified into two groups, those that are specified by the customer (external) and others that are due to intrinsic limitations of the processing equipment (internal). The external constraints comprise the tolerances on the circuit performance and the demand deadlines for the ICs to be delivered. The internal constraints, on the hand, are equipment and technology related, such as processing times of various steps, capacity and mean time between failures of fabrication equipment, etc. The aim of the fabrication line is to satisfy the given constraints while maximizing profit. Although the process and circuit are designed such that the resulting circuit performance for the fabricated ICs satisfy the constraints, process fluctuations inherent to VLSI fabrication processes cause the performance of a percentage of the ICs to violate the constraints. These process fluctuations are caused by material parameter variations, errors in setting the process control parameters (viz. temperatures, implant doses, etc.), degradation in the quality of the processing equipment and non-uniform processing condition. All of these variations manifest themselves similarly in the IC fabrication process and are collectively called *process disturbances*. The total profit is affected by the different fabrication/assembly costs and the revenue from the *acceptable* chips (Shyamsundar 1986a). The process disturbances cause the circuit performance of the ICs to deviate from their designed values resulting in yield losses and depleting the total profit.

To minimize the total costs (or maximize the total profit), it is necessary to minimize the effects of the process disturbances on the total costs. One of the salient parameters which affect total profit is the fabrication yield. However, to optimally schedule the lots for fabrication, it is necessary to have minimal yield variations from lot to lot. A stable or robust process minimizes the number of lots to be processed, thus decreasing the manufacturing costs. The mechanisms for profit loss can be countered by designing the process and the circuit for each individual IC type such that the parameters that strongly affect the profit are least sensitive to the process disturbances. During the wafer fabrication process the lot being processed is monitored via in-process measurements performed on pilot wafers and test structures. Depending on the deviations of the in the measured values from their designed values, the process is diagnosed and controlled to suit the necessary goals. Together, the design, diagnosis and control activities form a CAM system aimed at maximizing the total profit of the fabrication facility (see Figure 1).

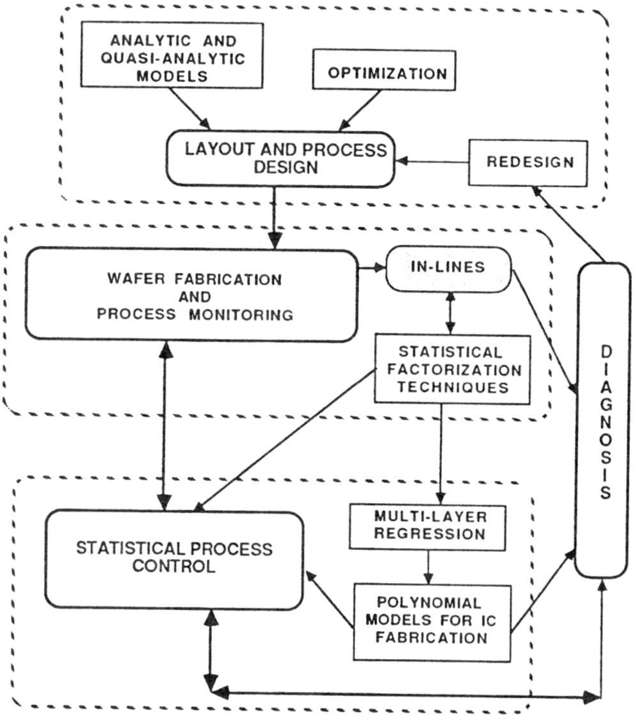

Figure 1: CMU-CAM System Capabilities

THE CMU-CAM SYSTEM

Present day VLSI manufacturing systems rely heavily on Computer Aided Manufacturing Systems that perform a number of *passive* functions, such as downloading process recipes, acquiring large samples of data from wafer probes and postprocessing the acquired data. However, the data is postprocessed without making full statistical inferences about the process from which the data is gathered. We, on the other hand, are developing a CAM system that will perform not only the passive functions presently in use but also a number of active functions like process monitoring, diagnosis and control based on the gathered data. This system which is being developed at the Carnegie Mellon University, henceforth referred to as the CMU-CAM system, aims at integrating the different phases of IC fabrication, namely design (process and layout), wafer processing, testing, diagnosis and control. These phases, which are presently treated as independent, will then form a set of strongly interacting subsystems that will aim at maximizing the profit from the fabrication line (Figure 1).

The first component in our CAM system deals with design for manufacturability. Its objective is to develop a design that is relatively insensitive to the random disturbances affecting the IC manufacturing process. The design problem can be formulated as determining a set of layout and process control parameters which will minimize the effect of the disturbances while keeping the circuit performance within the *acceptability region*. The acceptability region is the union of all points in the space of circuit performance that satisfy the external constraints. The optimization can be performed via *design centering* techniques (Director 1977, Brayton 1980) as well as quality improvement techniques using *signal to noise ratios* (Taguchi 1984, Box 1986). Data gathered by the CAM system together with statistical process and circuit simulators enable the characterization of the fabrication line which plays a crucial role in design and synthesis of the ICs. The design may also involve the manufacturing of test lots, data from which can be used to fine tune the design.

During the fabrication process, a number of *in-line* measurements are made (viz. oxide thickness, sheet resistance, critical dimensions). These measurements are made on pilot wafers as well as on test structures that reside on production wafers. The data from these measurements will then enable the system to determine the *process trajectory*, i.e. whether the lot being fabricated is encountering the expected process conditions for which it was designed. Based on the knowledge of the process trajectory, the process can be diagnosed to determine the causes of the process trajectory fluctuations. The statistical process control subsystem can then be invoked to use the results of the diagnosis and the data gathered during the fabrication steps to bring back the process to its normal trajectory. Diagnosis also plays an important role in correcting and updating the process controls from lot to lot. This is especially useful during initial stages of manufacturing an IC when the process is not completely stable and the lot to lot yield variations are high. In this case the process is said to be ascending the *learning curve*.

In the remaining sections of this paper, we develop the statistical process control subsystem and its interaction with the other components of the CAM system. In the first section we introduce the two forms of control that are available to the IC manufacturer. Next, we briefly discuss the objective function used by the CMU-CAM system for control. Efficient modeling techniques suited for process control are developed in the following section. In this section, we introduce the concepts of process *observability* and *controllability*. Finally, we describe the algorithms and concepts involved in statistical process control, specifically quality and feed forward control.

Statistical Process Control

In any fabrication line, two forms of control are possible, namely quantity control and quality control. A quantity control system aims at maximizing throughput, i.e., maximizing the number of wafers fabricated in a given period of time. Scheduling is very important since different process flows competing for the available resources are used in the fabrication line to produce different IC's. The quantity control approach works well for high volume manufacturing processes with a high yield but may not be entirely suitable for a newly designed VLSI fabrication process where the yield is varying from lot to lot and is typically much less than 100%. To counter the problem of low yield, quality control is necessary.

The aim of a quality control system is to maximize the profit over a certain period of time by maximizing yield. However, problems with regard to low yield may lead to changes in scheduling which is a part of quantity control. A typical situation is depicted

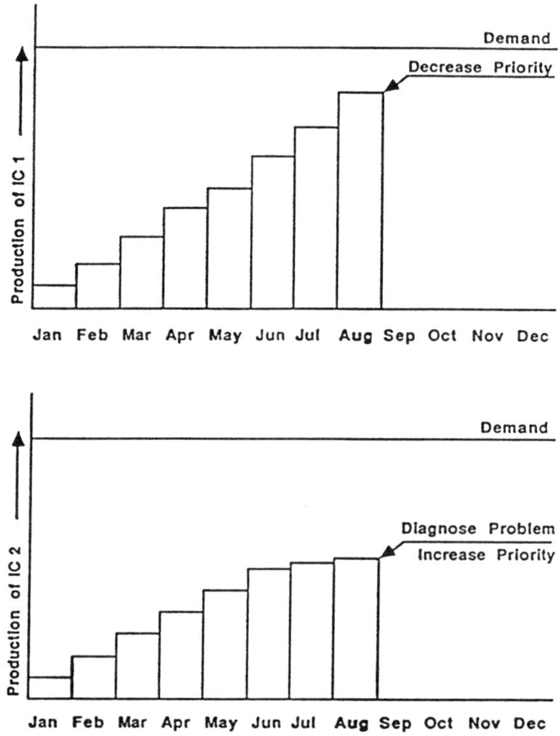

Figure 2: Manufacturing schedule in an IC Fabrication Line

in Figure 2 where production of the second IC is not proceeding at the required rate. This would necessitate a change in the fabrication schedule. We formulate the task of IC fabrication as one of profit maximization (Shyamsundar 1986a) and use a unified framework where both quantity and quality controls are performed simultaneously. Depending on the predicted profit, based on in-line measurements, four possible decisions can be made which minimize the total fabrication cost:

1. To continue processing the lot further without any modification to the process trajectory (the normal process). when the predicted fabrication yield is above its *threshold of acceptability*.

2. To correct the process trajectory by manipulating the future process parameters, so that the expected profit is set above the threshold of acceptability. This is done when the predicted profit fails to satisfy the previous criterion. If a feasible solution exists the lot is processed further with the new set of process parameters. The process of determining and manipulating the future process parameters is called *feed forward control*.

3. To *rework* a previous faulty process step. Rework is not possible for all process steps (e.g. implantation or thermal redistribution) especially where the step cannot be immediately observed via in-lines.

4. To reject the lot from further processing when the expected profit from the lot does not justify its further processing. This is where it is not cost effective to continue further processing or take corrective measures because of large process deviations from the normal trajectory.

In the IC fabrication process, yield maximization, as attempted by the quality control system, does not imply a profit maximization. This is especially true when a lot is rejected at an intermediate stage. Although the yield for that particular lot is zero it saves the cost of subsequent processing steps as well as freeing up the equipment which could then be used to produce a lot with a much higher yield and resulting in an increase in the total profit. Therefore, the outlook of the system must be from the viewpoint of profit maximization, as opposed to yield maximization, when trying to schedule, control and optimize the fabrication process.

At the start of the fabrication process, the lots are scheduled using production planning algorithms to maximize the profit. The algorithm determines an expected value of the fabrication yield based on prior knowledge of the process disturbances in the fabrication equipment, test runs, and information from the design phase. This value of the yield is then used by the planning algorithm to maximize the total profit. However, the process variations from lot to lot often lead to significant variations of the yield from the value expected, necessitating a modification of the plan after a lot with a significantly higher or lower yield has been processed. It is therefore imperative to monitor each lot undergoing the fabrication process and make quality control decisions which can then be reflected by rescheduling future lots if necessary. On the other hand, the quality control algorithm must be able to determine the effect of its decision on the total profit due to rescheduling and thereby determine the thresholds of acceptance and rejection. This information transfer between the quantity and quality control systems is to be carried out by a short term scheduler. The short term scheduler, as a part of the quantity control system, will determine whether a rescheduling is necessary as a result of quality control decisions, and if so, reschedule. At the same time the short term scheduler derives the expected change in the profit due to the quality control decisions, based on the success of previous lots, the time remaining to meet the demand and the present state of the lot being monitored. This information is then used by the quality control system to determine the thresholds of acceptance and rejection. Therefore, the quality control operations form the inner loop in the overall profit maximization problem.

PROFIT FUNCTION

In this section we will briefly discuss the objective function used by the CAM system together with the constraints that have to be satisfied by it. To develop the objective function, all ICs that are manufactured using the same recipe are classified into a single product group. Moreover, we assume that for each IC type the lower and upper bounds on the demand are specified. The objective function is the sum of the fabrication costs, the negative value of the revenue from the acceptable chips and the penalty for not meeting the demands associated with the different ICs (Shyamsundar 1986a). The aim of the fabrication process is to minimize this objective function.

The fabrication cost for each IC comprises the back end costs associated with testing, assembly and other back end processing and the front end costs associated with each wafer fabrication step. Penalties are assessed for those ICs whose lower bounds on the demands are not met. The objective of the fabrication line is therefore to fabricate at least as many lots of each IC as are required to meet the lower bounds on their demands within the specified due dates. However, since the resources are limited, the process flows for different ICs have to constantly compete with one another for resources. Therefore resources have to be allocated optimally so as to minimize the objective. This competition for resources and the various demands can be expressed in terms of three basic type of constraints: sequence constraints, resource constraints and timing constraints.

The sequence constraints are on the fabrication steps for each lot being fabricated. The constraints state that a particular fabrication step can be performed only when all the previous steps have been performed. The resource constraints state that a resource can be involved in one fabrication step at any time. The timing constraints are expressed in terms of the initiation time for the first lot of each IC type, which then forms an effective constraint on the entire production of that IC. Not all the ICs may be available at the starting time point and hence the timing constraints are necessary to specify when the production of a particular IC type can be scheduled. The number of timing constraints therefore equals the number of IC types. A detailed formulation of the profit function and the constraints can be found in (Shyamsundar 1987).

The individual components in the profit function vary considerably from line to line. For high volume lines with a good yield, the back end costs which are proportional to the number of good chips dominate, while in the case of a low yield the front end costs dominate. Moreover, the demand and the penalty associated with a ASIC fabrication line is very different from that of a high volume line. The ASIC production line tends to have a short turn around time with a much higher penalty for not meeting the demands than that of a high volume line. Clearly, an IC manufacturer needs to know the different costs, the penalty and the revenue per IC type to maximize the total profit from the fabrication line.

EFFICIENT MODELING FOR PROCESS CONTROL

During the design for manufacturability, the optimal value of the process controls are determined using knowledge of the process disturbances. Moreover, via simulations and test runs the expected distributions of the in-lines and the circuit performance are determined. The value process controls and the associated process disturbances for which the process is design constitute the *normal process trajectory*. During the fabrication of the lots it is necessary to determine how far a process is from its normal trajectory using in-line and circuit performance measurements. If the deviation is significant then the cause is determined via process diagnosis. In certain cases the deviation can be corrected by appropriate process control. During the design, diagnosis and control of the fabrication process the effect of the designables/controllables on the process trajectory is of importance. This motivates us to define two attributes of any fabrication process, *observability* and *controllability*. A process is said to be observable if the distribution of the IC performance can be determined in terms of the distributions of the in-line parameters. Similarly, a process is said to be controllable if the desired distribution of the output performance can be achieved by manipulating a set of process parameters. Both attributes are necessary for successful process control. In the remaining part of this section we will describe the method of deriving efficient models of the fabrication process which will render the process observable.

Let us represent the vector of layout and process control parameters by **L** and **P** respectively. Let **D**, a vector of continuous random variables, represent the disturbances. Then the in-lines, device parameters and the circuit performances, which depend on **D**, can be similarly represented as vectors of continuous random variables **U**, **X** and **Z** respectively. It is desirable that the models be in a form whereby the system is able to determine the in-lines that should be measured for process observability and the process controls that can be manipulated to control the process trajectory. This dictates that there exist explicit mappings from the space of observables and controllables to the IC performance. Moreover, the models must be accurate and also fairly efficient so that optimizations and statistical predictions can be made without incurring large CPU times.

For the purposes of design, the space of designable parameters is of a high dimensionality and also the models have to be valid over a large region of operation. In contrast, for control, only the salient in-lines need be observed, the process parameters to which the objective is sensitive is smaller, and the region over which these vary also tends to be relatively small. One of the ways of modeling the large variations of the designable parameters and the equivalent variations in the output performances accurately is via the use of statistical process simulator like **FABRICS** (Nassif 1984) in conjunction with a circuit simulator like **SPICE**. On the other hand, there exists a number of difficulties with the use of the simulators for process control. The simulators do not give an explicit relation between in-lines and IC performance, necessary from the viewpoint of observability. Moreover, they cannot be restarted at any step in the fabrication process which is necessary from the viewpoint of quality as well as feed forward control. These factors, together with the inherent low dimensionality of the input space, motivate us to derive an explicit two layer polynomial model, the first layer being a mapping from the space of process controls, layout and disturbances to the space of in-lines, and the second being a mapping from the space of in-lines to the space of device parameters and IC performances. In general, the resultant model is pseudo-two layered, since the second layer has to include all process parameters, layout parameters and disturbances that affect circuit performances and whose effects are not captured by any observable in-line. Standard linear regression is prohibitive in terms of space and time since the dimensionality of the input space is large, transformations are highly non-linear and the polynomial orders are not known a priori. Moreover the in-lines which serve as inputs to the second layer are intercorrelated. To alleviate these difficulties of high dimensionality and correlation, the problem is decomposed into smaller subproblems each of which is then solved independently. To solve the difficulty of unknown polynomial order, a method of multi-layer regression (Shyamsundar 1986b) is used which automatically determines the degree of the polynomials.

Decomposition

The set of observable in-lines are decomposed in two phases, *clustering* and *statistical factorization* (Mozumder 1986a). The aim of clustering is to partition the space of in-lines such that each group of in-lines is affected by a restricted set of process parameters and disturbances. These restricted sets can be used to model the respective in-line groups. This is achieved if the clustering is performed such that no two in-lines from two different clusters are correlated. Ideally, this would lead to clusters which do not have a common underlying factor, thus partitioning the space of process parameters and disturbances so that each in-line cluster depends on a unique subset of process parameters and disturbances. The clustering can be expressed mathematically as

$$\rho_{ij} < \rho_T \qquad \forall U_i \in G_p,\ U_j \in G_q \qquad \forall p \neq q \tag{1}$$

where

G_p, G_q are the pth and the qth clusters of in-lines and

ρ_{ij} is the multivariate correlation coefficient between in-lines U_i and U_j

The exact value of ρ_{T} depends on the confidence interval of the correlation coefficient and the error that can be tolerated due to misclassification. The clusters resulting from the use of this non-zero ρ_{T} are quasi-independent.

The problem of correlations amongst the in-lines within each cluster, which poses a problem for the second layer, can be resolved by determining a set of loosely correlated in-lines that would in effect be able to represent all other in-lines in that cluster. Principal component decomposition is applied whereby n in-lines in the cluster can be represented as a linear combination of n principal factors (Harman 1967). The decomposition can be expressed mathematically as

$$U_i = \lambda_{i1}F_1 + \cdots + \lambda_{ij}F_j + \cdots + \lambda_{in}F_n \tag{2}$$
$$i = 1,2, \ldots, n;$$

where U_i is the ith in-line in the cluster to be factorized, F_j is the jth principal factor and λ_{ij} corresponds to the factor loading of F_j on U_i. The factor loadings can be represented in terms of the eigen-values and eigen-vectors of the correlation matrix (Harman 1967). The resultant factor loadings are such that first factor accounts for the maximum variance in the cluster. All succeeding factors account for the maximum variance of the corresponding residual variance. The ratio of the ith eigen-value to the trace of the correlation matrix represents the fraction of the variance accounted for by the ith principal component. Therefore all those eigen-values which do not contribute significantly to the trace of the correlation matrix are deleted without loss of accuracy. Depending on the accuracy desired, the largest k out of n eigen-values are chosen. The desired accuracy can be expressed in terms of the fraction of the variance in the in-line cluster that can be accounted for by the chosen factors. In-lines within the cluster are then identified with these principal factors (Mozumder 1986b). These in-lines, known collectively as *factor in-lines*, then form a set of loosely correlated variables which are able to effectively represent the clusters to which they belong (Mozumder 1986a). If the initial set of in-lines are chosen such that all disturbances which affect the output performances affect at least one in-line in the set then the factor in-lines form the minimal set of observables. If the *joint probability density function* (*jpdf*) of the factor in-lines can be estimated from in-line measurements then the process is said to be observable.

In the two layer model, all factor in-lines in a cluster are represented in terms of process parameters and disturbances which govern the cluster. The non-factor in-lines and the device parameters/circuit performance are represented in terms of the factor in-lines and the process disturbances which affect the circuit performance but whose effects are not captured by the factor in-lines. The implementation details for the modeling algorithms can be found in (Mozumder 1986a).

STATISTICAL QUALITY CONTROL

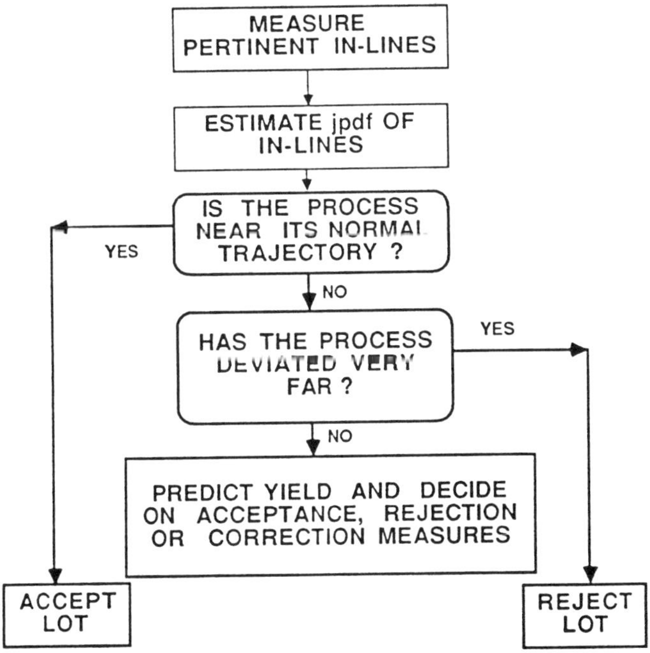

Figure 3: Quality Control of an IC Fabrication Line

The problem of statistical quality control can be viewed as determining bounds on the moments of the *jpdf* of the observable in-lines that will enable the system to accept, correct or reject the lot under fabrication. Let us denote the *jpdf*s corresponding to the vectors representing the disturbances (**D**), in-lines (**U**), device parameters (**X**) and circuit performance parameters (**Z**) by Ψ_D, Ψ_U, Ψ_X and Ψ_Z, respectively. Then the quality control decision making process can be expressed as a sequence of inference steps (see Figure3)

- To determine from the external constraints (i.e. the circuit parameters in terms of which the IC performance is specified), the in-lines which will render the process observable.

- To estimate the moments of Ψ_U from in-line measurements. The Ψ_U is then mapped to Ψ_Z, which in turn can be used for fabrication yield and profit prediction. This task can be viewed as a problem of *statistical hypothesis testing and estimation.*

- To determine the threshold of acceptance and rejection from the profit function via the short term scheduler and map these thresholds as tolerances on the moments of Ψ_Z, Ψ_U (for the problem of quality control), Ψ_P and Ψ_D (for the purposes of feed forward control).

- To make a quality control decision (accept, correct, rework or reject) based on the tolerances on the moments of the in-lines and the *jpdf* estimated from the in-line measurements after each salient step.

The first step consists of determining from the constraints and the objective, the in-lines that render the process observable. The models created are ideally suited for this purpose. The parameters to be observed are necessarily a subset of the factor in-lines and the process parameters and disturbances that form inputs to the second layer of the model. Since the models are represented in the form of polynomials, *dependency trees* which relate dependencies of the circuit performance parameters on the factor in-lines, process parameters and disturbances can be identified by parsing the second layer. Using the dependency trees, the factor in-lines, process parameters and disturbances in the second layer of the model are determined. However, all the parameters detected need not be monitored. It is necessary to monitor the ones that are inputs to the IC performance parameters in terms of which the external constraints are specified. The parameters to be monitored are called *observables* and are represented as a vector of continuous random variables **W**. The dependency trees can be extended similarly for feed forward control to determine the process parameters which affect the IC performance. These process parameters are collectively called *controllables*.

The next step in quality control involves estimating the *jpdf* of the in-lines of interest from measurements. The estimation procedure consists of two parts: determining the type of the distribution and the moments necessary to characterize it, and the sample size necessary to perform the estimation of these moments with the desired confidence. *Chi-square tests* (goodness of fit) show that most of the in-lines are normal while most of the remaining in-lines can be assumed to be log-normal. All the disturbances are assumed to be normally distributed and independent. The resultant *jpdf* in the n_w-dimensional **W** space, Ψ_W, can be represented in terms of a mean vector $E[\mathbf{W}] = (E[W_1], \cdots, E[W_{nw}])$ and a covariance matrix Σ_W, and is expressed as

$$\Psi_W = \frac{1}{[(2\Pi)^{n_w} \det \Sigma_W]^{1/2}} exp(-\frac{1}{2}[\mathbf{W}-E[\mathbf{W}]]^T \Sigma_W^{-1} [\mathbf{W}-E[\mathbf{W}]]) \tag{3}$$

The problem of estimating Ψ_W then reduces to the problem of estimating $E[\mathbf{W}]$ and Σ_W. Depending on the confidence level desired (which relates to the control accuracy) and the limitations on the number of test structures, the minimum sample sizes for estimating the mean, variance and covariance can be determined. The maximum of the three is the required sample size. Once Ψ_W and the constraints in the **Z** space are known, quality control decisions can be made either by determining Ψ_Z, the *jpdf* in the **Z** space, from Ψ_W, or by determining the acceptability region in the **W** space, $\mathbf{R_W}$, from the corresponding region in the **Z** space, $\mathbf{R_Z}$. The latter is chosen because mapping of *jpdf*s from one space to another is computationally intensive and has to done every time an in-line distribution is estimated. The inverse mapping is done via *simplicial approximation* with the desired accuracy. The details of the theory behind the simplicial approximation can be found in (Director 1977). The models serve the purpose of a fast forward mapping which enhances the efficiency of the simplicial approximation. The deviation of the process from its normal trajectory is determined by comparing *jpdf*s between the normal and the observed process while the constraints in conjunction with the process deviation enable the system to make one of the four quality control decisions.

The final step in quality control is to decide on acceptance, rejection or correction of the lot under observation. It is observed that the acceptability region in the in-line space is defined by a large number of constraint equations, even when the corresponding acceptability region in the output space is simple, thus making yield estimation in the **W** space computationally intensive. Therefore, it pays off to be able to determine inexpensive methods based on statistical inference that can identify the critical cases of acceptance and rejection based on the observed *jpdf* only. The procedure for rejection is based on identifying cases where the process deviation is large enough to be detected via a single in-line or a linear combination of them. Acceptance on the other hand is based on cases where the *jpdf*s are not significantly *different*. If, however, neither decision can be made with sufficient confidence then the total fabrication yield has to be predicted.

The parametric yield is estimated by integrating Ψ_W over the *acceptability region*, A_W derived using simplicial approximation, in the **W** space. Equivalently, yield prediction is also possible using the *jpdf* and an equivalent acceptability region and the parameters in the **Z** space. The fabrication yield can be expressed as

$$Y = \int_{A_Z} \Psi_Z \, dZ = \int_{A_W} \Psi_W \, dW$$

$$= \int_{-\infty}^{\infty} \cdots \int_{-\infty}^{\infty} \Psi_W \Phi(W) dW_1 \cdots dW_{n_w} \tag{4}$$

where $\Phi(W)$ is the indicator function and is defined as:

$$\Phi(W) = \begin{cases} 1, & \mathrm{I} \subset \mathrm{A_W} \\ 0, & \text{otherwise} \end{cases} \tag{5}$$

The next section deals with the criteria used to determine the critical cases for acceptance and rejection.

Rejection and Acceptance

The problem associated with the complexity of the indicator function and the resultant complexity in the yield estimation can be overcome by approximating the acceptability region as a hyperbox. The integrals can then be decoupled and the final yield can be approximated as the product of several partial yields. Since the various factor in-lines are

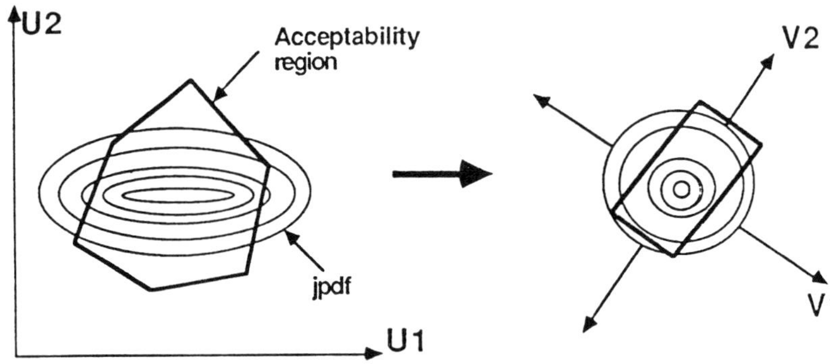

Figure 4: Generation of the Hyperbox

independent Σ_W and Σ_W^{-1} in Eqn.(3) are diagonal. Therefore, when the acceptability region is a hyperbox the yield, given by Eqn.(4), can be expressed as

$$Y = \int_{L_1}^{H_1} p(w_1)dw_1 \cdots \int_{L_{nw}}^{H_{nw}} p(w_{nw})dw_{nw} \tag{6}$$

where

L_i denotes the lower bound of the hyperbox in the ith dimension,

H_i denotes the upper bound of the hyperbox in the ith dimension,

$p(w_i)$ denotes the marginal distribution of W_i.

The orientation of the hyperbox is derived using a heuristic such that the hyperbox encloses the acceptability region while minimizing the enclosed volume. The largest face of the acceptability region, as given by the simplicial approximation, is aligned to one of the faces of the hyperbox. A resultant matrix **A** is derived which represents the rotation required to align the generated hyperbox along the coordinate axes, the details of which can be found in (Mozumder 1987). Given that the transformation of the axes is expressed as

$$\mathbf{V} = \mathbf{A}\mathbf{W} \tag{7}$$

the *jpdf* in the **V** space, Ψ_V, can be expressed as a multivariate normal distribution similar to Eqn.(3), where $\Sigma_V = \mathbf{A}\Sigma_W\mathbf{A}^T$ and $E[\mathbf{V}] = \mathbf{A}E[\mathbf{W}]$. However, even though **A** is unitary (since **A** represents pure rotation), Σ_V is not diagonal in general, which means that the decoupling cannot be achieved. The situation is remedied in the system by scaling the original space so that the covariance matrix is an identity. If the scaling matrix is denoted by **S**, where **S** is diagonal with $S_{ii} = 1/\sigma_{ii}$, then the resultant transformation can be expressed as $\mathbf{V} = \mathbf{QW}$ where $\mathbf{Q} = \mathbf{AS}$. The covariance matrix $\Sigma_V = \mathbf{Q}\Sigma_W\mathbf{Q}^T = \mathbf{I}$, and the Eqn.(6) can be expressed as a product of n_v integrals.

$$Y_V = \prod_{1}^{n_v} y_i \qquad \text{where} \quad y_i = \int_{L_i}^{H_i} p(v_i)d(v_i) \tag{8}$$

The limits of integration, L_i and H_i, correspond to the equations of the hyperbox in the transformed space. The criterion of rejection in terms of the *ith* partial yield can be written as

$$y_i \le \frac{Y_{rej}}{\prod_{j=1, j\neq i}^{n_v} y_j} \tag{9}$$

where Y_{rej} represents the rejection threshold.

Using Eqn.(9) the criterion for rejection can be expressed in terms of limits on the moments of the marginal distributions. Under the assumption that the nature of the distribution remains unchanged due to a change in the process trajectory the partial yields can be expressed as a difference of *error functions*. The error functions can then be approximated by a truncated power series in terms of the mean and the variance of each of the marginal distributions. Keeping one of the moments constant (at its nominal value) Eqn.(9) can be solved to find the limits on the other moment (Mozumder 1987). Therefore the rejection criterion is simplified into $2n_w$ comparisons.

Acceptance, on the other hand is based on the measure of a *difference* between the observed *jpdf* and that of the normal process, and its quantification in terms of the final fabrication yield. Nonparametric statistical techniques are applied to get a measure of the difference in attributes of scatter clouds which can then be mapped onto the change in fabrication yield. An example of such a technique for mean shift is the method of *Fisher's linear discriminant analysis*, where the sample means of the two scatter clouds are separated as far as possible with respect to their sample variance. The resulting distance is termed the *Mahalanobis distance* (Rao 1973) and is expressed as

$$\mathcal{M} = \Delta\mu^T \Sigma^{-1} \Delta\mu \tag{10}$$

where $\Delta\mu$ is the difference between the mean of the scatter clouds. The change in yield is then related to the distance by weighting the difference between the means with the yield sensitivity in the direction of the shift. This is equivalent to weighting the shift in mean along each dimension of **W** by the sensitivity of the yield to the corresponding observables. The *weighted Mahalanobis distance* is expressed as

$$\mathcal{M}_S = \Delta\mu^T \mathbf{Y_s}^T \Sigma^{-1} \mathbf{Y_s} \Delta\mu \tag{11}$$

where $\mathbf{Y_s}$ is a diagonal matrix with $y_{S_{ii}} = \partial Y/\partial W_i$. The weighted distance is used by the

CAM system as an appropriate measure of the yield fluctuations due to small changes in the mean. The method can be applied for any objective function given the sensitivity of the W_i to the objective. Methods of determining yield gradients have been described in (Styblinski 1986). If the yield change is smaller than the difference between the expected yield and the threshold of acceptance, the system decides on acceptance without any further steps.

It can be noted at this point that integration may be necessary in cases where the decision to accept or reject cannot be made with sufficient confidence using the inexpensive criteria. However, the dimensionality of the space of observables tends to be low, thereby making the integration computationally feasible. On the contrary, during design where the dimensionality of the space of designables is significantly higher, yield prediction by integration is infeasible.

FEED FORWARD CONTROL

Let the mapping of the IC performance from the device parameters and of the device parameters to the process and layout parameters be combined into a single set of models given by,

$$\mathbf{Z} = q(\mathbf{P},\mathbf{L},\mathbf{D}) \qquad (12)$$

The number of process control, disturbances and layout parameters affecting the IC performance of interest may, in general, be very large. To reduce the dimensionality of the problem, the circuit performance parameters are clustered, using the previously described method. Specifically, if $\mathbf{Z}_i$ refers to the *ith* group of circuit performance parameters and $\mathbf{P}_i$, $\mathbf{D}_i$, $\mathbf{L}_i$ refer to the *ith* group of input parameters, then any member of $\mathbf{Z}$ that is not in $\mathbf{Z}_i$ will not depend on any member of $\mathbf{P}_i$, $\mathbf{D}_i$, $\mathbf{L}_i$. Since, the input parameters are statistically independent of each other, the different groups of circuit performance parameters will in turn be statistically independent. Since the circuit performances have been partitioned into independent groups, the acceptability regions can be determined separately for each of these groups. The final acceptability region will be the union of these regions. Simplicial approximation can be used again to determine acceptability region for each group. Like modeling, this process is also an off-line activity and again has to be performed only once for a stable line.

Yield improvement using feed forward control

Feed forward control involves the modification of the process control parameters for the remaining fabrication steps to increase the final yield above an acceptable value. This is performed at each stage in the fabrication process where the process control parameters have to be modified, in order to increase the estimated yield iteratively. Whenever feed forward control is considered, it is necessary to *identify* the process i.e., the distributions of all the previous process disturbances should be determined from the results of the in-line measurements. Let the models relating the vector of in-lines, $\mathbf{U}$, to $\mathbf{P}$, $\mathbf{L}$ and $\mathbf{D}$, be given by

$$\mathbf{U} = F(\mathbf{P},\mathbf{L},\mathbf{D}) \tag{13}$$

Using these models, polynomial models are generated for the moments of the in-lines in terms of the process control parameters, layout parameters and the moments of the process disturbances. Let $\mathbf{U}_m$, $\mathbf{D}_m$ represent the vectors of moments of in-lines and moments of disturbances, respectively, that are necessary to characterize the process. Let

$$\mathbf{U}_m = F_m(\mathbf{P},\mathbf{L},\mathbf{D}_m) \tag{14}$$

represent the entire set of functions of interest to us.

Let $\mathbf{U}_m^0$ be the moments of the in-lines from the measurements taken. The process is identified by determining $\mathbf{D}_m^0$ such that,

$$||(\mathbf{H}_m)^T \cdot (\mathbf{U}_m^0 - f_m(\mathbf{P},\mathbf{L},\mathbf{D}_m^0))||_2 \tag{15}$$

is minimized, where $\mathbf{H}_m$ is a vector of weighting factors.

The models that relate the in-line measurements to the process control parameters and disturbances aid in process identification. The final yield should now be estimated. Let $\Psi_i(s_i)$ represent the *jpdf* of $\mathbf{S}_i$ and $\Phi_i(s_i)$ the indicator function for $\mathbf{S}_i$, analogous to Eqn.(5).

The *ith* partial yield, Y_i, can thus be given as the expected value of $\Phi_i(s_i)$ with respect to $\Psi_i(s_i)$,

$$Y_i = \int_{s_i} \Psi_i(s_i)\Phi_i(s_i)ds_i \tag{16}$$

If the number of partial yields is n, then the total yield is given by,

$$Y = \prod_{i=1}^{n} Y_i \tag{17}$$

The process control parameters to which the yield is maximally sensitive would have to be altered in order to increase the estimated yield above an acceptable value. By means of off-line computer simulations, the feed forward control system should arrive at a new vector of process parameters (by modifying some of the future process control parameters) such that the estimated yield, Y, is greater than the acceptable value of yield, $Y_{acceptable}$. It may not always be possible to determine such a vector of process control adjustments in which case the feed forward control procedure will fail and the fabrication process would have to be aborted. However, if the procedure succeeds, the new set of values for the future process control parameters, as determined by it, can be used for the future processing steps.

NUMERICAL EXAMPLES

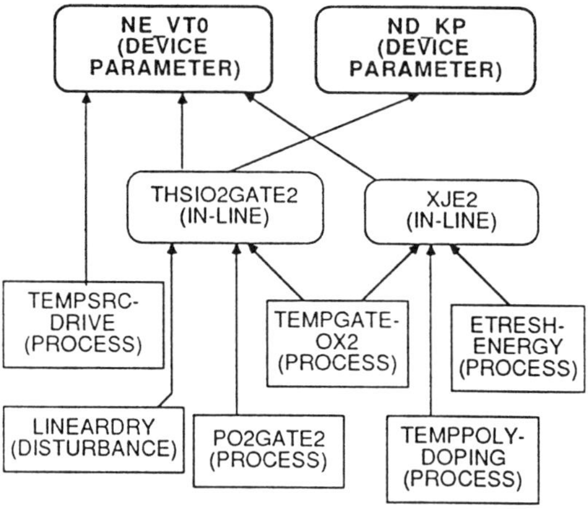

Figure 5: Example of a 2-layered Model

In this section we present computational examples to illustrate the software implementation of the various modeling and quality control algorithms that have been described in the previous sections. All the programs are written in C and run under the UNIX operating system on a VAX-11/785. **FABRICS** was tuned to a commercial 4-device **NMOS** process, the devices being strong depletion, strong enhancement, weak depletion and weak enhancement. A sample size of 200 was chosen and **FABRICS** and **SPICE** run to generate the data in form of vectors that could be used both by the problem decomposition as well as the regression routines.

In the problem decomposition phase the intra-cluster correlation matrix for each cluster was generated using the data from **FABRICS** and **SPICE**. Clustering was performed using a value of $\rho_T = 0.1$, and 10 distinct clusters were formed. Each cluster was then subjected to a principal component decomposition and the significant eigen-values, using a variance cutoff value of 0.975, were chosen to determine the factor loadings. Multi-layer regression was then used to create the process models. An example of the process model can be found in Figure 5. The device parameters modeled here are the zero bias threshold voltage of the strong enhancement device, NE_VTO and the transconductance of the strong depletion device ND-KP. They depend upon two factor in-lines, namely *thsio2gate2* (gate oxide thickness after the second gate oxidation step) and *xje2* (depth of the enhancement threshold implant after poly doping), and a process parameter *tempsourcedrive* (temperature of source drive in). The factor in-line *thsio2gate2* depends on two process parameters *tempgateox2* (temperature of second gate oxidation) and *po2gateox2* (partial pressure of oxygen in second gate oxidation), and one disturbance, *lineardry* (linear coefficient of dry oxidation), while *xje2* depends on three process parameters namely *ethreshenergy* (the energy of the strong threshold implant), *temppolydoping* (temperature of the poly doping step) and *tempgateox2* (temperature of the second gate oxidation step). The model is pseudo-two layer due to the inclusion of a process parameter in the second layer.

Using the models described we simulated the flow of a lot through the fabrication line where the means and variances of the process disturbances were changed from their nominal values to adversely affect the yield. A tolerance of 20% was allowed on NE-VT0 and 10% on ND-KP. The yield for the normal process was approximately 85%. The observable space consists of two factor in-lines, *thsio2gate2* and *xje2* and a process parameter *tempsourcedrive*. We evaluated the results using the quality control algorithms after the source drive in. Using the simulated data for the normal as well the perturbed process the algorithm for acceptance deduced a Mahalanobis distance of 0.0267. This would indicate a very small deviation in the process trajectory and hence would be a clear case of acceptance. However, to check the validity of our assumption we applied the algorithm for rejection via partial yields. The results were:
Partial yield due to thsio2gate2 = 79.3%
Partial yield due to xje2 = 98.3%
Partial yield due tempsourcedrive = 92.8%

Finally the fabrication yield was estimated using Eqn.(4) and found to be 71.0%

Both results confirm the decision made by the acceptance algorithm to accept the lot for further processing.

The CPU time required for the problem decomposition routines was 14 CPU minutes while the multi-layer regression for model building took 13.4 CPU seconds. 2.5 CPU minutes were required to create the simplicial approximation of the acceptability region and parsing the dependency trees. Finally, the system spent 22.8 seconds to make quality control decisions, including yield prediction. It is observed that most of the time spent by the system is in off-line activities and the on-line quality control decisions making algorithms require a short time. Moreover, the time required to create the models is negligible compared to the time that would have been required if the simulators were used for control.

CONCLUDING OBSERVATIONS

With the ever increasing demand for ASICs, low volume and short turn around times are becoming a characteristic of fabrication lines. CAM systems that perform passive functions are no longer capable of streamlining the fabrication lines towards producing high yield lots within the short time essential in meeting the customer demands. Moreover, it is no longer adequate to gather data from test structure at the end of the fabrication and diagnose the parametric and catastrophic process fluctuations. The reason for this is two-fold. Since in low volume production, every lot is critical to the highly constrained production schedule; it is imperative to monitor the process via in-line measurements and then control the process on-line. More importantly, since production lots are pipelined, it is desirable that any faulty process equipment be detected as early as possible. If the lot is diagnosed at the end of the process then all those lots which were in the pipeline will be subjected to the same process fluctuation until the time that the faulty equipment is identified and repaired. The exact number of lots will depend on the time interval between the faulty processing step and diagnosis. If the number of lots is large, this delay in diagnosis can lead to a significant yield loss and hence it is desirable that the delay be minimized. From the simple experiments that we have carried out on commercial fabrication processes it is evident that the proposed control algorithms are efficient enough in terms of CPU time and memory requirements to be used in commercial fabrication facilities. However, the efficiency is derived from having adequate process observability. Not only are the necessary in-lines to be measured, but the sample size should be large enough to allow statistical inferences with adequate confidence. This indicates a careful design of test structures as well as the use of automated measurement and data acquisition systems that will reduce the bottleneck during in-line measurements. The clustering and factorization algorithms provide the basis for the process observability; they identify the factor in-lines which form the minimal set of observables to be measured for adequate process monitoring. We are presently studying some of the existing commercial fabrication lines to check the feasibility of our algorithms. The data collected has been used to tune **FABRICS** for a number of advanced fabrication processes. Preliminary tests show that the control algorithms are robust and efficient enough to be applied in these processes. However, in most cases appropriate in-lines that are necessary to monitor the process are not measured with the desired accuracy thereby making on-line control infeasible. With the tightening of the performance tolerances and decrease of turn around times it is absolutely essential that appropriate in-lines be measured and the process trajectory monitored to successfully use the proposed CAM system.

ACKNOWLEDGEMENTS

We would like to acknowledge C. R. Shyamsundar for his contribution to the research presented in this paper.

REFERENCES

Box, G. E. P. 1986. *Studies in Quality Improvement: Signal to Noise Ratios, Performance Criteria and Statistical Analysis* (Tech. Rep. 11). Center for Quality and Productivity Improvement, University of Wisconsin-Madison,

Brayton, R. K., Director, S. W. and Hachtel, G. D. 1980. Yield Maximization and Worst-Case Design with Arbitrary Statistical Distributions. *IEEE Transactions on Circuits and Systems, CAS-27*, 756-764.

Director, S. W. and Hachtel, G D. 1977. The Simplicial Approximation to Design Centering. *IEEE Transactions on Circuits and Systems, CAS-24*, 363-372.

Harman, H. H. 1967. *Modern Factor Analysis*. Chicago. University of Chicago Press.

Mozumder, P. K., Shyamsundar, C. R. and Strojwas, A. J. 1986. Efficient Modeling Techniques for VLSI Process Control Applications. *In proceedings of 1986 IEEE*

International Symposium on Circuits and Systems.

Mozumder, P.K. 1986. *Statistical Quality Control of VLSI Fabrication Processes.* Master's thesis, Carnegie-Mellon University,

Mozumder, P.K. 1987. Statistical Quality Control of VLSI Fabrication Processes (PhD Proposal, Carnegie-Mellon University).

Nassif, S. R., Strojwas, A. J. and Director, S. W. 1984. FABRICS II: A Statistically Based IC Fabrication Process Simulator. *IEEE Transactions on Computer-Aided Design of Integrated Circuits and Systems, CAD-3*(1), 40-46.

Rao, C. R. 1973. *Linear Statistical Inference and its Applications.* New York: Wiley. 2nd. Edition.

Shyamsundar, C. R., Mozumder, P. K. and Strojwas, A. J. 1986. Cost Minimization Approach to VLSI Design and Manufacturing. *1986 IEEE International Symposium on Circuits and Systems Proceedings,* Vol. *1.*

Shyamsundar, C. R. 1986. *Adaptive Control of VLSI Fabrication Processes.* Master's thesis, Carnegie-Mellon University,

Shyamsundar, C. R. 1987. Profit Maximization Approach to VLSI Manufacturing (PhD Proposal, Carnegie-Mellon University).

Styblinski, M. A. 1986. Problems of Yield Gradient Estimation for Truncated Probability Density Functions. *IEEE Transactions on Computer-Aided Design of Integrated Circuit and Systems, CAD-5*(1), 30-38.

Taguchi, G. and Phadke, M. S. 1984. Quality Engineering Through Design Optimization. *IEEE Communication Society (Global Telecommunications Conference),* .

5 DEFECT-TOLERANT ARCHITECTURES

There has been a large amount of research worldwide devising fault-tolerant techniques which can be applied at the chip level and on developing ULSI chip architectures which can exploit such techniques. The reader is invited to consult Moore (1986), Jesshope and Moore (1986) and Saucier and Trilhe (1987) to see the range of techniques that have been proposed. The previous chapters of this book have illustrated the thesis that effective design of complex chips requires an intimate knowledge of the defect mechanisms and the way that these affect the yield and it may therefore be expected that successful exploitation of defect tolerance will only be obtained through such knowledge. In particular, defect-tolerant architectures usually exploit a high degree of regularity in order to have a large number of spares available to replace any defective parts, but at the same time rather complicated reconfiguration strategies may be required to overcome typical clustering of defects. Very great care is also needed in avoiding catastrophic defects which can affect the whole chip, such as power to ground shorts and this may involve considerations of circuit layout, fabrication processes, supply protection (Carley and Maly 1988) and, if necessary, even post-fabrication processing (e.g. microsurgery or discretionary metallization).

This chapter contains four papers which indicate recent thinking about how defect tolerance may be introduced into large area chips. Each paper goes to some lengths to show how the defect-tolerance strategy can react to the anticipated defect distributions.

Fuchs *et al.* (§5.1) look in some detail at three problems. They show how large regular arrays might be constructed from processing elements which do not have a conveniently rectangular layout, and they look at how the size of the processing elements can be traded for their performance and the number of elements required to satisfy an application. Finally they show how this granularity and the provision of spares can be chosen in the light of the expected defects in order to optimize the circuit yield.

Howells and Agarwal (§5.2) examine the ability of a tree architecture to withstand clustering of defects and show that this can only be successful if the spares are allocated in a way appropriate to each level of the hierarchy. Jones (§5.3) examines the yield of another tree hierarchy in which the processing elements are associative string processors. These have the additional very helpful property that useful computations can be obtained with different numbers of processors working in each branch.

It has often been suggested that rectangular arrays of processors with nearest neighbour interconnections could provide a richer interconnection for many applications than a string or tree architecture. Unfortunately such arrays pose severe obstacles to the reconfiguration needed to obtain defect tolerance. In the fourth section, Delgado-Frias *et al.* (§5.4) show that the problem can be solved for some applications which are well suited to a 2-D processor array but in which the computing resources need not be completely regular. They examine in some detail architectures to implement finite element analysis, low level image processing and semantic network computations.

REFERENCES

Carley L R and Maly W 1988 Circuit breaker for redundant IC systems *Proc. Custom Integrated Circuits Conf.*

Jesshope C R and Moore W R 1986 (eds) *Wafer Scale Integration* (Bristol: Adam Hilger)

Moore W R 1986 A review of fault-tolerant techniques for the enhancement of integrated circuit yield *Proc. IEEE* **74** (5) 684–98

Saucier G and Trilhe J 1987 (eds) *Wafer Scale Integration* (Amsterdam: North Holland)

5.1 The Impact of Parallel Architecture Granularity on Yield

Kent Fuchs, Ming-Feng Chang, Sy-Ken Kuo, Pinaki Mazumder and Craig Stunkel

INTRODUCTION

Design for on-chip parallel processing includes the definition and layout of individual processors in addition to the topological description of the interconnection network. Significant on-chip parallelism also requires large chip sizes, which rely on yield enhancement techniques to provide reasonable fabrication costs. Most design efforts utilizing architectural yield enhancement for parallel processing applications have focused on reconfiguration techniques for systolic type networks. Numerous architectural issues, which impact yield, have yet to be thoroughly examined. This paper introduces some of those architectural issues and examines them from a yield enhancement perspective. The issues examined include: (1) layout-based analysis of processor shapes and how specific shapes can be utilized to obtain compact layout for reconfigurable mesh networks; (2) performance and area (yield) trade offs for a variety of processor data path widths and memory requirements; and (3) an analysis of hierarchical redundancy in parallel architectures and the relationship to datapath width and local memory size. Three levels of redundancy structure are proposed with application to on-chip shared memory multiprocessors and reconfigurable mesh networks.

PROCESSOR GEOMETRIES AND LAYOUT

In order to minimize area and routing complexity, it is desirable that the processors should be close to each other and wiring length should be minimal. The main reason that parallel processor geometries are typically assumed to have a square aspect ratio is to distribute the delay uniformly across the horizontal and vertical directions of the chip. However, it is difficult to obtain a square layout from a natural rectangular or geometrically more complex layout. Aleliunas, Rosenberg (1982) and others, have investigated the problem of layout transformation. Most of the techniques proposed utilize the concept of grid transformation and are not always applicable to practical circuits. In this section of the paper, alternative shape geometries are investigated which form area efficient layouts on a chip or wafer to describe different array networks.

Polyomino cellular structures were originally proposed by von Neumann in connection with Cellular Automata and are utilized here as an alternative shape geometry to the square. Since a processing element is comprised of functional blocks, such as ALU, floating point units, memory, i/o buffer, etc., these functional blocks can be organized to describe the global shape of the processor similar to polyominoes (Golomb 1966).

The planar tessellation or tiling (i.e., covering the chip without overlapping) properties of these tiles are utilized in this section to construct layouts with maximum packing density and thereby enhanced yield. By applying linear transformations on the polyomino tiles, array networks can be mapped on the chip. Since the interconnections in a VLSI layout typically run either vertically or horizontally parallel to the edges of the chip, the tiles are oriented such that their edges are always parallel to the basis of the chip, represented by (x, y). The permissible orientations of a tile, τ_k, within a chip are due to the *reflections* about the x axis, the y axis and both x and y axes. The corresponding operators are denoted by π_x, π_y and π_z, respectively. The *rotation* about the z axis, orthogonal to the chip plane, is denoted by λ_z. The embedding of a repetitive tile structure is referred to here as a *mosaic*. For a VLSI implementation, such regularly structured layouts are easily constructed. These layouts are developed by replicating a singular type tile such that there is no gap between the successive tiles. Formally, a layout of τ_k under the embedding rule ϵ is called a *mosaic*, $\mathbf{M}_\epsilon(\tau_k)$, if and only if for every point within the layout, there exists an instance of the tile, τ_k denoted as $\tau_k{}^i$, such that for all $j \neq i$ $\tau_k{}^i \bigcap \tau_k{}^j = \phi$

Embedding algorithms

Using the notation of Mazumder (1987), two embedding algorithms are described in the following, which result in two alternative parallel processor structures with highly efficient layout, minimal interconnection and appropriate for yield enhancement through reconfiguration. The first algorithm constructs a mosaic layout which consists of tiles having the shape of an English alphabet L. The tile is formed by connecting three unit squares and is denoted by L_3, generally known as *triomino*. Hexomino tiles consisting of six squares and having P shape can also describe a similar mosaic layout. Layouts are recursively built according to the following:

Algorithm 1 The layout of L_3 describes a mosaic, $\mathbf{M}_{\epsilon 1}(L_3)$ under the following embedding rule (say $\epsilon 1$):

$$\mathbf{S}_n(L_3) = \begin{vmatrix} \pi_x \bullet \mathbf{S}_{n-1}(L_3) & \phi & \mathbf{S}_{n-1}(L_3) \\ \phi & \mathbf{S}(L_3) & \phi \\ \mathbf{S}_{n-1}(L_3) & \phi & \pi_y \bullet \mathbf{S}_{n-1}(L_3) \end{vmatrix}$$

where $\mathbf{S}_n(L_3)$ is the shape polynomial (Barnes 1982) of the layout having $n > 1$ *level of embedding*. $\mathbf{S}_1(L_3) = \mathbf{S}(L_3)$ is the shape polynomial of the triomino L_3.

An example is given to illustrate how the algorithm hierarchically generates the layout.

Example Assume $n = 3$ in Algorithm 1, then the mosaic layout $\mathbf{M}_{\epsilon 1}(L_3)$ will be given by

$$\mathbf{S}_3(L_3) = \begin{vmatrix} \pi_x \bullet \mathbf{S}_2(L_3) & \phi & \mathbf{S}_2(L_3) \\ \phi & \mathbf{S}(L_3) & \phi \\ \mathbf{S}_2(L_3) & \phi & \pi_y \bullet \mathbf{S}_2(L_3) \end{vmatrix}$$

The shape polynomial of the mosaic layout is given by:

$$\mathbf{S}_3(L_3) = \begin{bmatrix} \pi_x \bullet \begin{vmatrix} \pi_x & \phi & \iota \\ \phi & \iota & \phi \\ \iota & \phi & \pi_y \end{vmatrix} & \phi & \iota \bullet \begin{vmatrix} \pi_x & \phi & \iota \\ \phi & \iota & \phi \\ \iota & \phi & \pi_y \end{vmatrix} \\ \phi & \iota & \phi \\ \iota \bullet \begin{vmatrix} \pi_x & \phi & \iota \\ \phi & \iota & \phi \\ \iota & \phi & \pi_y \end{vmatrix} & \phi & \pi_y \bullet \begin{vmatrix} \pi_x & \phi & \iota \\ \phi & \iota & \phi \\ \iota & \phi & \pi_y \end{vmatrix} \end{bmatrix} \bullet \mathbf{S}(L_3)$$

$$
\mathbf{S}_3(\mathrm{L}_3) = \begin{bmatrix} \begin{pmatrix} \pi_x & \phi & \pi_z \\ \phi & \pi_x & \phi \\ \iota & \phi & \pi_x \end{pmatrix} & \phi & \begin{pmatrix} \pi_x & \phi & \iota \\ \phi & \iota & \phi \\ \iota & \phi & \pi_y \end{pmatrix} \\ \phi & \iota & \phi \\ \begin{pmatrix} \pi_x & \phi & \iota \\ \phi & \iota & \phi \\ \iota & \phi & \pi_y \end{pmatrix} & \phi & \begin{pmatrix} \pi_y & \phi & \pi_z \\ \phi & \pi_y & \phi \\ \iota & \phi & \pi_y \end{pmatrix} \end{bmatrix} \bullet \mathbf{S}(\mathrm{L}_3)
$$

The layout of $\mathbf{S}_3(\mathrm{L}_3)$ is shown in Figure 1. It can be seen that the diagonal (lower left to upper right) tiles are retained unaltered by the identity operator (ι), other tiles are rotated by the corresponding operator in the embedding matrix and the null operator (ϕ) indicates no occurrence of a tile in the corresponding position. In order to evaluate the importance of the above embedding algorithm, it is necessary to identify the interconnection network described by the mosaic layout. The graph described by $\mathbf{M}_{\epsilon 1}(\mathrm{L}_3)$ denotes communication geometries of square arrays with redundant cells as shown in Figure 2. In the embedding matrix the five non-peripheral diagonal operators (except ϕ) correspond to the location of redundant cells. It may be noted that the communication graph in Figure 2 corresponds to $\mathbf{S}_4(\mathrm{L}_3)$ and hence it has $5*4+1 = 21$ redundant cells corresponding to four $\mathbf{S}_3(\mathrm{L}_3)$ and the center $\mathbf{S}(\mathrm{L}_3)$.

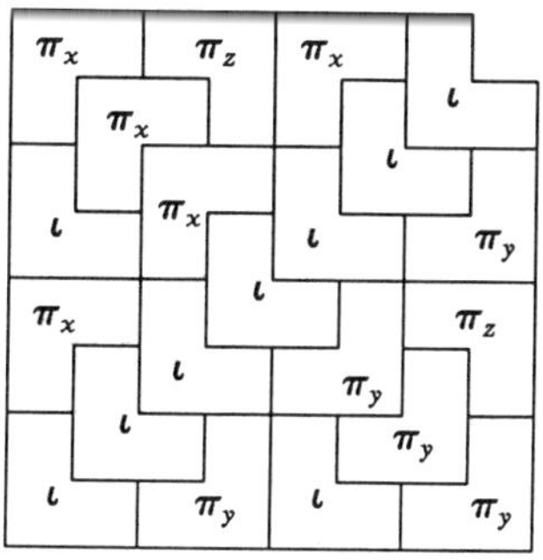

Figure 1. Layout obtained by Algorithm 1.

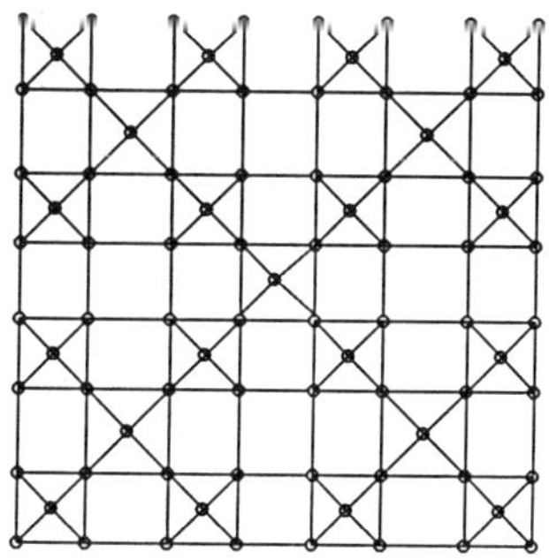

Figure 2. Reconfigurable mesh network.

Similar to the square meshes, the concept of tiling can also be used to describe hexagonal arrays as often used for systolic arrays. Figure 3 shows a mosaic layout made of I_2, generally known as *dominoes*. The n-th level of embedding consists of $4n-2$ dominoes, n dominoes in the top row, n dominoes in the bottom row, and $n-1$ dominoes in the left-most and rightmost columns. This describes an annular square region, having inner edge of length $n-1$ and outer edge of length n. The layout can be recursively built by the following algorithm:

Algorithm 2 The layout of I_2 describes a mosaic, $\mathbf{M}_{\epsilon 2}(\mathrm{I}_2)$ under the following embedding rule (say $\epsilon 2$):

$$
\mathbf{S}_n(\mathrm{I}_2) = \begin{vmatrix} \phi & \mathbf{S}(\mathrm{I}_{2n}) & \phi \\ \lambda_z \bullet \mathbf{S}(\mathrm{I}_{2n-2}) & \mathbf{S}_{n-1}(\mathrm{I}_2) & \lambda_z \bullet \mathbf{S}(\mathrm{I}_{2n-2}) \\ \phi & \mathbf{S}(\mathrm{I}_{2n}) & \phi \end{vmatrix}
$$

where $n > 1$ and $\mathbf{S}_1(\mathrm{I}_2) = \mathbf{S}(\mathrm{I}_2)$ is the shape geometry of I_2.

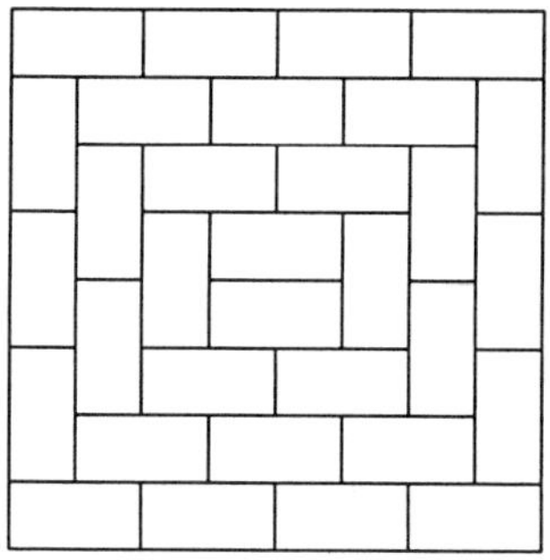

Figure 3. Layout obtained by Algorithm 2.

A number of other polyomino tiles also describe these array networks. The planar covering properties of these tiles have been pictorially represented by Figure 4 and from these geometrical arrangements the layout algorithms can be easily designed.

Figure 4. Mosaic layouts for hexagonal array network.

PROCESSOR ARCHITECTURE

Processor area

The approach employed here is to compare processor definitions on the basis of processing capability per chip area. A variety of data path widths have been utilized in parallel architectures, including single-bit and 32-bits (Hillis 1985). Processor area can be divided into two categories: logic *independent* of data path width d (such as most control logic, flag registers, memory address registers, etc.) and logic *dependent* on d. The logic area related to d can be further divided into areas roughly proportional to d, $\log d$, and $d \log d$. The area proportional to d corresponds to data buses, data registers, memory, fanout of control logic to the data path, etc. The area proportional to $\log d$ corresponds to addressing within a data word. Carry look-ahead often has logic area proportional to $d \log d$.

Given a particular design (*e.g.*, a 32 bit processor), the entire chip can be characterized in this manner, and this characterization can be used to predict the size of any other *similarly* designed processor of different granularity (*e.g.*, an 8 bit processor).

$$A(d) = \alpha_1 + \alpha_2 \log d + \alpha_3 d + \alpha_4 d \log d$$

For processors we have examined, $2 < \alpha_1/\alpha_3 < 25$ depending on the complexity of the control logic, and the size of the memory, with α_2 and α_4 typically much smaller than α_1 or α_3.

The effect on area due to multipliers, dividers, and floating point units can be determined by examining carry save adders. These units are typically designed to add or subtract a d length word multiple times in a cycle. Using carry save adders as a model, each carry save adder cell reduces three input bits to two output bits. If d_f words of length d are added, then it takes approximately $d_f d$ carry save adder cells to reduce the d_f words to two $d + d_f$ operands. These operands are then added with a conventional full adder. In addition, some of the logic for gating data into the carry save adders has area proportional to $\log d_f$ and d_f. Thus an important variable is the number of words to be added or subtracted per cycle for multiplies, d_f.

Considering d_f as a new variable for $A(d,d_f)$, the equation becomes:

$$A(d,d_f) = \alpha_1 + \alpha_2 \log d + \alpha_3 \log d_f + \alpha_4 d + \alpha_5 d_f + \alpha_6 d d_f + \alpha_7 d \log d \qquad [1]$$

This characterization of area is useful in estimating the effect of processor definition on yield. Equation [1] is utilized later in this paper for yield calculations.

Cycle time

The cycle time for a given processor architecture can also be modeled as a function of data path width d. The cycle time can generally be characterized by a constant overhead and a delay proportional to $\log d$. Control as an example contributes to the constant overhead. The delays proportional to $\log d$ result from factors such as the number of carry look-ahead levels required, and number of levels of fanout required to drive control lines to d width buses and register logic.

For multipliers, the cycle time is often determined by the number of carry save adder (CSA) levels. For $d > d_f$, this number of levels will not change as d is increased or decreased. d_f is therefore included in the equation for cycle time. Using hardware multipliers as an example, cycle time is a balance between the path delay through the full adder and the path delay through the CSA tree. Since each CSA level performs a 3 to 2 reduction in the number of data words, the number of CSA levels is proportional to $\log_{3/2} d_f$, and therefore the delay through the CSA tree is proportional to $\log d_f$. Other hardware multipliers place CSA levels and full adders in the same delay path. This results in a rough characterization of the maximum delay (the cycle time) as a function of the form:

$$CT(d,d_f) = \beta_1 + \beta_2 \log d + \beta_3 \log d_f \qquad [2]$$

Number of cycles

The number of cycles required to execute most operations will depend on d. It will also depend on the width of the data for the particular operation, D. For example, an integer addition will take $\lceil D/d \rceil$ cycles to complete for data width D and ALU width d. For a hardware multiply or floating point calculation, it is again beneficial to consider d_f, the number of words being added simultaneously during a cycle. The number of cycles typically required for a hardware multiply is

$$(\lceil D/d \rceil - 1)\left\lceil D/d_f \right\rceil + \lceil D/d \rceil$$

where the last $\lceil D/d \rceil$ represents the final summation by the full adder.

Various widths of data may be operated on during different parts of an algorithm or in different algorithms that represent a general workload for a processor. Assuming a variety of data widths, the number of cycles for a given processor description can be represented as

$$NC(d,d_f,D) = c_1 + \sum_{i=1}^{j} c_{2i}\left\lceil \frac{D_i}{d} \right\rceil + \sum_{i=j+1}^{j+k} c_{3i}\left\lceil \frac{D_i}{d} \right\rceil \left\lceil \frac{D_i}{d_f} \right\rceil \qquad [3]$$

where D_i for $i=1, \ldots, j$ represents the data widths for operations with complexity on the order of data width, and D_i for $i=j+1, \ldots, j+k$ represents data widths for operations such as multiply that have complexity of the order of the data width squared.

Modeling performance/area

The objective in formulating expressions for area and execution time is to obtain a model that will predict maximum performance per area.

$$\max PA(d,d_f) = \max \frac{\text{performance}}{\text{area}} = \max \frac{1}{A(d,d_f)CT(d,d_f)NC(d,d_f)}$$

Figure 5 is the graph of $PA(d,d_f)$ for parameters that describe a floating point processor executing an equal number of multiplies and adds. The y-axis is the *relative* values of $PA(d,d_f)$. For $d=32$, the decrease going from $d_f=16$ to $d_f=32$ indicates that the gain in performance for the floating point hardware does not compensate for the penalties of increased area and cycle time. Figure 6 shows a simple processor with no multiply or floating point hardware. This figure shows the effect of varying the mix between single bit logic operations and 32 bit addition operations. These results indicate that, depending on the mix of instructions, a processor data path width d between 1 and 32 is more efficient than either a bit-serial or 32 bit processor.

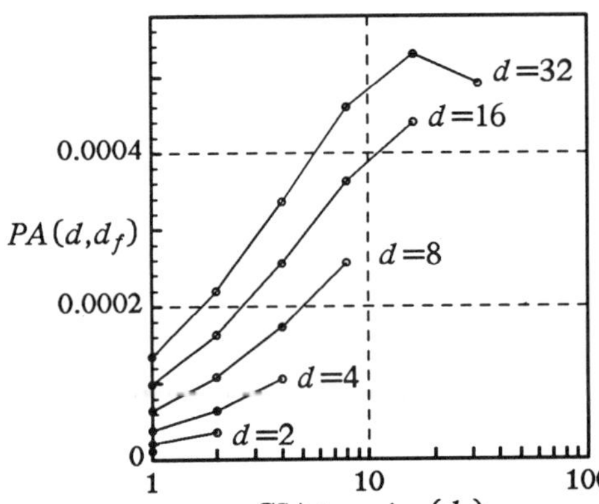

Figure 5. Varying d,d_f for floating point.

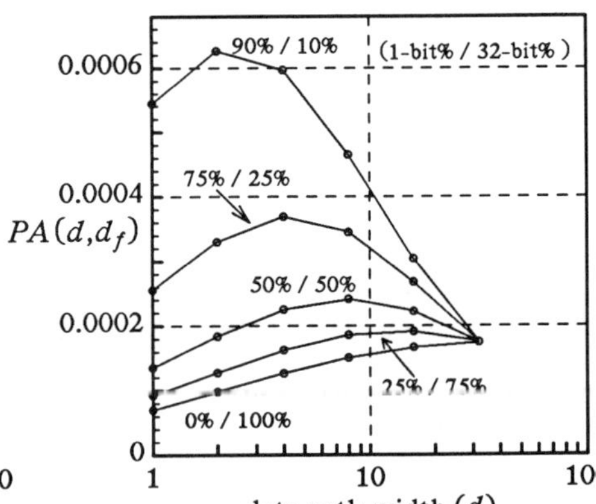

Figure 6. Varying operation granularity.

Application of the model to parallel processing

Several factors are important in modeling $PA(d,d_f)$ for an area consisting of multiple processors. As the total processing power of a chip or wafer increases, the total I/O bandwidth or the memory size must increase to maintain a balance between processor computation and I/O (Kung 1986). Both of these considerations potentially affect the area $A(d,d_f)$. Matrix multiplication and triangular factorization are used here to illustrate these considerations.

The number of I/O ports to and from a mesh array will be assumed to be fixed due to external interconnection limitations regardless of the number of individual processors in the mesh and their definitions. In performing a computation, a set of processors is said to be *balanced* if the computation time equals the I/O time. As presented by Kung (1986), an increase of processing power in a balanced set of processors with fixed I/O bandwidth necessitates an increase in memory capacity. For the case of matrix multiplication or triangular factorization, any relative increase p in processing power with no I/O rate increase for the mesh must be accompanied by a relative memory increase of p^2 to maintain the balance. Using the data in Figure 5, assume a balanced mesh for $d=16$ and $d_f=4$. The relative memory increase required for a $d=32$ and $d_f=4$ processor is:

$$p^2 = \left[\frac{PA(32,4)}{PA(16,4)}\right]^2 \approx \left[\frac{0.00034}{0.00026}\right]^2 \approx 1.71 \qquad [4]$$

This required memory affects the area calculation $A(d,d_f)$, and therefore the $PA(d,d_f)$ calculation. The optimum 32-bit processor has p^2 times as much memory as the 16-bit processor. As $PA(d,d_f)$ changes with memory requirements, p^2 changes accordingly. The optimum processor can be found graphically or analytically (a quadratic equation). To illustrate the process a graphical solution is given. The $\alpha_4 d$ term in equation [1] can be expanded into

$$(\alpha_{4a} + \alpha_{4m} w(d,d_f))d = \alpha_{4a} d + A_m(d,d_f)$$

where $w(d,d_f)$ represents the number of words in the processor memory and $A_m(d,d_f)$ represents the approximate space of the processor memory. For Figure 5, $\alpha_1=10$, $\alpha_2=.1$, $\alpha_3=.05$, $\alpha_4=.5$, $\alpha_5=.01$, $\alpha_6=.05$. Assume that initially α_4 splits into $\alpha_{4a}=.2$ and $\alpha_{4m} w(d,d_f)=.3$. For this assumption,

$$A_m(32,4) = 0.291\ A(32,4), \quad \text{and} \quad A_m(16,4) = 0.221\ A(16,4)$$

Setting $A(32,4) = A(16,4)$ (equal *total* area for meshes), $A_m(32,4) = 1.32\ A_m(16,4)$.

Since $p^2 = 1.71$, $A_m(32,4)$ must be increased to keep the processor computation time versus I/O time balanced. Figure 7 illustrates the solution, for which $p^2 \approx 1.5$ ($p \approx 1.22$), indicating that the 32-bit processor is still the best choice for matrix multiplication or triangular factorization, and that for arrays of equal total area, arrays of 32-bit processors will complete these algorithms 22% faster than arrays of 16-bit processors. The memory area in the 32-bit processor was increased by 23% to balance the computation time versus I/O time.

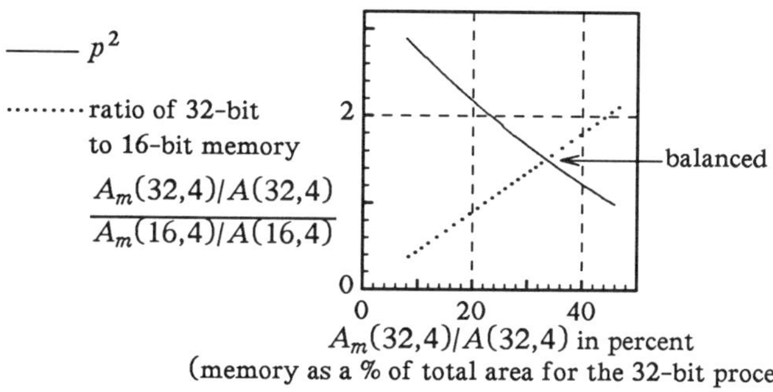

Figure 7. Finding the optimum ratio of memory area to total processor area.

PARALLEL PROCESSOR ARCHITECTURE YIELD

By designing with a hierarchy of redundancy, flexibility in reconfiguration strategies can be provided. Three levels of redundancy are proposed in this section of the paper with application to the parallel processor architectures previously discussed. An alternative analysis has been performed by Huang and Daughton (1986) who analyzed the relationship between yield and redundancy in wafer scale circuits with two levels of hierarchy.

A typical PE can be divided into four functional blocks, namely the control unit (CU), arithmetic logic unit (ALU), floating point unit (FPU) and local memory (LM). Due to layout and design differences the defect densities of these modules and their reconfiguration capabilities may be different. Therefore, the relative size of each block may also affect yield. The effect of these four functional blocks on the yield of the system is considered in the analysis of this section. Our analysis uses the simplifying assumption that every module has a Poisson defect distribution. It should be noted that this assumption has been shown not to be valid in some cases, particularly due to clustering of defects and an increased defect density near the perimeter of the wafer (Stapper *et al* 1983).

Redundancy within the PE

The PE organization under consideration consists of two groups of functional blocks as shown in Figure 8. The group N contains blocks without redundancy. Its total area is A_N, and its yield Y_N can be determined directly by its area. The other group R contains blocks with redundancy. Assume there are m blocks, and each block i contains n_i repeated elements. If we allocate r_i spare elements for block i, then the yield for each block and the whole group is given as:

$$Y_{R_i} = \sum_{j=0}^{r_i} C_j^{n_i+r_i}(1-Y_{e_i})^j {Y_{e_i}}^{n_i+r_i-j} \quad , \qquad Y_R = \prod_{i=1}^{i=m} Y_{R_i} \qquad [5]$$

Shared spare functional blocks between PEs

The second level of redundancy is at the functional block level, i.e., spare functional blocks can be shared by a group of PEs as shown in Figure 9. Each PE again consists of an unrepairable part N and a repairable part R. Assume there are n PEs and r spare repairable parts shared by these n PEs, then the yield for each PE is given as:

$$Y_{PE} = Y_N Y_R' \quad , \qquad Y_R' = Y_R + (1-Y_R)\sum_{i=0}^{n-1} F_i \sum_{j=1}^{r} f\left(\frac{j}{i+1}\right) R_j \qquad [6]$$

where F_i is the probability that given one PE's R block is faulty and i out of the other $n-1$ PEs need to be repaired, and R_j is the probability that j out of the r spare blocks are functional.

$$F_i = C_i^{n-1}[Y_N(1-Y_R)]^i[1-Y_N(1-Y_R)]^{n-1-i} \quad , \qquad R_j = C_j^r Y_R^j (1-Y_R)^{r-j}$$

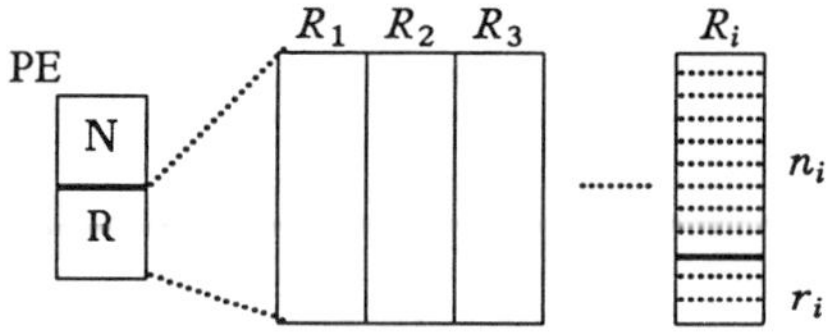

N : functional blocks without redundancy.
R : functional blocks with redundancy.
Figure 8. Redundancy within functional blocks.

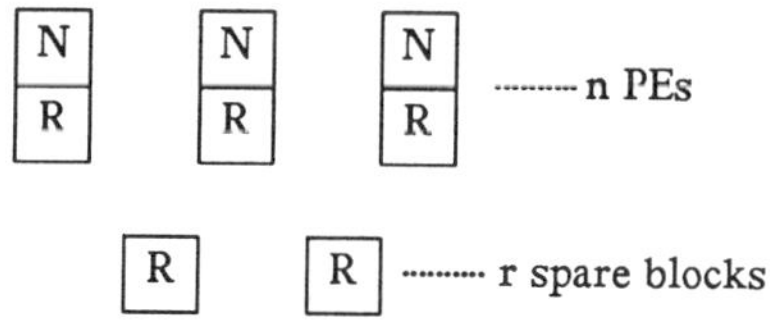

N : unrepairable.

R : repairable.

Figure 9. Shared spares between functional blocks.

$$f(\frac{j}{i+1}) = 1 \text{ (if } j \geqslant i+1) , \quad \text{or } \frac{j}{i+1} \text{ (if } j < i+1)$$

PE replacement

If a faulty PE cannot be repaired by using redundancy within functional blocks or by the shared spare functional blocks, then the faulty PE is a candidate for replacement by a spare PE. The analysis scenario for PE level replacement is shown in Figure 10. There are two groups of PEs. Assume there are N_1 PEs in group 1, N_2 PEs in group 2, and M total PEs are required from these two groups to make the system functional. In this case, the replacement between the two groups is at the PE level and its yield is given as follows,

$$\text{Chip Yield} = \sum_{i=0}^{i=N_1} P_i \sum_{j=M-i}^{j=N_2} Q_j , \quad P_i = C_i^{N_1} Y_{PE_1}^i (1-Y_{PE_1})^{N_1-i} , \quad Q_j = C_j^{N_2} Y_{PE_2}^j (1-Y_{PE_2})^{N_2-j}$$

where Y_{PE_1} and Y_{PE_2} are the PEs yield for group 1 and group 2 respectively, which can be derived from equation [6].

Hierarchical yield analysis for shared memory and mesh architectures

The yield for two types of parallel architectures is analyzed in the following. The first is the fault-tolerant mesh of Figures 1 and 2. The other is a shared memory multiprocessor with a single bus with its layout configuration as shown in Figure 11. For the fault-tolerant mesh, in addition to PE level replacement, we also allow reconfiguration within each PE's local memory. For the shared memory multiprocessor system, all three levels of hierarchy are utilized. Our goal is to maximize the absolute yield, which is defined to be the chip yield per unit area.

$$\text{Absolute Yield} = \frac{\text{Chip Yield}}{\text{Chip Area}}$$

The processor without redundancy has memory of area A_m and yield Y_m with additional functional blocks of total area A_N and yield Y_N. If memory repair is allowed through block replacement and the memory is divided into n_i blocks, and its r_i redundant blocks, then the

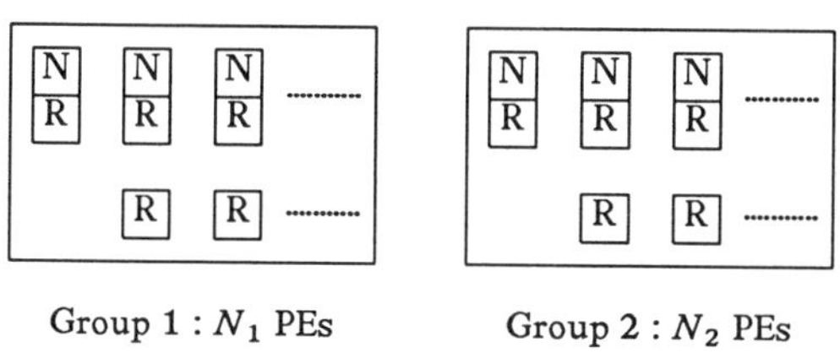

Figure 10. PE level replacement.

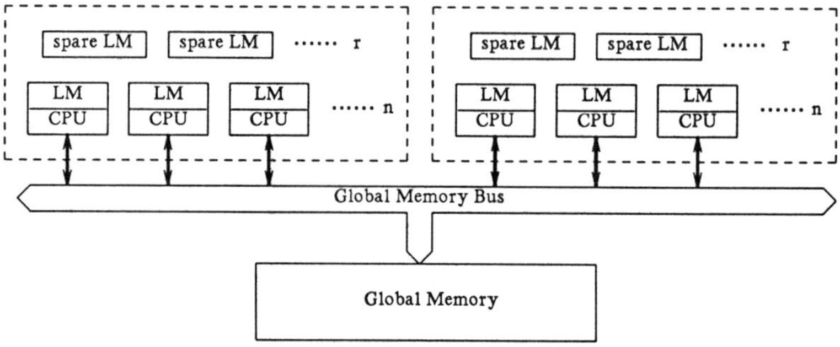

Figure 11. Shared memory multiprocessor with single bus.

area and yield for the processor with redundancy within local memory is

$$A_{PE} = A_N + A_m \frac{n_i + r_i}{n_i} \;, \qquad Y_{PE} = Y_N \sum_{j=0}^{r_i} C_j^{n_i+r_i} (Y_m^{\frac{1}{n_i}})^{n_i+r_i-j} (1 - Y_m^{\frac{1}{n_i}})^j$$

Since the layout of the reconfigurable mesh of Figure 2 is recursively defined, its yield can also be recursively defined. Let Y_i denote the yield of $S_i(L_3)$,

$$Y_i = Y_{i-1}^4 + 4Y_{i-1}^3 Y_{PE} C_{i-1}$$

where Y_{PE} is the yield of a single processor (i.e., Y_1), and C_{i-1} is the possibility that $S_{i-1}(L_3)$ is faulty and can be repaired by one spare processor which is added in the center of $S_i(L_3)$.

$$C_1 = 1 - Y_{PE} \;, \qquad C_2 = 4Y_{PE}^3 (1 - Y_{PE})^2$$

$$C_3 = 3Y_2^3 Y_{PE}^3 (1 - Y_{PE})^2 + (1 - Y_{PE}) Y_2^3 C_2 + 3Y_2^2 Y_{PE} C_2^2 \;, \qquad C_i \approx 0 \quad \text{for } i \geqslant 4$$

The area of each functional block for different data path widths is given in Table 1. The defect density in these examples was chosen to be $0.04/mm^2$. An example curve showing the relationship between Y/A and memory redundancy for different processor sizes and different mesh sizes is in Figure 12. The corresponding area for each case is shown in Table 2.

Table 1. Area(mm^2) of functional blocks.

d	MEM	CU	ALU	FPU
8	2.4	2.95	1.6	3.28
16	4.8	3.10	3.2	12.96
32	9.6	3.25	6.4	51.52

Table 2. Area(mm^2) of redundant meshes.

$d=8\ S_4(L_3)$	$d=16\ S_3(L_3)$	$d=32\ S_2(L_3)$
870	505	354

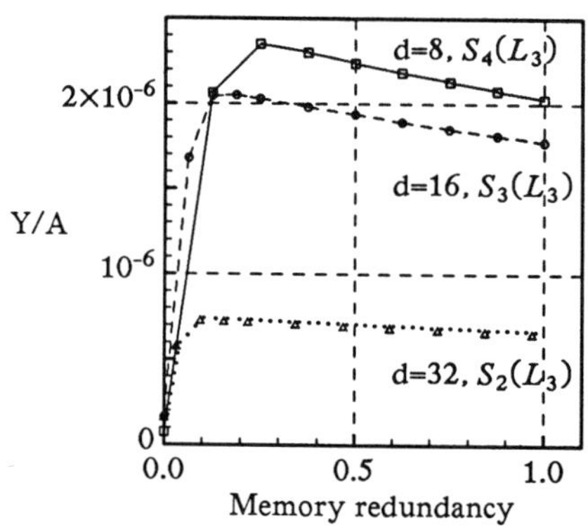

Figure 12. Absolute yield vs. memory redundancy for fault-tolerant meshes.

The yield for the shared memory system architecture of Figure 11 can be determined by calculating the processor, bus and memory area in conjunction with the previous yield equations for the reconfigurable mesh. Without considering the interconnect overhead or the global memory, the total area for all the processors is given by

$$A = 2(A_N + A_m \frac{n_i + r_i}{n_i} \frac{n+r}{n})n$$

Example curves showing the relationship between Y/A and various levels of redundancy are in Figure 13-14. Figure 13 illustrates the efficiency of redundancy within memory blocks compared to shared spare memory blocks with the condition that each group has ten 16-bit PEs and eight total PEs are required. In Figure 14 maximum yield is achieved with moderate memory redundancy.

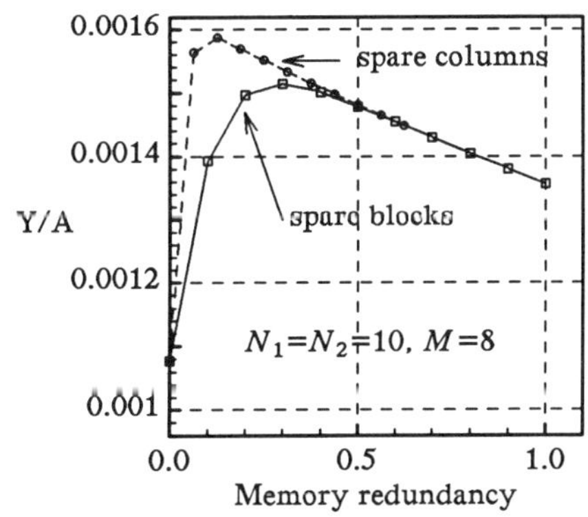

Figure 13. Absolute yield vs. memory redundancy for a shared memory system.

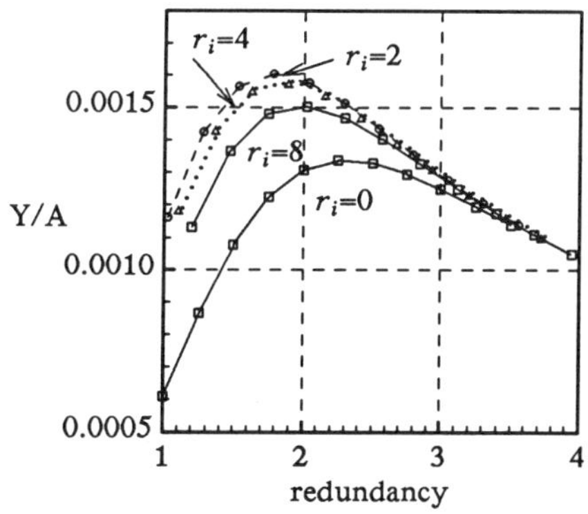

Figure 14. Absolute yield vs. PE redundancy for different memory redundancy in a shared memory system.

SUMMARY

A yield-based analysis was presented in this paper for processor architectures utilized for on-chip parallel processing. Individual processor geometries were examined and several shapes were shown to result in compact layouts implementing reconfigurable mesh interconnection networks. An examination of performance and hierarchical yield was made for a variety of processor data path widths and memory requirements.

ACKNOWLEDGMENTS

This research was supported in part by a Faculty Incentives for Excellence Award from Digital Equipment Corporation and in part by the Semiconductor Research Corporation under Contract 86-12-109.

REFERENCES

Aleliunas R and Rosenberg A L 1982 On embedding rectangular grids in square grids, *IEEE Transactions on Computers* **C-31** 907-913

Barnes F W 1982 Algebraic theory of brick packing 2, *Discrete Mathematics* **42** 129-144

Golomb S W 1966 Tiling with polyominoes, *Journal of Combinatorial Theory* **2** 280-296

Hillis W D 1985 *The Connection Machine* (Cambridge, MA: The MIT Press)

Huang J S T and Daughton J M 1986 Yield optimization in wafer scale circuits with hierarchical redundancies *Integration* **4** 43-51

Kung H T 1986 Memory requirements for balanced computer architectures, *Proc. of 13th Annuanl Symposium on Computer Architecture* 49-54

Kuo S Y and Fuchs W K 1987 Efficient spare allocation for reconfigurable arrays *IEEE Design and Test* **4** 24-31

Mazumder P 1987 Planar decomposition for quadtree data structures *Computer Vision, Graphics, and Image Processing* **38** 258-274

Stapper C H, Armstrong F M and Saji K 1983 Integrated circuit yield statistics, *Proceedings of the IEEE* **71** 453-470.

5.2 A Cluster-Proof Scheme for Improving Yield and Reliability of Large Area Binary Tree Architectures

Michael Howells and Vinod Agarwal

INTRODUCTION

The binary tree has been shown to be a useful architecture for concurrent processing (Mead and Conway 1980). High performance searching engines (Bentley and Kung 1979), dictionary machines (Somani and Agarwal 1985), and other specialized Very Large Scale Integration (VLSI) multiprocessor systems have been designed based on the binary tree architecture. The major drawback of the binary tree architecture, like most other multiprocessor architectures, is that the complexity and size of the system increases the likelyhood of failure of the hardware. Thus many researchers have proposed VLSI implementations of binary tree architectures which incorporate redundancy and permit reconfiguration of the system in response to failures (Raghavendra *et al* 1984, Rosenberg 1983, Hassan and Agarwal 1986, Lowrie and Fuchs 1985). These reconfiguration schemes have focused mainly on improving the reliability of the system.

With the trend towards large area devices and Wafer Scale Integration (WSI), much larger binary tree architectures may be integrated on a single device. Unfortunately, with this increase in area has come a similar increase in fabrication related defects which has caused very low device yields. Reconfiguration schemes used to improve reliability do not seem suited to the problem of yield enhancement due to the clustering characteristic of defects. In this paper a reconfiguration scheme which enhances both the yield and the reliability of large area binary tree architectures is briefly described. The goal of this work is the design of large binary tree architectures of eight to twelve levels having hundreds or thousands of small (0.05 square centimeter) processing nodes integrated on a single large area chip. For more details about this work, please consult either Howells and Agarwal (1987) or Howells (1987).

PREVIOUS WORK

Hassan and Agarwal (1986) proposed a scheme for designing a fault-tolerant binary tree architecture which achieves a good increase in system reliability for a relatively low cost in redundant hardware. Their approach involves subdividing the entire binary tree into modules that can each tolerate one fault. A typical module is a two level subtree comprising four nodes, one of which is a spare. If any one of the nodes in such a module becomes faulty, the remaining three are capable of reconfiguring a working two level subtree module. The Modular approach has the advantage of localizing all aspects of fault-tolerance within each module, thus greatly simplifying the necessary control algorithms. In addition, although 33% redundancy in nodes is required for a two level subtree module, the number and complexity of the redundant links and switches is lower than for other approaches. The layout of the approach is also shown to be $O(N)$, which is optimal. Unfortunately the Modular approach is not suitable for yield enhancement of large area binary trees because the modules cannot tolerate multiple faults which can be expected to occur due to defect clustering.

Lowrie and Fuchs (1985) proposed the SOFT approach for a fault-tolerant binary tree architecture which attempts to alleviate the problem of multiple faults occurring within a local region of the tree. The spare nodes are placed at the leaf level of the binary tree and redundant links and switches are provided to allow faulty nodes to be replaced by any unused spare at the base of the subtree which is rooted at themselves. All nodes on the path between the spare node and the faulty node have their I/O links reconfigured so that one of the sons of each node replaces the node itself. Due to the increased number of spares available for replacing any particular faulty node, the reliability of the SOFT binary tree is much better than that realized with the Modular approach, even with only 25% redundant nodes. The layout of the SOFT binary tree is also shown to be $O(N)$. The increase in reliability is at the cost of increased complexity of the reconfiguration hardware and control algorithms over those of the Modular approach. Nonetheless the SOFT approach does appear to be suitable for yield enhancement as the sharing of spares makes the scheme more tolerant of multiple faults caused by clustered defects.

DEFINING THE PROBLEM

In order to improve both the yield and reliability of large area binary tree architectures, the faults caused by both clustered defects and uniformly distributed failures must be tolerable. This problem is compounded when the entire system is integrated on a single large area chip rather than subdividing it into a number of smaller, less complex and less defect-prone chips.

Existing approaches appear to offer solutions for dealing with faults due to uniformly distributed failures or defects, but Stapper (1986a) has shown that most defects tend to be clustered. The effect of this clustering of defects on a VLSI implementation of a binary tree architecture will depend on the nature of the clusters as they relate to the size and layout of the nodes in the tree. Recently Stapper (1986b) presented an equation which allows the yield of a group of physically adjacent chips, some of which are spares, to be computed. If Stapper's equation is used to compute the yield of a group of four adjacent nodes, one of which is a spare node, each of area 0.05 square centimeters, then it can be shown that for reasonable values of the clustering parameter α, the yield of such a group decreases due to the effect of defect clustering. Hence, for architectures with small node areas, defect clustering is a major concern.

If clustered defects are a concern, then the layout of the architecture will play a role in deciding what parts of the structure are susceptible to clusters. Consider a cluster of defects mapped onto a binary tree architecture having small nodes and a layout similar to that of the $O(N)$ H-tree layout in which three level, H-shaped subtrees are recursively embedded. Due to this layout technique, interconnected nodes in the lower levels of binary tree are physically closer to each other than at the higher levels of the tree. In fact, the interconnection distance between nodes doubles every two levels as one ascends the binary tree from the leaf level. In order to keep the costs of redundant sparing reasonable, spares can usually only be made available to a closely interconnected family of nodes. A cluster of defects will only be a problem if it affects a number of nodes in the same family, which, due to the properties of the H-tree layout, will only occur at the lower levels of the binary tree. Thus, for an H-tree layout of a binary tree having small nodes, only the reconfiguration hardware of the lower levels of the tree needs to be able to tolerate clustered faults.

THE CLUSTER-PROOF SCHEME

The Cluster-Proof scheme, proposed in Howells (1987), exploits these layout characteristics in order to choose the optimal point to place the spare processing nodes having the least cost on performance and the greatest benefit for yield enhancement. The point chosen is k levels from the bottom of the n level binary tree, the tree being cut into an upper subtree and many lower subtrees as shown in figure 1. The level chosen represents the point at which an average defect cluster is determined to no longer affect two interconnected nodes at that level due to the interconnection distance between them. Sets of spares are placed adjacent to the root of each of the lower subtrees and redundant links and switches are provided so that these spares may be used to replace faulty nodes within either the upper subtree or their respective lower subtree. The upper subtree, for which defects and failures can be considered uniformly distributed, is implemented using a switch and interconnect scheme similar to the SOFT approach, but with the spares being supplied from the lower subtrees. Within the lower subtrees clustered faults can be tolerated, as global sparing is used to allow any pattern of faulty nodes to occur. Global sparing, which is usually considered to be impractical due to high costs, may be achieved at reasonable cost within the lower subtrees due to the short interconnection distances between the nodes in a lower subtree.

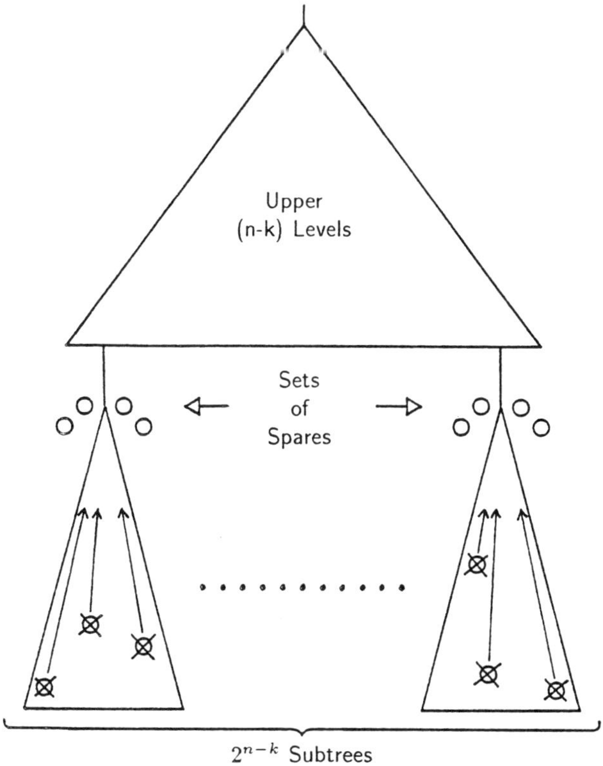

Figure 1. Overview of the Cluster-Proof Approach ©IEEE 1987

As the Cluster-Proof scheme assumes the use of an upper subtree based on a slightly modified existing approach, only the reconfiguration hardware of a lower subtree will be discussed. The reconfiguration hardware of a lower subtree comprises a set of spares at the top of the subtree, one redundant link per node in the subtree, and a number of different switch sets. This is shown in figure 2. On the I/O link between each pair of interconnected nodes is an isolation switch (IS) which permits the I/O links to be re-routed to the spares, via a redundant link (R-link), if either node becomes faulty. The R-links are routed to the top of the tree, where they are grouped into a bus (R-bus) which is isolated from the R-links by a set of switches (LBS). Each spare has a switch array (SBS) which permits the I/O links of the spare to connect to any of the redundant links in the R-bus so as to allow the spare to replace any faulty node. With this hardware, a Cluster-Proof subtree can tolerate faults which occur in the nodes, the spares, the I/O links, the isolation switches, and the R-links, assuming that the R-bus, the LBS and the switch arrays of the spares are known to be good. Faults in the redundant links or in the isolation switches are treated as if both nodes using this switch or link were faulty. Faults in the upper subtree are reconfigured as described in Lowrie and Fuchs (1985), with the exception that the spares are provided by the lower subtrees. It must be assumed that mechanisms for fault-detection, fault-diagnosis and recovery exist and are free of faults. The fault-detection and diagnosis would probably be implemented using built-in self-test (BIST) hardware localized within each subtree.

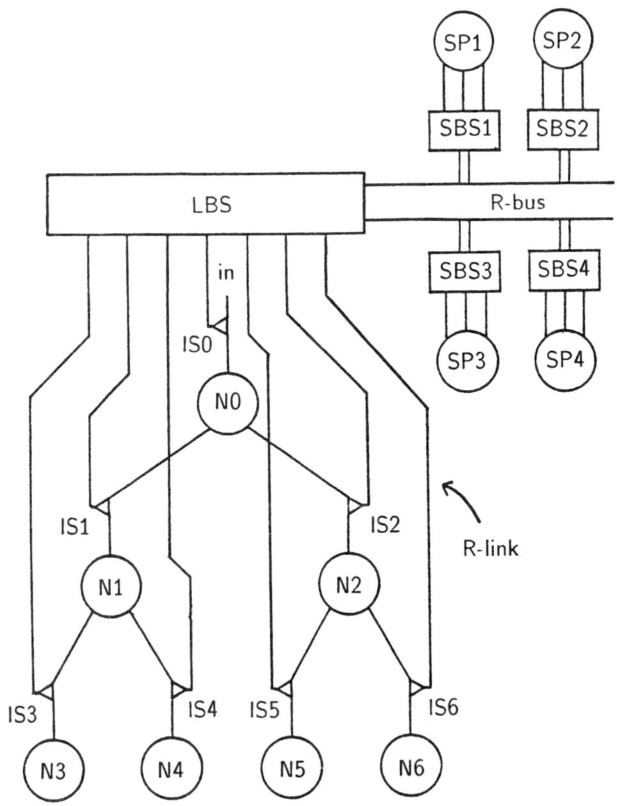

Figure 2. A k=3 Cluster-Proof Lower Subtree ©IEEE 1987

The layout of the Cluster-Proof binary tree can be done using an optimal $O(N)$ H-tree layout. To layout the lower subtree as an H-tree, the spare nodes are placed in a row adjacent to the root node of the subtree by widening the central channel of the subtree block. The isolation switches are incorporated into the nodes and the switch arrays are incorporated into the spares. The redundant links coming from each isolation switch force a widening of the channels, but do not cause any additional complexity as the routing of these links is parallel to the existing links. The layout of a $k = 5$ lower subtree is shown in figure 3. If the entire lower subtree block is then considered as a single large node and the approach being used for the upper subtree also uses an H-tree layout, then it follows that the layout of the entire Cluster-Proof binary tree may be an $O(N)$ H-tree layout.

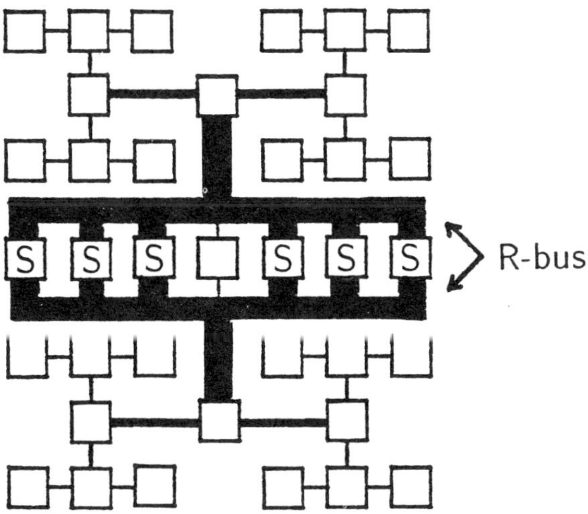

Figure 3. Layout of a k=5 Lower Subtree ©IEEE 1987

EVALUATION

The yield and reliability of the Cluster-Proof scheme was computed and compared with the yield and reliability realized using the SOFT and Modular schemes. The comparison was done for eight, ten and twelve level binary tree architectures at different defect densities and failure rates. For the Cluster-Proof scheme the level chosen for the spares was set at half the total number of levels in the tree ($k = n/2$). It was observed that as the size of the lower subtrees increases, the number of spares required to achieve a reasonable yield or reliability decreases, due to better sharing of the spares. The yield was calculated using an analytic approach based on Stapper (1986b) and also using a Monte-Carlo style defect simulation. In these calculations a number of assumptions were made and hence the computed yield values are not absolute, although they are fine for comparison. Both methods gave similar results and, as can be seen in figure 4, the yield of the Cluster-Proof scheme is much greater than the yield of the other schemes. In addition, the yield results for the Cluster-Proof scheme remain high as the size of the tree and the corresponding lower subtree are increased, while the yield results for the other schemes get much worse with increasing tree size. The results of analytic calculation of the reliability also show that the Cluster-Proof scheme gives greater reliability than the other schemes.

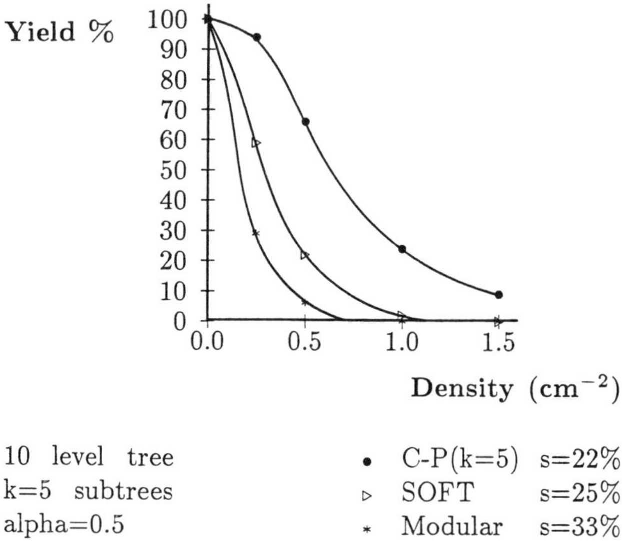

Figure 4. Simulated Yields for 10 Level Trees. ©IEEE 1987

To contrast the costs of the Cluster-Proof tree with those of the SOFT and Modular trees, the amount of redundant hardware for each of these three schemes was compared. Three different measures of redundancy were used, based on the costs of the redundant nodes, switches and redundant links in the scheme. Table 1 summarizes the results of the redundancy calculations. In general the Cluster-Proof scheme uses more switches and redundant links than either of the other two schemes but fewer spares. As the cost of the redundant links and switches is much less than the cost of the redundant nodes, the savings in nodes can be considered to offset the increases in switches and links. Thus the Cluster-Proof scheme can be considered to have redundancy costs comparable to those of the SOFT approach and the Modular approach.

Tree Size	Scheme	Nodes	Switches	Links
8-level	Modular	85	1705	289
	SOFT	64	2240	255
	C-P (k=4.s=25%)	64	2956	255
10-level	Modular	341	6852	1157
	SOFT	256	8960	1023
	C-P (k=5.s=22%)	224	17912	1767
12-level	Modular	1365	27305	4628
	SOFT	1024	35840	4095
	C-P (k=6.s=19%)	768	112688	12159

Table 1. Redundancy Results for Various Schemes. ©IEEE 1987

CONCLUSIONS

The problem of designing a reconfiguration scheme suitable for both yield and reliability improvement has been examined. The importance of clustered defects on the yield of binary tree architectures having small nodes and the role of layout in determining the mapping of these defects onto the structure of the architecture have been discussed. A reconfiguration scheme for both yield and reliability improvement of VLSI implementations of large area binary tree architectures has been developed and evaluated. The scheme gives much better yield and reliability than other existing schemes at a comparable cost in redundancy, while using an optimal $O(N)$ H-tree layout.

REFERENCES

Bentley J and Kung H T 1979 A tree machine for searching problems, Proceedings of the International Conference on Parallel Processing 257-266

Hassan A S and Agarwal V K 1986 A modular approach to fault-tolerant binary tree architectures, IEEE Transactions on Computers 35 356-361

Howells M C 1987 A cluster-proof approach to yield enhancement of large area binary tree architectures, Master of Engineering Thesis - McGill University

Howells M C and Agarwal V K 1987 Yield and reliability enhancement of large area binary tree architectures, Proceedings of the International Symposium on Fault-Tolerant Computing 290-295

Lowrie M B and Fuchs W K 1985 Dynamic reconfiguration in tree architectures, Technical Report - University of Illinois

Mead C A and Conway L A 1980 Hierarchically organized machines, Introduction to VLSI Systems 8.4

Raghavendra C S, Avizienis A and Ercegovac M D 1984 Fault-tolerance in binary tree architectures, IEEE Transactions on Computers 33 568-572

Rosenberg A 1983 The Diogenes approach to testable fault-tolerant arrays of processors, IEEE Transactions on Computers 32 902-910

Somani A K and Agarwal V K 1985 An efficient unsorted VLSI dictionary machine, IEEE Transactions on Computers 34 841-852

Stapper C H 1986a On yield, fault distributions and clustering of particles, IBM Journal of Research and Development 30 326-338

Stapper C H 1986b Yield statistics for large area ICs, Proceedings of the International Solid State Circuits Conference 168-169

5.3 Influence of Defects and Defect Clustering on the Silicon-Efficiency of a Hierarchically-Organized Wafer-Scale Parallel Processor

Simon Jones

INTRODUCTION

In recent years, many researchers both in academia and industry have been investigating the design and development of wafer-scale parallel processing computer architectures. A key issue in the successful implementation of such architectures is the design of a suitable interconnection strategy that can support fault-tolerance and provide adequate signal distribution.

A hierarchically-organised tree interconnection strategy has been shown to be a particularly promising structure for WSI implementation (Jones and Lea 1986). The tree interconnection strategy (Figure 1) provides fault-tolerance through 'lopping off' faulty branches. Furthermore, high-speed communication can be achieved by zoning clocks within sub-trees and utilising the 'logarithmic' point-to-point communication distance of a tree.

There has been significant analysis of the fault-tolerance of tree structures using mathematical models (eg. Fourman 1986, Rosenberg 1986). This paper offers an alternative view, looking at the Silicon-efficiency of a particular tree structure using actual circuit layouts under a range of defect densities and clustering parameters.

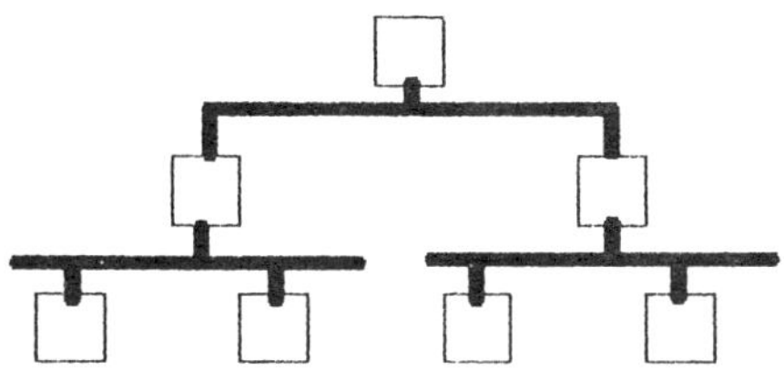

Figure 1. Hierarchical interconnection strategy

WASP

WASP is a wafer-scale implementation of an associative string processor (ASP). An ASP is a distributed-logic memory fine-grain parallel processing computational structure. An ASP comprises a linear string of identical associative processing elements (APEs) that operate in parallel to achieve high-speed cost-effective structured data processing (Figure 2).

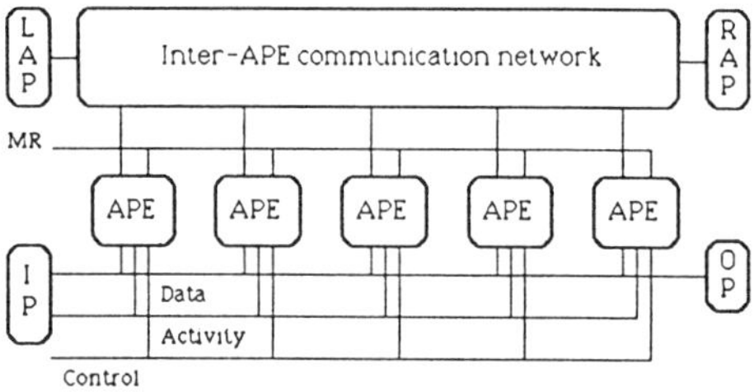

Figure 2. ASP structure

Each APE is connected to an inter-APE communication network, that runs in parallel with the APE string. All APEs are connected to common data, activity and control busses and share a common feedback line (MR). These busses and the MR line are supported by an external string controller which also maintains the left (LAP) and right (RAP) activity ports of the inter-APE communication network. Each APE comprises a data register, an activity register and a word-parallel comparator (implemented through an associative memory) together with logic for local arithmetic processing and communication with other APEs in the string.

A segment of an associative processor can be cost-effectively implemented with an associative parallel processor (APP) as shown in Figure 3.

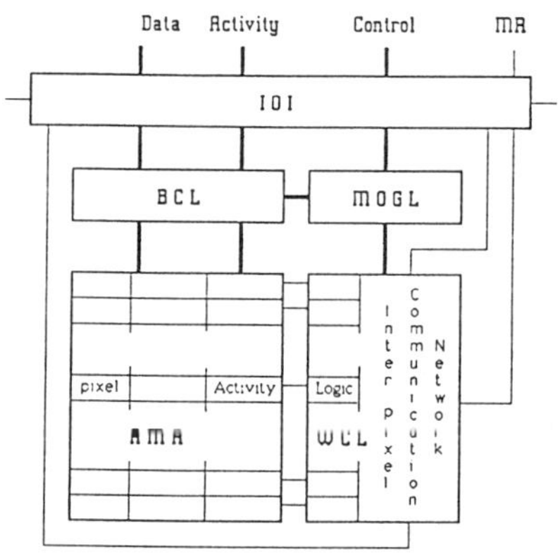

Figure 3. APP structure

Wasp image processor

The furthest developed WASP variant to date, is a version targeted at the support of image processing applications (WASP IP) (Lea 1986, Jones, Warren and Lea 1986).

WASP IP implements a complete 8192-pixel patch processing module, comprising 4 main elements (Figure 4).

(1) Wafer interface (WI) modules providing IO to the outside world.

(2) RAM modules for storing image patches.

(3) CC modules for controlling the APP modules and interchanging processed patches in the APP with other patches held in the RAM modules.

(4) APP modules for the parallel processing of image patches.

In operation, a 2-dimensional 8192-pixel image patch (eg. 512 x 16) is mapped into the associative string. Thus a 512 x 512 video frame could be processed by sweeping the 512 x 16 patch across the frame. The RAM modules support multiple patch storage and patch transfers between the RAM buffers and external frame stores can be overlapped with patch processing within the APP modules to achieve high-speed patch processing.

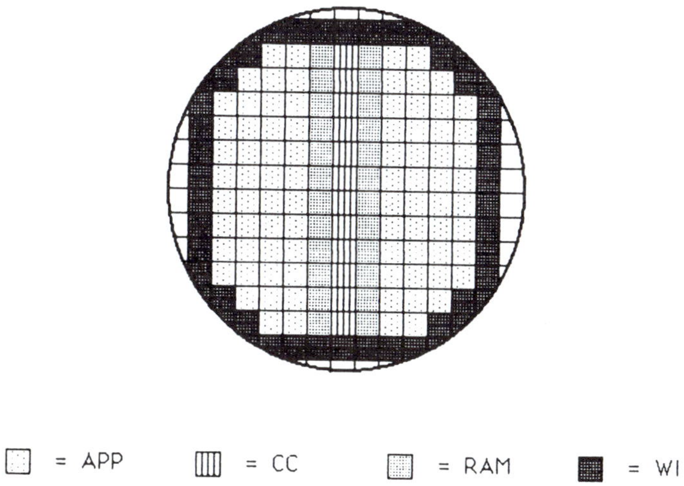

Figure 4. WASP image processing module

WASP fault-tolerance

WASP implements a hierarchical interconnection strategy with the WI modules being the root of the tree, the branches are the CC blocks and the twigs RAM blocks, with the leaves being the APP blocks. Faulty groups of APEs, APP modules or even complete branches are bypassed by connecting their LAP and RAP ports appropriately. Since all APEs can be used, regardless of their location, the WASP fault-tolerance strategy is effectively an 'n from m' strategy.

The most yield-efficient approach would be to include every working APE into the associative string (Figure 5), in practice the overhead of including LAP/RAP bypass logic in each APE proves ineffective. Hedge and Lea (1986) demonstrate that by organising the APEs into sub-strings (viz. groups of APEs) and bypassing faulty sub-strings (even if only 1 APE in the sub-string is faulty) a more silicon cost-effective strategy will result. This mechanism is illustrated in Figure 6, where the sub-string bypassing is implemented in the 'row links'.

In the same paper Hedge and Lea also investigate the design tradeoffs with different sub-string length for WASP. They determine that with 64-APE APPs a sub-string length of 4 (viz. 4-APE 'blocks') provides the best tradeoff.

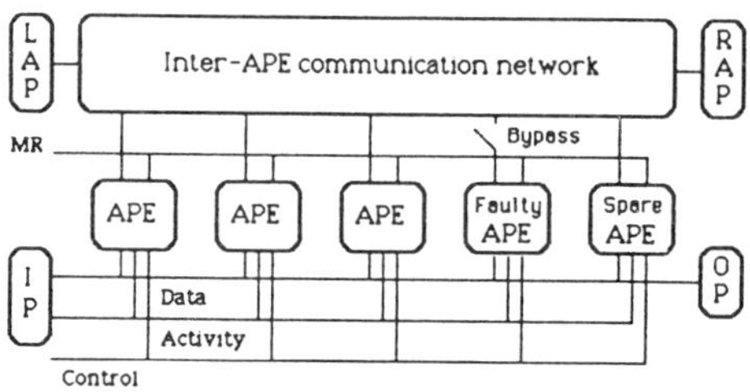

Figure 5. Ideal ASP fault-tolerance

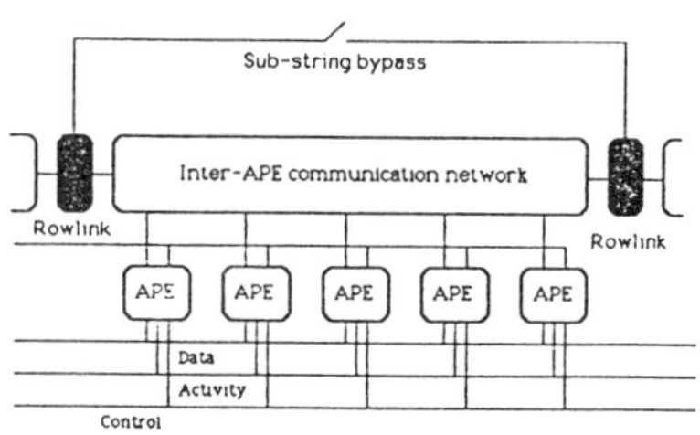

Figure 6. ASP by-pass strategy

EXPERIMENTATION

Objective

The objective of the experimentation is to determine the silicon-efficiency of the WASP tree interconnection strategy under a range of defect densities and defect clustering parameters.

APE circuitry

A WASP APE comprises a 37-bit CAM word plus an APE slice.

The relevant areas of the APE are

CAM word: 71928 um^2 WCL area: 72000um^2

Giving an APE area of: 143928um^2

The APEs are grouped into 4-APE sub-strings, which neglecting sub-string bypass logic area (estimated at 4% overhead in Hedge and Lea 1986) gives a sub-string area of

$$0.57571mm^2$$

Assumptions

In the experimentation, the following assumptions have been made

(1) The yield model used is a commonly occurring one based on a negative binomial distribution where expected yield is given by

$$Y = (1+DA/x)^{-x}$$

Where A = silicon area
D = average defect density
x = clustering parameter

(2) Expected average defect density (D) range is 0/02-0.08d/mm^2.

(3) The expected clustering parameter (x) range is as in (Carlson and Neugebauer 1987).

0.5-4

To model no defect clustering, an x value of

1 000 000

is used.

Wafer-scale metrics

The following wafer-scale metrics based on previous work by the author (Jones and Lea 1986) are used to evaluate the silicon-efficiency of the WASP tree.

(1) Available storage (AS) this measures the average number of APEs expected to be working on the wafer and is defined as

AS = Y(APE) x N(wafer)

where Y(APE) is the probability of a single APE being functional and N(wafer) is the number of APEs on the wafer.

(2) Harvest (H) measures the proportion of working APEs that need to be utilised to achieve the desired number of usable APEs.

Harvest is defined as

H = DS / AS

where AS is as defined above and DS is the desired storage (viz. 8192-APEs in the WASP image processing device).

Results

Clustering	Defect densities (d/mm^2) .02	.04	.06	.08
.5	.99	.98	.97	.96
1	.99	.98	.97	.96
2	.99	.98	.97	.96
4	.99	.98	.97	.96
1000000	.99	.98	.97	.95

Figure 7. ASP sub-string yield.

Clustering	Defect densities (d/mm^2) .02	.04	.06	.08
.5	12402	12265	12132	12003
1	12401	12262	12125	11992
2	12401	12260	12122	11986
4	12401	12259	12120	11983
1000000	12400	12258	12118	11979

Figure 8. WASP ASP yield.

Clustering	Defect densities (d/mm²) .02	.04	.06	.08
.5	66.05%	66.79%	67.52%	68.25%
1	66.06%	66.81%	67.56%	68.31%
2	66.06%	66.82%	67.58%	68.35%
4	66.06%	66.82%	67.59%	68.37%
1000000	66.06%	66.83%	67.60%	68.38%

Figure 9. WASP harvest requirements.

CONCLUSIONS

The small size of the 4-APE blocks means that their yield is relatively high even with quite severe defect densities.

It is this high-intrinsic yield that gives the encouraging figures in Table 2. Indeed, even with random defects and $0.08d/mm^2$ over 11900 APEs can be expected to yield on an 'average WASP' wafer.

These figures means that a harvest of 65-70% (viz. 65-70% of all WI-CC-RAM-APE controller paths need to be functional to generate an 8192 APE wafer 50% of the time) must be attained.

From Figure 2, it can be seen that clustering does not affect the number of working APEs attained and may not be a major concern in terms of APE yield. However, this does not mean that WASP is insensitive to clustering. The APE-string controllers and the WI and CC blocks comprise irregular circuitry to realise microcontrollers. For such irregular structures, replication is needed for fault-tolerance. It is here in these circuits that any clustering may cause significant yield hazards. Further work needs to be performed to fully determine clustering in WASP. Two WASP test chips are currently being processed to determine this.

ACKNOWLEDGEMENTS

The Author would like to thank all the members of the Brunel University WASP team for their encouragement of this research. Thanks are due to Professor J.J. O'Reilly of UCNW Bangor for his support for my work.

REFERENCES

Jones S.R., Lea R.M. 1986, 'Interconnection Strategies for the WASP device', Wafer-Scale Integration (C.R. Jesshope and W.R. Moore Eds), pp 82-91 Adam Hilger.

Fourman M.P. 1986, 'Redundancy Strategies for WSI' Wafer-Scale Integration (C.R. Jesshope and W.R. Moore Eds), pp 72-

81, Adam Hilger.

Rosenberg A.L. 1986, 'Graph-theoretic Approaches to Fault-tolerant WSI Processor Arrays', Wafer-Scale Integration (C.R. Jesshope and W.R. Moore Eds), pp 10-23, Adam Hilger.

Lea R.M. 1986, 'A WSI Image Processing Module' Wafer-Scale Integration, (G. Saucier, J. Trilhe, Eds), pp 43-58, North-Holland.

Jones S.R., Warren K.D., Lea R.M. 1986, 'Engineering an Image Processing Module in WSI', Silicon Design Conference, pp 449-458, EDA Ltd, 31/33 High-Holborn London WC1V 6BD.

Hedge S.J., Lea R.M. 1986, 'Intra-module fault-tolerant strategies for the WASP device', IFIP workshop on Wafer-Scale integration (Grenoble University, France), pp G3.1-G3.10.

Carlson R.O., Neugebauer C.A. 1987, 'Future trends in Wafer Scale Integration', Proceedings of the IEEE, pp 1741-1752.

5.4 High Harvest Approaches for 2-D Arrays

José Delgado-Frias, Will Moore and John Trotter

INTRODUCTION

Numerous schemes have been proposed for incorporating fault–tolerance into a two–dimensional (2–D) array so as to avoid the defects that can be expected in wafer scale integration (Moore 1986a,b). The typical approach is to fabricate a two–dimensional array of cells along with appropriate switches which allow a smaller array to be configured from the working cells. This often results in a *low harvest* of the working cells (i.e. in the final configured array many of the fault–free cells are not used). This arises because to construct an array of a fixed size, the number of spare cells must allow for the worst expected defect density. Mapping a regular two–dimensional array requires a complicated switching network that has an overhead in time, area, and configuring cost. Also complex techniques are used to map the complete array; but this still does not harvest all the fault–free processors. The usual clustering of defects adds to the problem by introducing longer worst case paths between working cells.

In this paper we examine three applications for two–dimensional arrays which do not necessarily require a completely regular array nor an array of a fixed size. Due to these features we can employ fault–tolerant schemes which produce a higher harvest of the fault–free cells.

WSI Technology

Wafer–scale integration (WSI) technology offers the potential for improving the reliability and the performance of whole wafer integrated systems. Improved reliability is expected because the total number of pins will be reduced, mechanical connections between the different packaging levels being the primary contributor to system failure rates. The performance (particularly the speed) can be improved by having very short interconnection lines between the system's elements. Other potential advantages include: lower power (fewer pad drivers), smaller system volume and lower system cost (some manufacturing steps are eliminated or reduced in number).

However WSI yield losses, which may be due to gross or to random defects, are the major limitation on WSI technology success. Gross defects usually originate from manufacturing errors and cause a large portion of the wafer to have no functional chip. Random or *spot* defects may be caused by defects in photolithography, process quality, or starting silicon material. In a well controlled process the random defects are far more frequent than the gross defects. Other possible WSI problems are: whole wafer testing with the restricted number of external connections and the higher cost of.reconfiguring the wafer.

Previous Approaches

Typical approaches that have been proposed for fault–tolerance of 2–D arrays are:

Row/Column redundancy. In this approach spare rows and/or columns are used. Rows or columns with faulty cells (even with a single faulty cell) are replaced by the fault–free spares which would otherwise be inactive. This scheme is the most popular one for adding redundancy to memory chips. The row/column approach has been also used in two–dimensional arrays (Bentley and Jesshope 1986). In general approaches based on this type of redundancy have simple hardware and require no extra software at run time. However they can tolerate only a small number of defects.

Restructurable VLSI. The restructurable VLSI (RVLSI) approach to fault tolerance is based on laser cutting and welding technology. RVLSI research which is aimed at implementing systems with complexities approaching a million gates is being conducted at MIT Lincoln Laboratory (Raffel 1986). The approach is intended for modular architectures that can be arranged as an array of cells embedded in a regular interconnection matrix. The interconnection matrix is a fixed pattern of two metallization layer buses which have 30 to 60 lines each. Not all the lines have the same length; the shortest available line is used to communicate between two cells. A laser is used to form interconnections between the two levels which are separated by a special dielectric, amorphous silicon, at specific link sites. The same laser can be used to disconnect a wire link by removing the metal on either level. A 16–point×16–bit Fast Fourier Transform (FFT) 2–D array has been implemented (Rhodes 1986).

Triple Modular Redundancy (TMR). The TMR is a masking redundancy approach where a unit is triplicated and a voter scheme is used to poll the outputs and masks the faulty ones. There must be at least two good outputs; thus yield requirements may be higher than could be obtained with today's WSI technology. The TMR approach was implemented by Trilogy in ECL technology (Peltzer 1983). In such an implementation triplicated modules were separated on the wafer in order to provide less sensitivity to clustered defects. However, delays were introduced as a consequence of the long lines required to communicate the three modules with their voter circuit.

Complex Switch Networks for 2–D Arrays. The switching fault–tolerant approach uses standby switches to bypass faulty modules and reconfigure a two–dimensional array onto a faulty grid. The Configurable Highly Parallel (CHiP) processor (Hedlund and Snyder 1982) is essentially a 2–D array but the inclusion of extensive switching between cells allows it to be configured in a variety of architectures (such as trees, linear arrays, etc). An application of the CHiP approach has been developed on the Wafer Scale Systolic Processor (WASP). In this implementation the wafer is partitioned into 3×4 cell blocks to permit the formation of 2×2 subarrays. The predicted cell utilization was only 30 percent for cell yields of 67 percent.

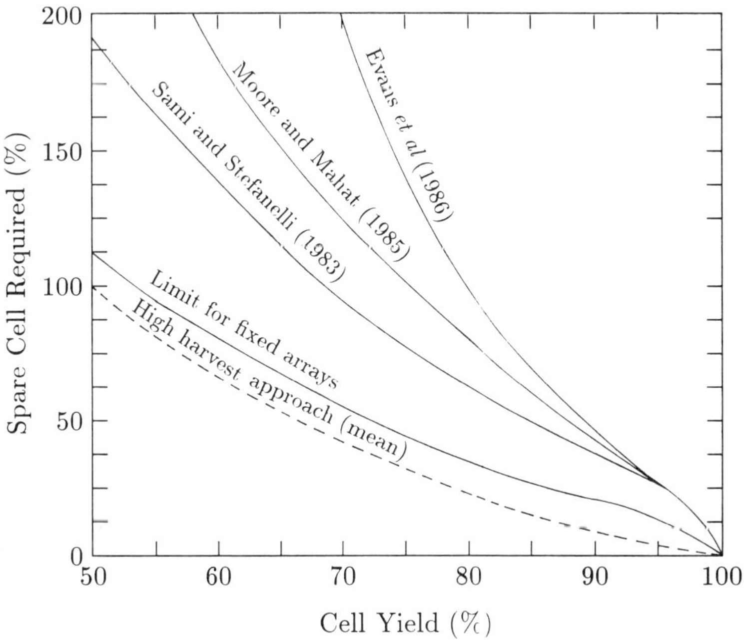

Figure 1. Percentage spare cells required vs cell yield for 90% yield of a 16×16 array cells.

Switching strategies may be optimized for mapping two–dimensional array. Figure 1 compares the effectiveness of the schemes proposed by Evans *et al* (1986), Moore and Mahat (1985) and Sami and Stefanelli (1983) in terms of the percentage of spare cells required in order to configure a 16×16 array of cells with a 90% confidence. The lower curves, representing a more effective harvest of the working cells, are obtained through progressively more complicated switching networks which introduce their own problems of reliability and of the difficulty of test and reconfiguration. The limiting case shown in Figure 1 assumes that all the working cells can be connected into the 16×16 regular array required. It is clear that the requirement to configure the circuit into a *regular* array of cells causes a considerable complication in the design and still results in the loss of a sizeable proportion of the working cells.

A further problem is associated with the requirement in these examples to configure a working circuit with a fixed target *size* (number of cells). The bottom curve in Figure 1 shows the *mean* number of spare cells needed to compensate for the yield loss if *all* the working cells can be used irrespective of how many turn out to be working. Other penalties in designing to a fixed target size include the fact that the design must cater for the worst expected yield loss included the problem of local defect clustering as well as batch to batch variations.

These problems imply that there may be substantial advantages in the designs which use elementary fault–tolerant switching but obtain near 100% harvest of working cells by allowing less regular arrays of variable target size.

High Harvest Approaches

The goal of the high harvest (HH) approach is to use most of the available computational elements in order to produce the largest possible two-dimensional array. Advantages of this scheme include: a high density of active cells, a simple fault-tolerant scheme, and a lower sensitivity to clustered defects. The density of active cells is higher than the above WSI approaches since a higher percentage of the fault-free cells are used. The fault-tolerant schemes for the HH approach are simpler than other approaches because it does not attempt to produce a perfectly regular array. This feature decreases both the amount of hardware involved in the reconfiguring process and the complexity of the fault-tolerant schemes. Defect clustering represents one of the most difficult problems for fault-tolerant schemes. The level of redundancy usually has to be chosen to cope with not simply the worst case defect density but also the worst cluster. The high harvest approaches are less sensitive to this problem since the loss of cells affects only the area where the faulty cells are located; and the overall system performance is less affected.

Three high harvest approaches are presented below for three common applications. The proposed architectures are also supported by robust algorithms. The first approach is based on time redundancy and its application is to the solving of boundary value differential equations. In this scheme the time to solve a problem depends on the average number of faults (or yield). The second approach is called granularity reduction and is aimed at image processing applications. In this scheme the time to solve a problem does not change but the granularity does. The third scheme is applied to semantic networks which are used in artificial intelligence for manipulating knowledge. In this scheme knowledge trees are mapped into a faulty array. The solution time and the amount of knowledge are only slightly affected by the module yield.

A TIME REDUNDANCY APPROACH

Boundary value differential equation computations are widely used in a number of scientific and engineering fields such as physics, mechanical engineering, and electrical engineering. Interest in finite difference methods (Young and Kincaid 1986) for solving differential equations has been increasing steadily. In such methods the derivatives in a differential equation are approximated by difference quotients that may be obtained from a truncated Taylor series. Finite difference methods become more attractive with the advent of specialized VLSI and WSI hardware components. A customized finite difference oriented multiprocessor system may now be built economically.

In boundary value differential equation computations (finite difference algorithms) each element needs only the values of its neighbours to the north, east, south and west in order to compute its own value (Young and Kincaid 1986). This feature leads to a simple but efficient communication scheme. There are no long communication line requirements that cause delays and require larger transistors to drive the larger capacitances.

The BVE Architecture

A possible architecture (Delgado-Frias and Green 1987) for the boundary value engine (BVE) is an array of processing elements (PEs) as shown in Figure 2. Each PE computes an element of the boundary value differential equation system. The

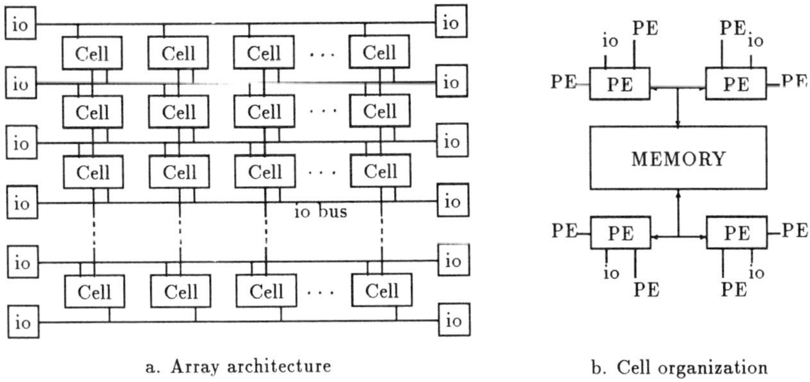

Figure 2. Boundary value engine architecture.

computation of each element is performed by a PE according to a program that specifies the relationship among values of the nearest neighbours. Such a relationship is a function of the boundary values and the differential equation being solved. Values among nearest neighbour PEs are passed by means of a dedicated line or through memory.

A cell in the BVE architecture consists of a memory and the four PEs attached to it. PEs in the same cell have communication paths through memory while PEs in different cells exchange data by means of dedicated lines. Since latency, the time between issuing a memory request and getting the response, represents a major difficulty in some multiprocessor systems (Arvind and Iannucci 1983), large global memory is avoided. Instead local memories are distributed in the array so that short access time and delays can be achieved. A program and four sets of data are stored in a two-port memory module that is shared by two PEs in each port. Data can be read by any PE in the same cell but data values can be changed only by the PE in charge of that element.

Each processing element (PE) is composed of three **Wa**fer-scale **S**erial **P**rocessors (WaSPs) (Delgado-Frias and Contreras 1987). Figure 3 shows the BVE processing element structure. Each WaSP has the same configuration; this feature lowers the design and implementation effort. Each WaSP performs one of the three functional tasks: computational, memory address computation, and I/O instructions which includes PE to PE communication.

Fault-tolerant Scheme

In the time redundancy approach most of the fault-free modules are used. Faulty modules are replaced by a good module in time; i.e. the job of a faulty module is shared by a good module or a set of modules. Previous approaches (Sami and Stefanelli 1984) to time redundancy have slowed down the system by a factor of two in order to accomodate the extra time required for time redundancy, left some fault-free PEs idle, and required a PE yield of at least 50 percent.

In this approach each PE has a three to one WaSP redundancy where only one fault-free WaSP is needed to have a functional PE; any faulty WaSP is rendered inactive. This approach is not a triple modular redundancy (TMR) scheme; where at

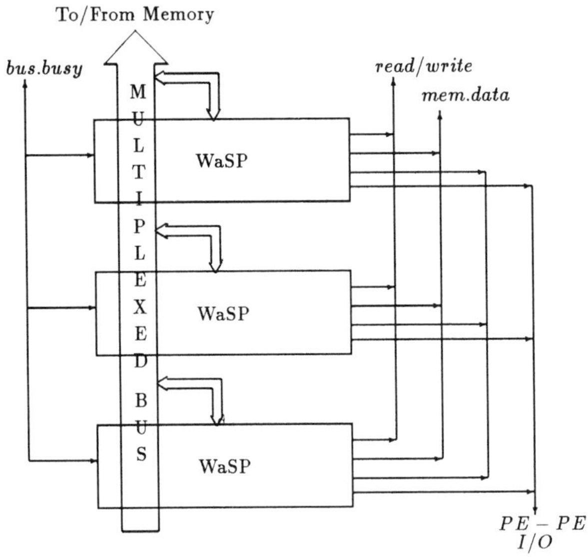

Figure 3. BVE processing element structure.

least two modules must work. Each WaSP contributes to the PE speed. WaSPs in a PE are connected to the same buses and modules, so they are able to perform any functional PE task. The WaSP has a small area and short communication paths that lead to higher yield and permit faster clocks. In order to avoid a large area overhead and complex communication paths the WaSP modules have no built–in redundancy.

Because of variations in the number of working WaSPs, some PEs will be slower than the average PE. However, the effects on computation time of slow PEs are relatively small. A slow PE is surrounded by faster PEs that are computing values closer to the final ones at faster rate; thus at the next iteration the slow PE will have better data to process and the overall convergence is not greatly affected. The defect clustering effects on the system performance are not as critical as they are for hardware redundancy, since clusters slow down only those portions of the system but the remaining part of it executes the task at normal speed.

The degree to which WSI time redundancy scheme affects the total computation time depends on the WaSP yield as shown in the graph of Figure 4. This graph was a result of several array simulations; a random number generator being used to assign WaSP defects. It can be observed that increasing the WaSP yield from 30% to 60% reduces the execution time significantly. But a further increase in the yield from 60% to 85% has little effect on the execution time. Using today's technology a yield between 60–85% may be expected for small sized WaSPs.

The Boundary Value Engine (BVE) combines a set of features; such as local communication, cell regularity, and fault–tolerance; that make it suitable for WSI implementation. The time redundancy scheme for a yield between 65–85% is an excellent approach for the BVE architecture since silicon real estate is efficiently used, time penalties are relatively small, and less concern in defect clustering is imposed on design. As feature sizes continue to shrink, the architecture can easily be extended to

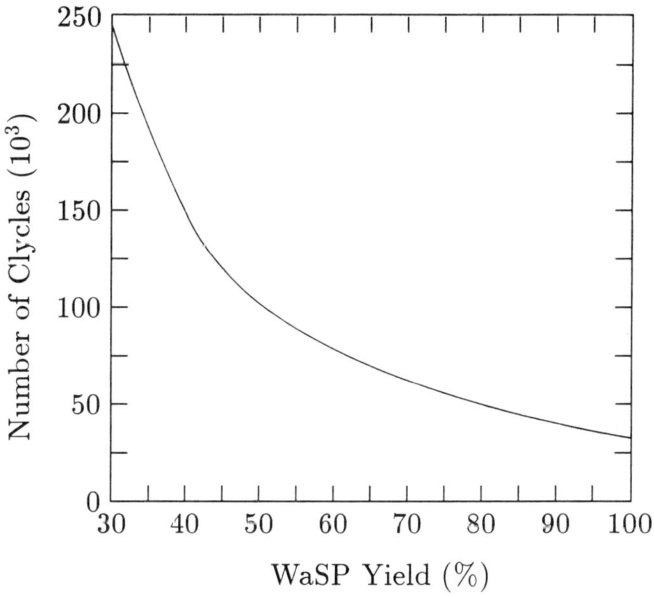

Figure 4. Performance/WaSP–yield graph.

accomodate more computational cells. This time redundancy approach to 2–D WSI arrays offers advantages of: 1) simpler hardware, since faulty computational elements are not replaced; 2) efficient use of silicon real estate, all functional PEs are used; and 3) defect clustering is easier to handle, those areas where defect clusters are present will have slow PEs, but computations can still be carried out.

A GRANULARITY REDUCTION APPROACH

An example of granularity reduction system to enhance yield is our development of MESH. The MESH system hardware is initially a two–dimensional array of processors. Its main application is the solution of the minimum energy type of problem, such as low and intermediate level image processing operations (Blake and Zisserman 1987). The MESH architecture can be used as a front end on an image processing system, providing the processing power for such algorithms. The idea of minimum energy is analogous to a thin rubber sheet constrained at certain points. The rubber sheet takes up the surface of minimum energy once it is released.

MESH System

The MESH system is similar to many image processing arrays where areas of the image are mapped onto each processor. The system combines hardware tolerance to fabrication errors with algorithmic fault tolerance which exploits the robust nature of the minimum energy type of algorithms. MESH will run at constant speed irrespective of whether there are defects or not. As the number of defects increases the system loses its resolution and the granularity of the solution decreases. Potentially all the

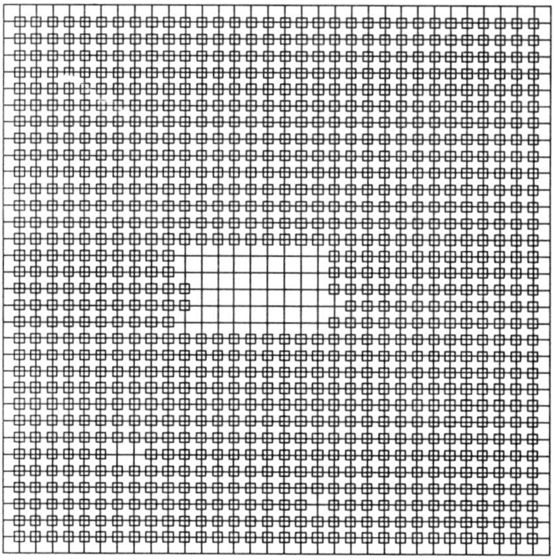

Figure 5. MESH processor array with faulty PEs ignored.

available processing elements can be used so the harvest and the computational power of the system remain high.

Fault–tolerant Scheme

The MESH system uses a simple fault–tolerant strategy which is to ignore the faulty processors and connect their horizontal and vertical neighbours together, effectively removing the defective processor from the array while maintaining a nearest neighbour interconnection. This leaves voids in the array where the faulty processors are. These voids could become quite large in the presence of clusters of faults as illustrated in Figure 5. If the problem were to be mapped onto the array at this stage, parts of the problem would not be covered by any of the processors. The resources of the array are distorted to cover the defective areas by spreading the available processors more evenly over the problem space. The inherent robustness in the minimum energy algorithms allows them to be executed on this type of irregular hardware.

MESH reconfigures the conceptual positions of its processors by applying the principle of minimum energy to the array with holes in. Processors move to positions as shown in Figure 6. To map data into the MESH each processor takes data from the area around its virtual position which was determined during reconfiguration. Data are mapped out in similar way, each processor outputting to the appropriate area on the output grid.

This system provides the user with a robust architecture which can stand manufacturing defects. In overcoming these defects MESH uses a novel combination of hardware fault–tolerance and the robustness in the algorithms that will be executed. As the yield decreases, the resolution with which the problem is solved reduces. The speed in the system remains constant. The MESH system attempts to harvest all of the working processors and put them to work. MESH provides a high harvest yield–enhanced system for minimum energy algorithms.

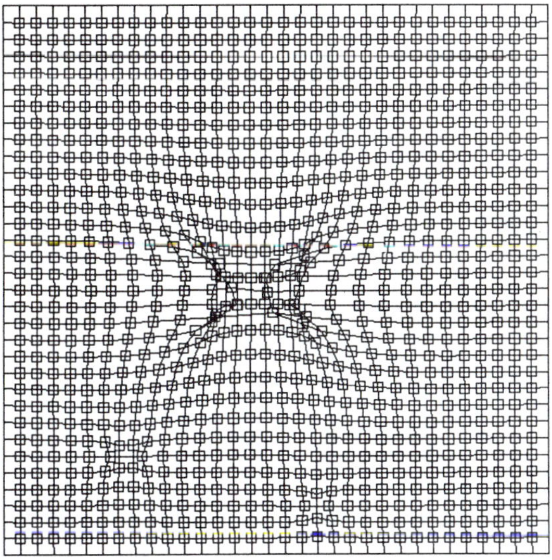

Figure 6. The virtual positions of processors after reconfiguration.

A SEMANTIC NETWORK APPROACH

This approach, as the two previous ones, is oriented towards an specific application which, in this case, is AI semantic networks. Many artificial intelligence applications require some sort of knowledge manipulation which demand a great amount of computational power. Knowledge-oriented systems, such as semantic network architectures, provide an efficient way for manipulation and representation of knowledge.

Semantic Network System

The semantic network system architecture is a two-dimensional array with nearest neighbour interconnection (NNI). The two-dimensional arrays are the most common architecture for semantic network applications (Delgado-Frias and Moore 1987) since it provides the cell connectivity required for this application. In semantic networks knowledge is stored in the nodes and links. Figure 7 illustrates the approach, nodes represent concepts (or things) and links represent relationships between concepts. In the proposed architecture each node and link is manipulated by a processing element. The PE has a bit-serial execution unit that is able to perform logic operations. Each PE has several registers where concepts and/or relationships (nodes or links) may be stored. This feature allows each PE to manipulate several knowledge items.

Fault-Tolerant Scheme

Non-deterministic and distributed algorithms are used in the AI knowledge oriented applications. They are non-deterministic because there is no unique path from initial state to the goal. Several alternatives are searched in parallel since the distributed nature of the algorithms allows it. These features are important in the fault-tolerant design.

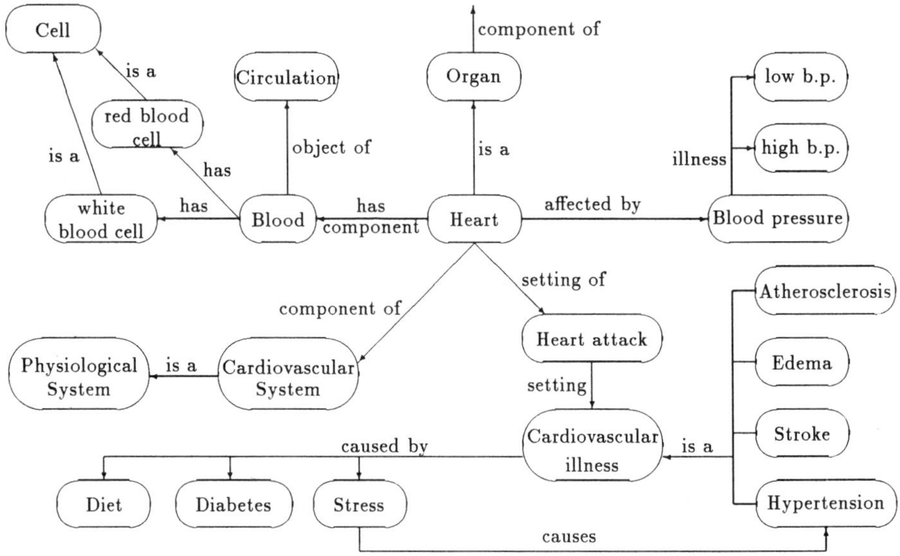

Figure 7. A portion of a knowledge base semantic network.

In this non-deterministic system several paths can be searched simultaneously. Thus, several independent processes may be running at a given time. Such processes need not finish at the same time, in fact they usually do not. Synchronization is only needed between nearby PEs in order to communicate results and/or variable names. Clock skew problems are avoided through this minimum synchronization requirement.

When loading the knowledge base into the system, PEs will store nodes or links of the semantic network. Once those PEs have used all their storage capacity they are full and nothing else can be stored in them. Faulty PEs are always set full; no access therefore can be made to them. When there is a need to store more links than the number of available neighbours, one or several PEs may be used as buffers in order to supply the required number. This approach is similar to the fanout scheme in digital electronics. Buffers introduce a small delay that will have a minimum effect on the overall system performance.

Small random defects may affect only one or few registers in the PE. In this case such a PE will be able to store fewer links or nodes than the good PEs. Some PEs will not use all their registers to store knowledge. Figure 8 shows part of an array with a semantic network mapped onto it. Here it can be observed how the PE's registers are linked to each other; it also shows some faulty and semi-faulty PEs.

Having a semantic network machine implemented in WSI technology will result in a faster, a more reliable and perhaps a cheaper system.

CONCLUSIONS

We have presented three novel approaches to achieve a high harvest of processing elements in irregular two-dimensional arrays. The approaches are application dependent but cover some very important areas. The HH hardware approaches are possible because of the robust algorithms that are used.

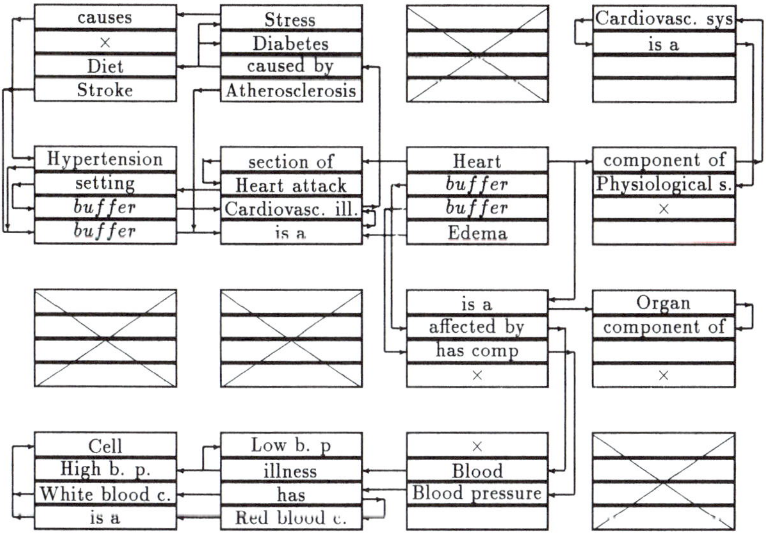

Figure 8. The semantic network mapped onto a 2–D array.

We believe that new approaches to 2–D WSI arrays should necessarily reflect their applications if they are to be reasonably efficient. High harvest schemes are appropriate for applications where high computational power, parallelism and speed are required. The need for robust software algorithms might seem to be a disadvantage but such algorithms are frequently used anyway even with a perfect architecture, for example some elements in the boundary value problem are computed at different rates in a uni–processor implementation in order to reduce the computational requirements, images commonly require noise filtering and some PEs in the semantic network may be filled by another branch. A feature of these HH approaches is that the fault–tolerance strategy is simple and requires little extra hardware.

ACKNOWLEDGEMENT

The authors gratefully acknowledge the financial support of the UK Alvey Programme (Project ARCH 019) and the SERC.

REFERENCES

Arvind and Iannucci R A 1983 A Critique of Multiprocessing von Neumann Style, *10th Int. Symp. on Computer Architecture* 426–436

Bentley L and Jesshope C R 1986 The Implementation of a Two Dimensional Redundancy Scheme in a Wafer Scale High-speed Disk Memory, *Wafer Scale Integration* Jesshope C R and Moore W R Eds. (Bristol, UK: Adam Hilger)

Blake A and Zisserman A 1987 *Visual Reconstruction* (Cambridge, MA: MIT Press)

Delgado-Frias J G and Contreras I 1987 WASP: A WSI Bit-Serial Processor, *CompEuro '87: VLSI and Computers* 164–167

Delgado-Frias J G and Green D M 1987 A Multi-processor WSI Boundary Value Architecture, *Int. Conf. on Computer Design* Port Chester, NY

Delgado-Frias J G and Moore W R 1987 Semantic Network Processing Systems, *Tech. Report OUEL 1677/87* University of Oxford

Evans R A, McCanny J V, and Wood K W 1986 Wafer Scale Integration Based on Self-organization, *Wafer Scale Integration* Jesshope C R and Moore W R Eds. (Bristol, UK: Adam Hilger)

Hedlund K S and Snyder L 1982 Wafer Scale Integration of Configurable Highly Parallel (CHiP) Processor, *Proc. Int. Conf. on Parallel Processing* 262–264

Moore W R and Mahat R 1985 Fault-tolerant Communications for Wafer Scale Integration of a Processor Array, *Microelectronics and Reliability* **25** (February) 291–294

Moore W R 1986a A Review of Fault-Tolerant Techniques for the Enhancement of Integrated Circuit Yield, *Proc. of the IEEE* **74** (May) 684–698

Moore W R 1986b A Critical Review of Fault-Tolerant Chips and WSI, *Wafer Scale Integration* Jesshope C R and Moore W R Eds. (Bristol, UK: Adam Hilger)

Peltzer D L 1983 Wafer-Scale Integration: The Limits of VLSI?, *VLSI Design* **4** (May) 43–47

Raffel J I 1986 The RVLSI Approach to Wafer Scale Integration, *Wafer Scale Integration* Jesshope C R and Moore W R Eds. (Bristol, UK: Adam Hilger)

Rhodes F M 1986 Applications of the RVLSI to Signal Processing, *Wafer Scale Integration* Jesshope C R and Moore W R Eds. (Bristol, UK: Adam Hilger)

Sami M G and Stefanelli R 1983 Reconfigurable Architectures for VLSI Processing Arrays, *National Computer Conference* 565–577

Sami M G and Stefanelli R 1984 Fault-Tolerance of VLSI Processing Arrays: The Time Redundancy Approach, *Proc. of the 1984 Real-Time System Symposium* 200–207

Young D M and Kincaid D R 1986 A Tutorial on Finite Difference Methods and Ordering of Mesh Points, *Proc. of the Fall Joint Computer Conference* 556–559

6 YIELD PREDICTION WITH DEFECT TOLERANCE

If the realistic prediction of yield is a difficult task for conventional VLSI circuits, then it is doubly difficult for defect-tolerant circuits. The extra complexity arises primarily because the statistical defect data have to be combined not just with typical circuit layout information but also with a system-level interpretation of the failure effects. At the same time information is required about defect statistics and their variations at the replaceable cell level and not merely at the chip level.

In the first section, Bolouri and Lea (§6.1) look at the problem from an experimental point of view. They examine the results of testing several batches of a particular VLSI chip to see how much information can be obtained about spatial distributions of defects which can then be applied to a wafer-scale design. To the extent that the VLSI chip implements very similar logic to the WSI design, they avoid the problems of having to analyse defect size distributions. The approach is simple and easy to implement but cannot generate information at the sub-chip level.

Lewis and McCabe (§6.2) describe a test chip aimed at characterizing these lower-level defect statistics. The chip is based on previous proposals for very low level defect tolerance and provisional results appear to confirm the feasibility of such designs as well as their ability to give the low level statistical data required.

Wang *et al.* (§6.3) attempt an analytical solution to the prediction of the yield of a defect-tolerant systolic array and obtain a tractable solution because their design is completely regular (albeit with a two-level hierarchy) and because they ignore the effects of defect clustering at the higher level.

The final section in this chapter, by Ramacher (§6.4), presents a system-level perspective on defect-tolerant VLSI/WSI. Again a very simple yield model is used with a single cluster parameter and a simple redundancy scheme, but the contention is that this is adequate for making system level decisions. Indeed the contribution points out that the yield is only one element in the complex task of cost analysis.

6.1 ULSI and WSI Yield Estimation: An Empirical Approach

Hamid Bolouri and Mike Lea

INTRODUCTION

Wafer-scale integration promises smaller, more reliable and higher-performance systems (McKirdy et al 1986). However, as Sumerling (1986) has shown, the cost of ULSI/WSI based systems will depend crucially on the device yields. In order to achieve the necessary yield levels, ULSI/WSI architectures tend to employ complex fault-tolerance strategies which do not lend themselves easily to mathematical yield modelling techniques. The paper reports on an empirical technique being investigated at Brunel University which aims to bypass the difficulties in ULSI/WSI yield estimation through automatic process characterisation, simulated defect map generation and yield estimation. The emphasis of this approach is on using readily available chip yield information from the target process to estimate expected WSI device yields at any given time.

In the paper, the Brunel WASP (WSI Associative String Processor) device (Lea 1986), is taken as an example of a WSI architecture targetted for fabrication using standard CMOS technology. Currently, 2 WASP devices are being designed, a full-wafer WASP device and a smaller ULSI version. For simplicity, only the full-wafer device will be considered here. Figure 1 shows the floor plan of the full-wafer WASP device.

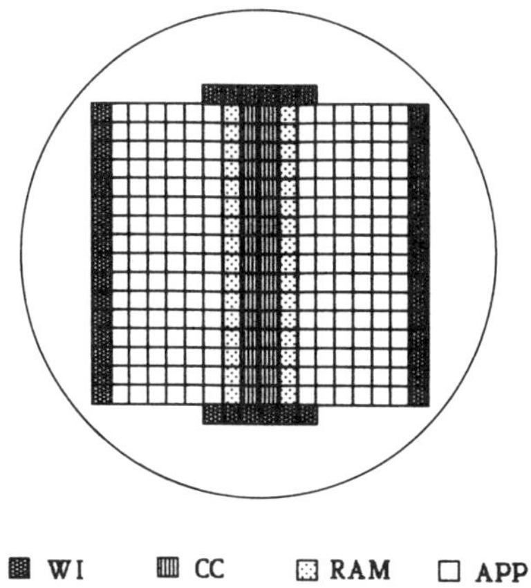

Figure 1, the WASP device

Figure 2 illustrates the way in which the WASP modules form a fault-tolerance hierarchy. The Associative Parallel Processor (APP) modules at the bottom of the hierarchy are only accessible if both the Communication and Control (CC) and the RAM modules at the node directly above them are functional. The WASP Wafer Interface (WI) must be functional in order for the device to be accessible. The major circuit blocks within each module are designed to be fault-tolerant (intra-module fault-tolerance). Inter-module fault-tolerance is provided in the form of spare nodes at each level of the hierarchy.

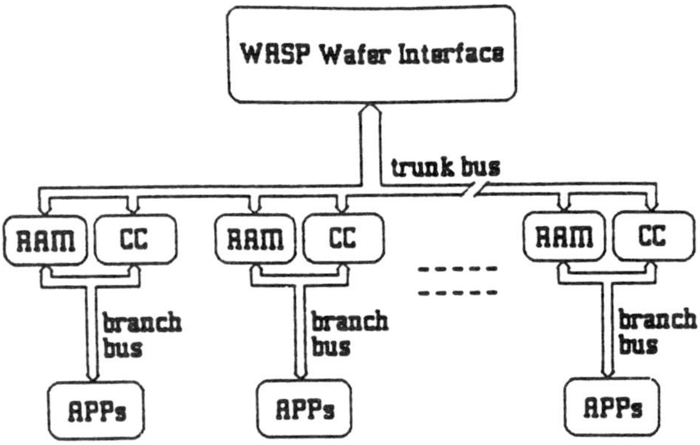

Figure 2, the WASP fault-tolerance hierarchy

THE ULSI/WSI YIELD ESTIMATION PROBLEM

The mathematical yield models proposed by Koren and Pradhan (1987), Franzon (1986) and Mangir (1984) all assume that the distribution of defects within a fault-tolerant processor array is random while the number of defects in the processor array is calculated using a clustering formula. Clearly, this assumption is more likely to be true in the case of intra-module fault-tolerance and for module sizes up to VLSI device sizes. In the case of inter-module fault-tolerance, the validity of the above assumption becomes doubtful. For instance, the array of CC modules in the WASP device spans the diameter of the wafer. Yield observations reported in this paper and elsewhere (eg. Perloff et al 1978) suggest that there are significant variations in defect density across such large areas.

The influence of defect clustering on WSI device yields

This section attempts to illustrate the importance of including inter-module defect density variations in estimates of device yields by taking the WASP device as an example.

The graphs of Figure 3 are calculated by assuming module failures are uniformly distributed. The yield of the WI modules is not included in the calculations since the effect of WI yields of less than one would be a simple scaling of the Y axis (WASP yield). The CC and RAM module yields are combined together because functionality of a CC-RAM node depends on both modules being functional.

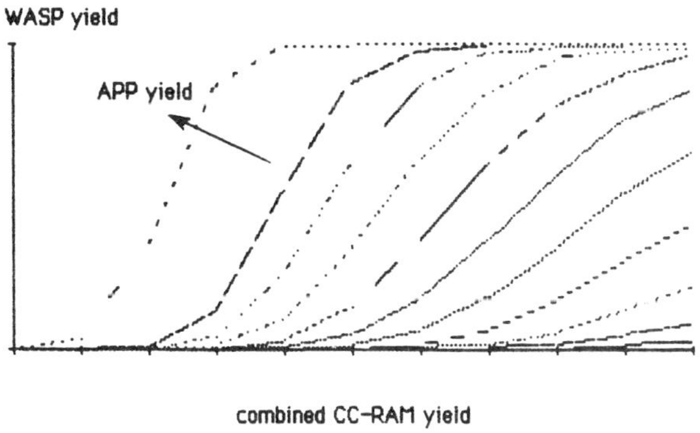

Figure 3, estimated WASP yield assuming uniform failure distribution

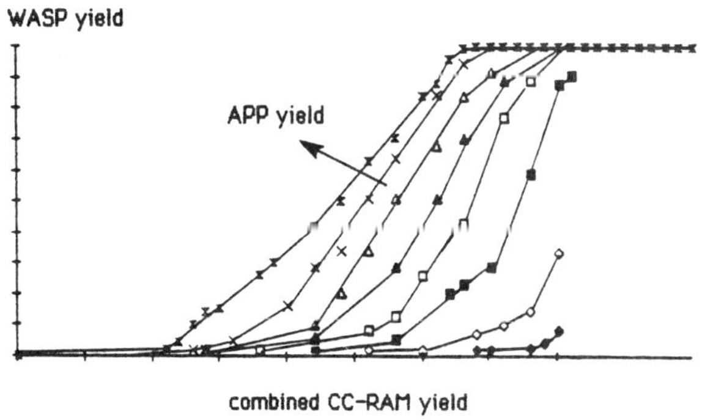

Figure 4, simulated WASP yield assuming non-uniform module failures

In contrast to Figure 3, the graphs of Figure 4 were derived by superimposing observed cross-wafer yield variations in a particular batch of chips, onto the floor-plan of the WASP device. For each chip site on the wafer, an average yield was calculated for the batch. The WASP floor-plan was superimposed onto the chip grid such that each chip site corresponded to a WASP module. Note that this step implies scaling the chip size to the size of the WASP modules and results in several WASP modules being placed at the very edges of the wafer where the yield is very low. In practice, the WASP device would not occupy this low yielding region and would therefore have better yields. The yield of each WASP module was calculated by weighting the nominal module yield with the yield of the chip site corresponding to the location of the module. Where the position of a WASP module did not correspond to a valid chip site (such as the wafer test sites), a weight was derived by a local averaging of neighbouring chip site yields. To determine whether a module was good or faulty, a random number between zero and one was generated. If the random number was smaller than the module yield then the module was deemed to be good. The WASP device was deemed to be functional if sufficient good modules existed at each level of the hierarchy. For each module yield assignment, WASP yield was averaged over a minimum of 100 simulations. This process was repeated for a range of nominal module yields to produce the curves in Figure 4. Note that the module yields given in the figures are average yields (ie. average module yield after the weighting process).

Comparison of the graphs in Figures 3 and 4 points to the effects of inter-module defect density variations on device yields. Firstly, the maximum possible yield of the modules is less than 1 due to the presence of several zero yield sites in the batch. Secondly, there is a distinct difference between the 2 Figures in the estimated WASP yield for lower values of APP yield. In particular, low APP yields give unacceptably low WASP yields when cross wafer yield variations are taken into account. Thirdly, the graphs of Figure 4 show a significant increase in sensitivity to the yield of the CC-RAM module pairs. This is important because the CC modules are potentially the least regular circuit blocks in the WASP device, and therefore, the least amenable to yield improvement through intra-module fault-tolerance. The remedy could be in increasing the amount of inter-module redundancy for the CC-RAM pairs.

THE EMPIRICAL APPROACH

The graphs of Figure 4 are useful as a pointer to the dangers of ignoring inter-module yield variations. However, major design decisions such as the amount of inter-module CC-RAM pair redundancy can not be made on the basis of yield observations from a single batch. On the other hand, any observed ULSI/WSI device yield characteristics could be significant if they persist in yield simulations based on several batches over a period of time. The aim of the empirical approach proposed here is to provide a fully computerised route for the estimation of WSI device yields using only chip yield information from the day to day output of the target process. It is important that the proposed technique be fully automatic and based on readily available data so as not to interfere with the running of the fabrication plant. Furthermore, because the yield estimates are generated automatically, repeated yield estimates can be made over a period of time with no additional manpower overheads. The following sections describe an experiment carried out to establish the feasibility of the proposed approach.

Data acquisition and analysis

Photographs of a hundred wafers from three CMOS batches (kindly supplied by Plessey (Caswell)) were used as the source data for the experiment. A spot of red ink on a chip indicated that the chip had failed probe testing. For the purposes of the experiment, the coordinates of good and faulty chips were digitised by hand. The resulting chip yield maps were analysed and characterised with two objectives:

(i) to compare the observed chip yield patterns with simulated yield patterns for a uniform distribution of defects, in order to highlight the presence of defect clusters

(ii) to produce references for measuring the degree of correlation between the generated defect maps and the original observations.

The following yield characterisation measures were chosen:

(i) vertical yield: wafers are superimposed on each other and the number of good chips in each location is recorded. The map of the number of good chips per wafer location is used to extract information regarding radial and angular variations in yield (Lagerberg, 1978)

(ii) good/faulty neighbour tests: for each good (or faulty) chip on each wafer, the number of good (or faulty) neighbours are recorded, in order to quantify the degree of defect clustering.

Figure 5 compares the characteristics of the observed chip yield data with the characteristics of 1000 simulated yield maps with uniform defect distribution and the same overall chip yield. Clearly, the observed good and faulty chip distributions are not due to a uniform defect distribution. Indeed, an annulus of lower defect density at half radius is clearly observed and distinct angular variation is evident.

```
          o o - o o o
        o - - + T + - - o
      o # + + + + - # # - o
    o # + + + + - + - - - - o
    - - + + + + + + + - # # -
  - + # + - + + - + # + - # -
  - T + + + + - T - # - - - T
  - + + + + + + + + + - + - -
  - + + # + + - + + + + - + o
    o # + + I I I I
      o - + + + + + + + - -
        o - + + T - - - o
            o o o o o
```

´o´ denotes zero yield regions

´#´ denotes areas where yield is within +/- 10% of batch average

´+´ denotes areas where yield is greater than 110% of batch average

´-´ denotes areas where yield is less than 90 % of batch average

a vertical yield contour map

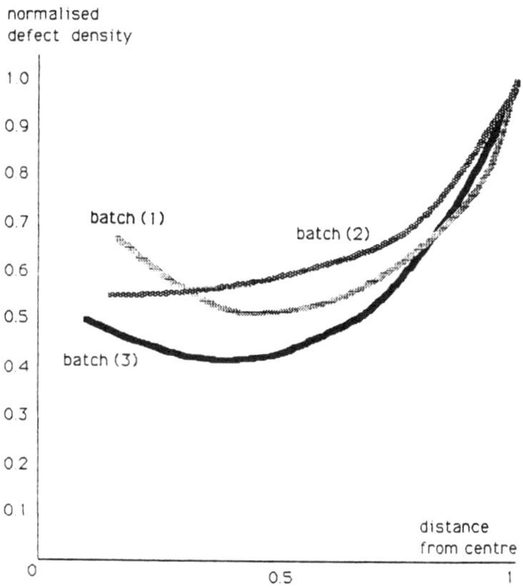

observed radial variations in yield for three batches

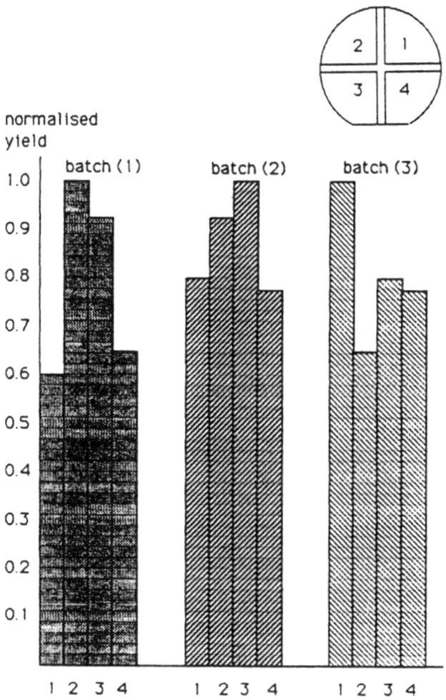

observed angular variations in yield for three batches

Figure 5, non-uniformities in the observed yield data

An interesting observation is the difference between the observed neighbour count profiles of the batches, and the neighbour count profiles of simulated random defect wafers. Figure 6 compares the neighbour count profiles of batch (2) with the neighbour count profile of simulated random defect maps with the same Poisson defect density. There is clearly a very significant difference between the two profiles. Random defect neighbour count profiles were generated for a range of defect densities. Figure 7 compares the most closely matching profiles with the actual batch profile. The difference between the defect density estimated using Poisson statistics and the defect densities which most closely reproduce the observed characteristics is about 25%. This is a reflection of the degree of defect clustering in the process, causing the effective defect density to be lower than that predicted by Poisson statistics.

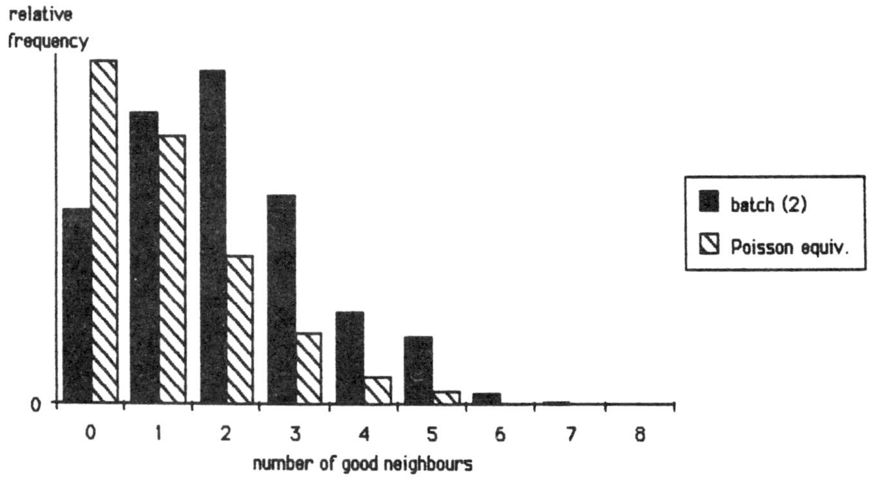

Figure 6, neighbour count profiles of actual and random defect wafers

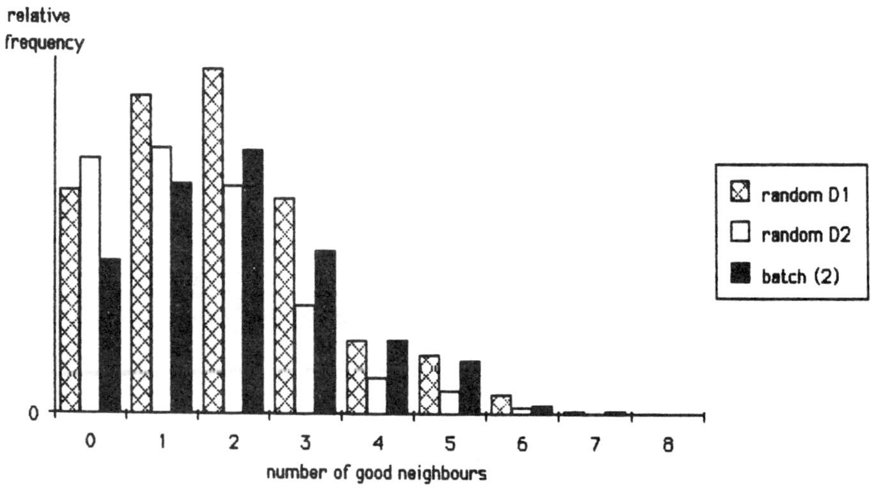

Figure 7, random defect profiles most closely matching actual data

Generation of defect maps

Three batches have been characterised. For each batch, the vertical yield map was divided into a number of regions of homogenous yield. For each region (i), an average defect density (D_i) was calculated from the vertical yield map. The number of defects in each region on a simulated wafer were sampled from a Poisson distribution with mean D_i. The sampled number of defects were then randomly distributed within the region. Because the generated defects are distributed randomly within each region, it is possible to generate an almost infinite set of non-identical defect maps with the same characteristics. Indeed, it is necessary to average device yields over many such defect maps in order to reduce the effect of fluctuations in estimated device yields caused by the random positioning of the defects.

Three different approaches to defining regions of homogenous yield in the vertical yield maps were investigated. Firstly, a program was written which used the general cross product statistic (Upton and Fingleton 1985) to dynamically search for areas of homogenous yield. At a 95% confidence level in the homogeneity of the regions, the result was a simple partitioning of the vertical yield maps into high and low yielding regions. For higher levels of confidence, the number of regions quickly increased. The second approach was to divide the vertical yield map into 4 quarters and to partition each quarter into a number of chip-wide annular regions. The reason for this choice was the clear radial and angular variations identified previously. In the third approach, each chip site was considered to be a separate region.

Defect maps generated using the 3 approaches described above were superimposed onto a grid of chip locations corresponding to the original observations. The resulting chip yield maps were used to test the neighbour count characteristics of the generated defect distributions against the original observations. While the results from all 3 approaches correlate well, the 1 chip per region technique appears to offer the closest match to the actual batch profile, as illustrated in Figure 8.

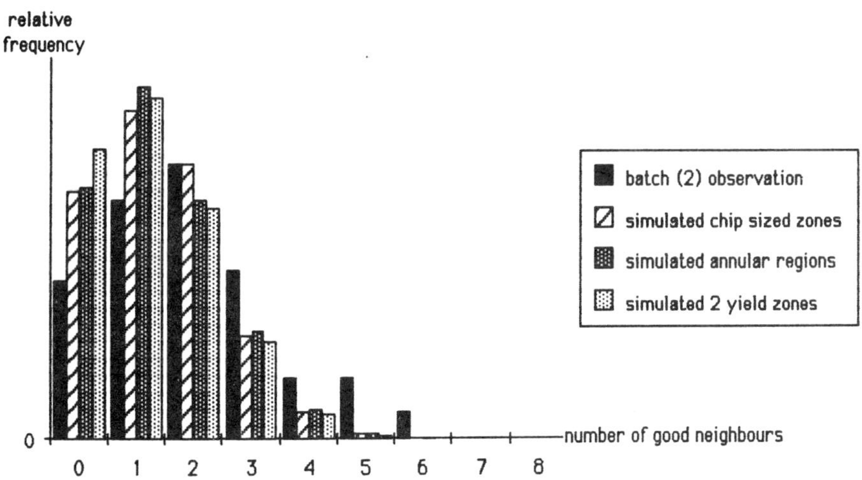

Figure 8, comparison of neighbour count profiles of the simulated defect maps with original data

module floor-plans become more clearly defined it is possible to make finer decisions based on where the defects are within a module.
In the case of the 3 batches analysed earlier, it was not possible to apply the generated defect maps to the WASP architecture because the analysed wafers were smaller than the WASP wafer size.

CONCLUSIONS

The paper has highlighted the need for a WSI yield estimation technique which takes large scale variations in defect density into account. A technique for the automatic generation of defect maps from readily available chip yield information has been described. A Pascal program which uses chip yield maps (recorded by automatic test equipment at probe test) as its input and produces defect maps as its output has been developed. The advantage of this approach is that it does not rely on costly process analysis techniques or on complex mathematical models of yield. Device yield is estimated by superimposing the generated defect maps onto the device floor-plan. A shortcoming of this approach is that any small scale defect clustering occurring within the fabricated chips is not detected, and therefore, not characterised.

ACKNOWLEDGEMENTS

The authors would like to thank Dr. G.W.Sumerling of Plessey Research (Caswell) for providing the raw yield data, and also for many stimulating discussions. We would also like to thank the WASP research team, and the U.K. SERC and the Alvey Commission for their support.

REFERENCES

Franzon P 1986 Yield Modeling for Fault Tolerant VLSI Arrays, Proc. Int. Workshop on Systolic Arrays, Oxford University

Koren I and Pradhan D K 1987 Modeling the Effect of Redundancy on Yield and Performance of VLSI Systems, IEEE Transactions on Computers C-36 344-355

Lagerberg J P L 1978 Vertical Yield: An Aid in Characterising the Process Yield of Integrated Cicuits, Proc. ESSCC 167-168

Lea R M 1986 A WSI Image Processing Module, in Wafer Scale Integration, Eds. Saucier G and Trilhe J, North-Holland 43-58

Mangir T E 1984 Sources of Failure and Yield Improvement for VLSI and Restructurable Interconnects for RVLSI and WSI: Part 1 - Sources of Failure and Yield Improvement for VLSI, Proc. IEEE 6 667-669

McKirdy R D et al 1986 Comparative Appraisal of the Implementation of an Image Processing Module Using ULSI and WSI Associative String Processors, Proc. Int. Workshop on Systolic Arrays, Oxford University

Perloff D S et al 1981 Microelectronic Test Chips in Integrated Circuit Manufacturing, Solid State Technology September 1981 75-80

Sumerling G W 1986 Cost Models for ULSI and WSI, presented at the 3rd Silicon Design Conference, London

Upton G and Fingleton B 1985 Spatial Data Analysis by Example, Volume 1 Point Pattern and Quantitative Data, John Wiley and sons, chapter 3

6.2 A Fault-Tolerant Testchip for Gathering Wafer Defect Data

Anthony Lewis and Andrew McCabe

INTRODUCTION

The optimal design of circuit architectures which are to be tolerant of fabrication defects requires a good understanding of the density and distribution of defects across a wafer. The extent to which defects tend to cluster is an especially important parameter for many fine grain architectures such as bit level systolic arrays.

Yield data obtained by testing conventional production chips provides data on which chips were defect free. It provides little or no information on the number of defects within a faulty chip or their spatial distribution. The effectiveness of fine grain fault tolerance schemes cannot be properly evaluated unless information on the expected defect count and defect distribution within chip-sized units is known.

In this paper we describe the design of a specialist test chip for gathering fine-grain defect data. Defects within this chip can be pinpointed in space, thus allowing high resolution wafer defect maps to be obtained. The chip itself contains fault tolerant structures, so that it can continue to provide useful data when defects are present.

DEFECT-TOLERANCE IN FINE-GRAIN LOGIC ARRAYS

The regular structure of systolic arrays (especially at the bit level) offers the systolic chip designer the opportunity to provide a degree of tolerance to processing defects within his design. This can be accomplised by adding redundant processors, bypass wiring and control circuitry so that defective processors can be isolated and the spares utilised in their place. In this way, the design and manufacture of array-based chips which are substantially larger than current state-of-the art would appear feasible. This follows naturally from the now widespread use of redundancy within large memories as a means of obtaining economic manufacturing yields. Many examples of fault-tolerance techniques for array architectures have been proposed (McCabe et al 1985, McCanny and McWhirter 1983). A few examples of techniques for 2-D arrays are shown in Figure 1.

When designing defect-tolerance into an array-based circuit, many design decisions which will affect the effectiveness of the defect-tolerance must be taken. These include fundamental decisions such as choice of redundancy and reconfiguration strategy, and detailed design decisions such as size of the smallest circuit block to be bypassed.

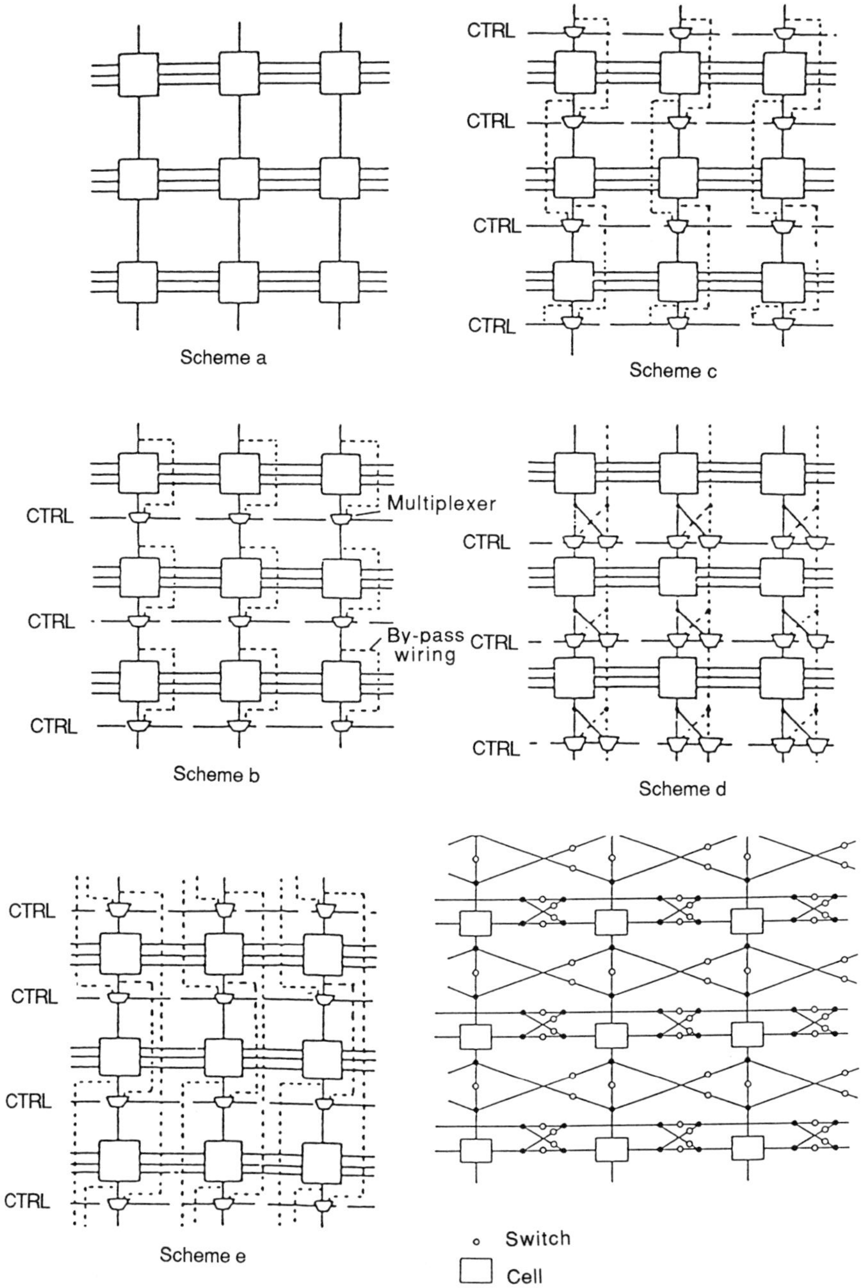

Figure 1: (a) 2-D array without fault-tolerance, (b-f) Examples of fault-tolerance techniques for 2-D arrays

The relative merits of different defect-tolerance strategies must therefore be evaluated by calculating parameters such as expected yield improvement vs area overhead cost. Figure 2, for example, shows a range of yield vs area overhead curves for a typical bit-level systolic array design. The points on each curve show the effect of providing different numbers of spare processors, and each curve corresponds to a different design option. It can be seen that design option "a" offers the best yield vs area performance. In addition it is clear that if too few spares are provided chip yield is significnatly compromised. Too many spares implies an unnecessarily high overhead for little extra benefit, which will reduce the cost-effectiveness of the approach. It may also add unnecessarily to the critical circuit area and thereby cause yield to worsen.

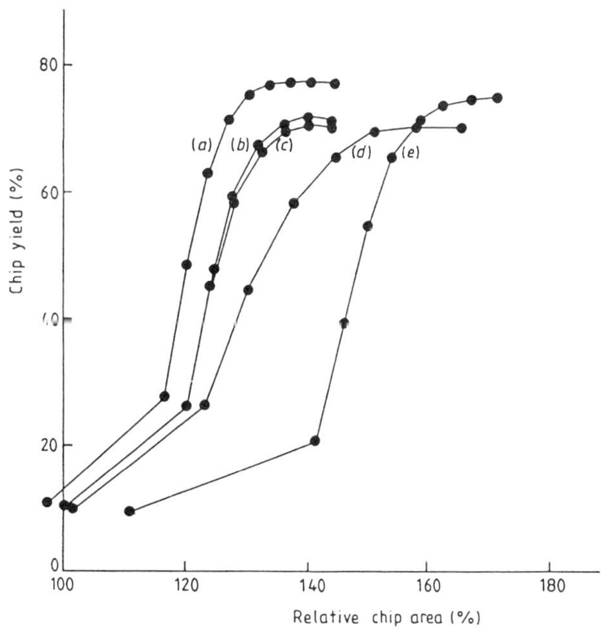

Figure 2: Predicted yield of 32 x 32 inner product processor array with original chip yield of 10%. (a) Off-line coefficient loading, (b) on-line loading (vertical buses), (c) as (b) but repair limited to maximum of one bypass stage fault, (d) on-line loading (horizontal buses, double columns), (e) on-line loading (horizontal registers)

Calculations such as these provide important guidance for the optimisation of the defect-tolerance strategy for each individual design. Clearly a realistic model of the expected defect density and distribution is a necessary prerequisite. To date, most work in this area has assumed a uniform distribution of point-defects within a chip-sized unit, if not across the entire wafer. In practice defect sized and defect distribution can be expected to vary, and defect clustering may occur. Many proposed defect-tolerance strategies for fine-grain arrays require bypass wiring which is physically close to a defective cell to be fully functional. These strategies are potentially sensitive to defect clustering, and yield predictions derived from a uniform model may be misleading. A non-uniform defect model, such as the Generalised Binomical Model, is to be preferred but the selection, development and charactisation of such requires detailed knowledge of the spatial distribution of defects on representative wafers.

ROLE OF TEST CHIP

Conventional testing of integrated circuits provides data on which die on a wafer under test are defect free, it does not provide information on the number and distribution of defects within the chip sized unit. The test chip described in this paper was designed to enable defects within a chip to be pinpointed in space, thus allowing high-resolution wafer defect maps to be obtained. It was designed to be sensitive to typical processing defects and to offer good fault diagnosability. Fault-tolerance within the design allows it to continue to provide useful data when defects are present. The data obtained from testing a selection of wafers will allow the validity of the uniform model to be assessed, and the alternative models to account for clustering to be developed if necessary.

The importance of bypass techniques as a means of achieving fault tolerance in regular arrays has been discussed. It is important that an efficient means of diagnosing faulty area within the array and reconfiguring the array appropriately to exclude these faults are established to maximise the potential of these techniques. The use of similar bypassing techniques is central to the operation of this chip and it will therefore provide a means of evaluating suitable diagnosis and reconfiguration strategies.

A simple logic array incorporating fault tolerant features forms the basis of this design. This provides a reasonable degree of area resolution (not as high as a memory type structure however) whilst maintaining design simplicity and is also representative of the logic structures found on typical MOS VLSI circuits. It also factilitates fault location without the need for extensive peripheral circuitry which itself would be prone to defects. A Sequential Access Memory (SAM) was designed and fabricated in conjunction with this chip, this offers a higher area resolution data but is not representative of VLSI logic structures and does not allow reconfiguration strategies to be investigated.

DESCRIPTION OF TEST CHIP

The Fault Tolerant Logic Array (FTLA) is formed by a 16 x 16 two dimensional array of identical cells as show in Figure 3. Each cell has

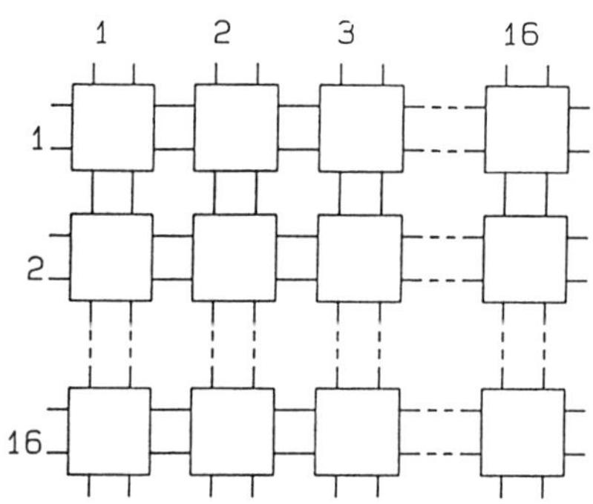

Figure 3: Logic array

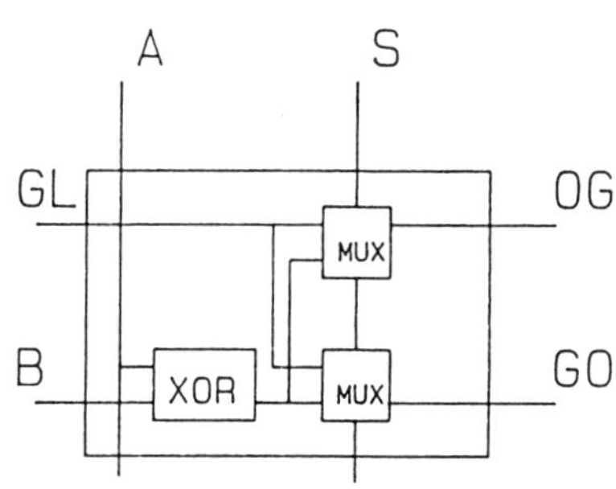

Figure 4: Array cell

two horizontal inputs (GL,B), two vertical inputs (A, S) and two horizontal outputs (OG, GO) as shown in Figure 4. The cell consists of a two input exclusivie OR (XOR) gate and two 2:1 multiplexors. The XOR gate is fed by one vertical input (A) and one horizontal input (B). The second horizontal input into the cell (GL) is a bypass wire. The array is partitioned into rows. Along each row the outputs of one cell form the input of the next. There is no communication between the cells vertically but the inputs A and S are broadcast to all cells fo a column. Figure 5 shows the construction of a row.

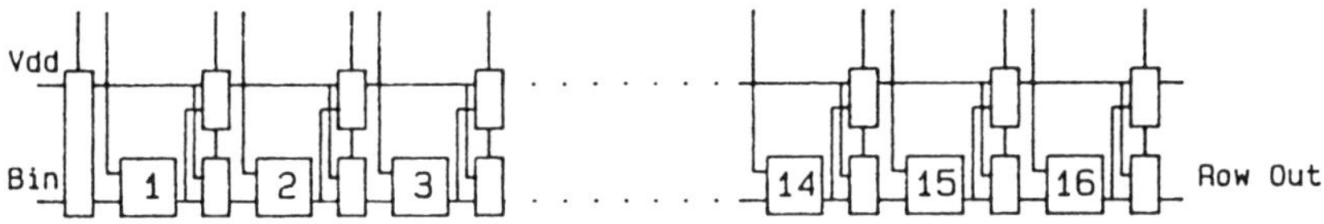

Figure 5: Row diagram

The array forms a flexible network of XOR logic and bypass circuitry controlled by multiplexors. Approriate use of the multiplexors, the vertical select lines, S, and the bypass wires allows any number of XOR gates along a row to be selected or bypassed, as dictated by test requirements. A column of multiplexors is added to the left input of each row to allow the first XOR gate to be bypassed if required. All sixteen rows are served by a common input bus, Bin, and have independent outputs as showm.

The logical design of the XOR gate has not been minimised thereby increasing its physical layout area and ensuring a high active to peripheral circuit area ratio. This is done so that cell area to bypass area ratio is more representative of a typical bit level array design. In addition it ensures that defects are more likely to occur in the XOR gate, where they can be tolerated by using the multiplexor to bypass the fault cell, than in the bypass circuitry and control logic where the existence of defects is potentially most damaging.

Small discrete areas of logic can therefore be isolated and tested independently. The resolution is defined by the smallest area in which defects can be isolated which is of the order of the size of the unit cell i.e. approx. 0.015 mm^2. If an isolated area of logic is found to be faulty, due to the existence of a defect, then its position on the wafer is known. Further tests on that row can take place provided that this area is avoided in subsequent tests by reconfiguring the row appropriately.

The design is still susceptible to defects occuring on the control circuitry, which may cause a gross failure of part or all of a die, and defects on the bypass circuitry which may seriously limit the information gathering potential in the locality of the defect.

TEST STRATEGIES AND PROCEDURES

An appropriate test strategy must take into account a number of conflicting requirements. The number and form of the test vectors should be sufficient to exercise the device fully at a functional level and also at a transistor level to ensure full defect coverage. In addition, in order to function effectively as a defect monitor the tests should be suitably specific to ensure maximum locatability thereby allowing the positions of defects to be pin-pointed precisely. On the other hand it is essential that test times are kept within acceptable limits and the form of the results is suitable to facilitate efficient analysis. Certain fault conditions impose limitations on acheiving these objectives, for example, the existence of certain faults may limit the testability of other areas of the device.

The array is tested on a row by row basis and the test strategy adopted is based on locating single or multiple struck at faults on the input, output and control signals associated with the functional blocks as outlined in Figures 4 and 5. It is assumed that the faults are time-independant i.e they are permanent faults, and that there is negligible coupling between cells.

The test strategy is designed to address the requirements of full fault coverage and locatability whilst minimising test times by applying a series of relatively short efficient tests to each row. This technique achieves full coverage without having to test each of the 2^{16} possible row configurations. Each test involves configuring a row appropriately to isolate a specific area of logic and testing it independently. For example, a particular test may require that cells 1 and 2 are selected whilst the rest of the cells within that row are bypassed. This is achieved by applying the correct control signals to the multiplexor select inputs, S. Therefore, with the exception of those tests designed to test the integrity of global paths, each test involves testing a linear array of cascaded XOR cells linked via multiplexors.

Each configuration is subjected to a series of four test vectors designed to fully exercise the inputs of those XOR gates which are selected. For example, the test vectors for a single isolated XOR cell with inputs designated A and B [(A,B)] are (0 0) (1 0) (0 1) and (1 1). The test algorithms ensure that for any level of cascade the test vectors are such that each XOR cell within the array has its inputs exercised fully. Figure 6 outlines the test route.

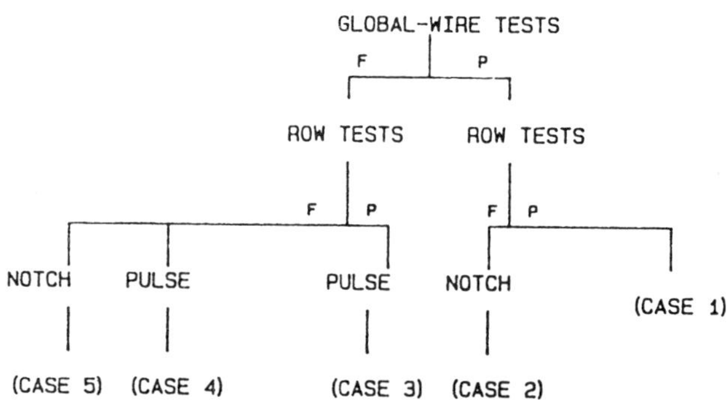

Figure 6: Test strategy

The existence of a fault free global-wire path along the entire length of a row greatly simplifies subsequent testing and analysis. To this end the first series of tests i.e. GLOBAL WIRE TESTS aim to test the integrity of a global wire path along the row. All sixteen XOR cells along the row are bypassed and the input to the row toggled and the output examined. The second set of tests, ROW TESTS, select all XOR elements in a row thereby producing a linear array of sixteen cascaded XOR elements and appplies the appropriate test vectors.

The main group of tests are those designated NOTCH TESTS and PULSE TESTS in Figure 6. These isolate and test smaller area of logic. The first group of NOTCH TESTS select each XOR element in turn whilst simultaneously bypassing all the other elements in the row. Figure 7 shows the data path in a typical NOTCH TEST which has selected the second cell only and bypassed the rest. It can be seen that in addition to the XOR gate of cell 2 being checked for stuck at faults on its inputs and output that the upper multiplexor on cell 2 and lower multiplexor on cell 1 are also checked for certain stuck at faults. Each cell in the row is tested in this way thereby building up the fault coverage of that row. A second group of NOTCH TESTS then selects two adjacent cells in turn and bypasses the rest. These tests are designed to test that logic not covered during the initial set of NOTCH TESTS, namely, the lower inputs to the lower multiplexors, thereby ensuring maximum fault coverage at a functional level.

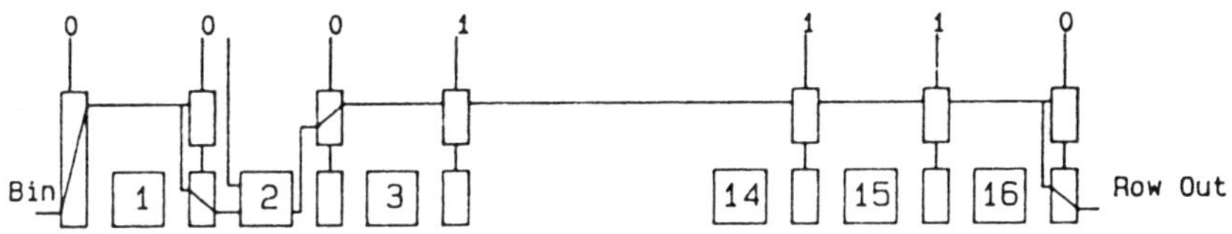

Figure 7: Notch test

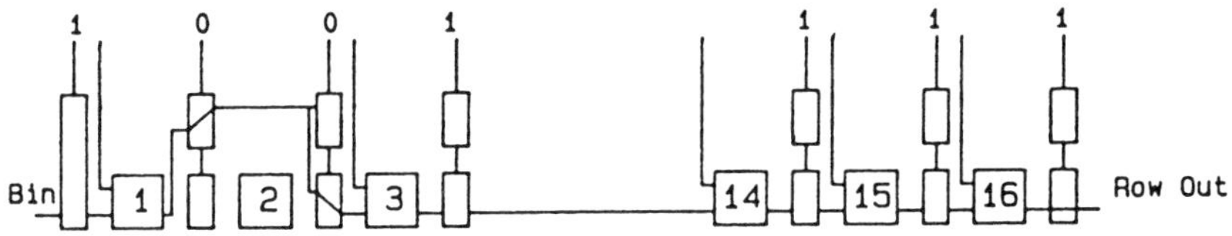

Figure 8: Pulse test

The PULSE TESTS form the compliment of the NOTCH TESTS. Each XOR cell is bypassed in turn whilst simultaneously selecting all other elements of the row. Figure 8 shows the data path in a typical PULSE TEST which has bypassed cell 2.

At the base of the tree depicted in Figure 6 a number of cases are listed which correspond to the following conclusions which may be drawn from the preceeding tests:

CASE 1 XOR gates OK Global wires OK

CASE 2 Global wire OK - Locate positions of all defects on XOR row

CASE 3 XOR gates OK - Locate positions of all defects on Global wire

CASE 4 Locate single fault on XOR row unless there is an associated local fault on the Global wire

CASE 5 Locate single fault on Global wire unless there is an associated local fault on XOR row

By analysing the results from these tests appropriately the positions of defects within the array can be pinpointed. In this way the flexibility of the reconfigurable nature of the array is utilised to maximum advantage. However certain fault conditions impose limitations on the usefulness of this technique. For example, a fault which affects an XOR gate and all the surrounding bypass circuitry may prevent any path through the row being configured. In this instance other faults on the row are masked out and the whole row has to be written off as faulty. In addition the present group of tests cannot resolve certain other faults and pin-point their positions automatically. For example, the location of multiple faults on an XOR row in the presence of a single defect on the global wire of that row. In such circumstances the test strategy applies an exhaustive sequence of tests and lists successful configurations. The software cannot analyse ths list automatically at present but if a path through the faults can be configured it may be interpreted manually if required.

RESULTS AND DISCUSSION

The data obtained from this chip is in a form which can be processed to construct bit-fault maps of a wafer pin-pointing areas affected by defects. In addition, it can be processed to calculate more precise defect density values and to investigate the form and effects of clustering due to process variations. Such information is valuable to the Design Engineer investigating fault tolerant systems and may prove helpful to the Process Engineer who requires higher resolution defect data than that offered by conventional data bases. It should be noted that defect densities and distributions are characteristic of particular processing lines and therefore results obtained from one may differ markedly from another.

Figure 9 is an example showing bit-fault maps of four die from the same wafer designed and fabricated in 3 μm -bulk CMOS technology. An "X" indicates a fault within an XOR cell whilst a "B" indicates a fault on the bypass wire. Figure 9 (i) shows a single isolated defect within the XOR gate of a cell, such a defect is avoided by routing a path around the affected area via the global wire provided the local global wire is itself free from defects (which in this particular instant it is). Figure 9 (ii) demonstrates a typical case were reconfiguration is not possible. Here the question marks indicate that the two rows contains a number of faults whose positions cannot be located exactly. For example, a single defect may affect the function of an XOR gate and the associated bypass circuitry, in which case an exhaustive sequence of tests would be unable to find a suitable path through the faulty rows and the whole row has to be written off as faulty. In this instance the routines were unable to reconfigure the rows successfully indicating a defect distribution of this type. Figures 9 (iii) and (iv) show die in which a slightly higher density of defects is apparent and where a degree of localised clustering is discernable. In both of these cases full reconfiguration is possible despite the relatively high density of defects. It was also found that a significant number of die had either a defect density so high or a catastrophic defect on a control wire that the present level of fault tolerance was insufficient to enable any of the sixteen rows to be configured.

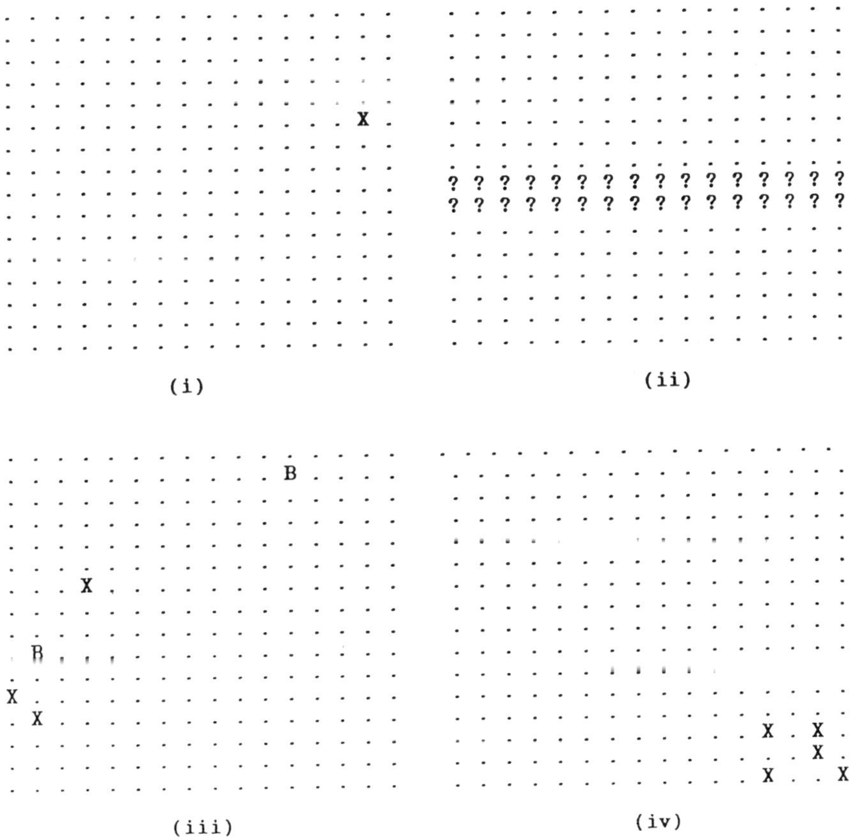

Figure 9: Bit fault maps

Although only a small sample has so far been tested and predictions based on one-off processing runs are to be viewed cautiously initial results have been encouraging and it would appear as if the fault tolerant structures and reconfiguration techniques as described in this paper enable the yield of reconfigurable arrays to be enhanced significantly.

The results obtained from this test chip will be used to evaluate the most efficient way to implement fault tolerance in a regular array chip have provided a valuable insight into the most effective diagnosis and bypass strategies in reconfigurable arrays. for example, producing suitable algorithms for generating appropriate test vectors and reconfiguration codes. It is likely that commercial applications of such systems will be required to posses efficient self testing and reconfiguration capabilities where exhaustive testing is inappropriate and short effective algorithms such as these are required.

ACKNOWLEDGEMENTS

This work was carried out as part of the Alvey Wafer Scale Collaboration. The support of the Alvey Directorate and the Alvey WSI team is appreciated and acknowledged.

REFERENCES

McCabe A P H, Moore W R and Bawa V, 1985; "Fault-tolerance in a large bit-level systolic array", Proceedings in 1985 Workshop on Wafer Scale Integration

McCanny J V and McWhirter J G, 1983; "Yield Enhancement of bit-level systolic array chip using fault-tolerant techniques"; Electronic letters, 19, pp 525-527

6.3 Yield Analysis of WSI Systolic Arrays with Two-Level Redundancy

Minghsien Wang, Michal Cutler and Stephen Su

INTRODUCTION

Systolic arrays are good candidates for implementation in wafer scale integration (WSI) due to their regular structure. But fault-tolerant redundancy techniques must be employed to enhance the yield of WSI systolic arrays to a reasonable level. Without redundant components in WSI systolic arrays, the yield would approach zero. In order to evaluate the effectiveness of various fault-tolerant redundancy techniques, a suitable yield estimation model is required.

A simple yield model for integrated circuits using a Poisson distribution has been proposed (Hofstein and Heiman 1963). The probability of having k defects in a chip of area A is $Pr\{k, A\} = \exp(-DA)\,(DA)^k / k!$, where D is the average defects density. Therefore the probability of having a good chip is $Pr\{0, A\} = \exp(-DA)$. Murphy (1964) recognized that the defect density may vary from chip to chip on a wafer and from wafer to wafer. Therefore the defect density is a random variable with probability density function f(D). The yield of a chip with area A is

$$Y = \int_0^{\infty} \exp(-DA)\, f(D)\, dD.$$

According to the experimental data of manufacturing processes, Stapper (1973) found that a Gamma distribution was practical for f(D) and the number of defects on a wafer followed the general negative binomial distribution (or Polya-Eggenberger distribution). The probability of a wafer having k defects is

$$P_W(k) = [\,\Gamma(k+\alpha) / k!\,\Gamma(\alpha)\,]\,\{\,(\lambda/\alpha)^k / [\,1 + (\lambda/\alpha)\,]^{k+\alpha}\,\} \qquad (1)$$

where α is a parameter related to the variation of defect density and λ is the average defect density on a wafer. $\Gamma(k)$ is the Gamma function of k. It has a mean and variance of λ and $\lambda\,[\,1 + (\lambda/\alpha)\,]$. The parameters λ and α can be easily determined from the manufacturing data. Stapper also gave the yield of a chip with area A as

$$Y = [\,1 + A\,(\sigma^2 / D)\,]^{-(D/\sigma)}\ ,$$

where D is the mean defect density and σ is the variance of the defect density.

Mangir and Avizienis (1982) presented a yield model for VLSI array processors with redundant cells. They introduced the concept of *bypass coverage* which is the probability that faulty cells can be successfully bypassed and replaced by redundant ones. They also considered the effect of the density of interconnection wires on yield improvement. Koren and Pradhan (1987) gave a slightly different yield model for fault-tolerant array processors. The array processors in their model consist of several types of elements. Each type of elements includes spares. The yield is the probability that for each type of elements, there are enough spares to replace the faulty ones. They also considered the effect of imperfect production testing in the yield model.

TWO-LEVEL REDUNDANCY SCHEME FOR MESH ARRAYS

In general, manufacturing defects of WSI/VLSI devices can be classified into two categories:
(1) *distributed defects* - the defects which are distributed uniformly on a wafer,
(2) *clustered defects* - the defects which are clustered together in a portion of a wafer.

A **two-level redundancy** scheme for mesh systolic arrays has been proposed (Wang et al. 1987) to tolerate manufacturing defects and operational failures.. A mesh array is partitioned into several identical subarrays (blocks). Spare cells (processing elements) are added into each block as *the local redundancy*. The local redundancy is effective for the distributed defects. But, it is not sufficient to overcome clustered defects. Some blocks in the wafer are assigned as *the global redundancy* for tolerating failure of blocks. The global redundancy is effective for the clustered defects. Switches and routing wires for reconfiguration are added between cells and blocks according to the selected reconfiguration scheme. The area of switches and routing wires between two cells (or blocks) is referred as *the cell (or block) connecting channel*. Fig. 1a shows a 3 by 2 block which contains 2 spare cells and a spare cell connecting channel as the local redundancy. Fig. 1b is an array consisting of 4 by 3 blocks, which contains 3 spare blocks and 2 spare block connecting channels as the global redundancy. Vertical wires (see Fig. 2) are used to connect cells vertically. In the yield analysis, these vertical wires are considered as part of the cell. Connecting channels are used to enable horizontal connections between cells or blocks on adjacent columns.

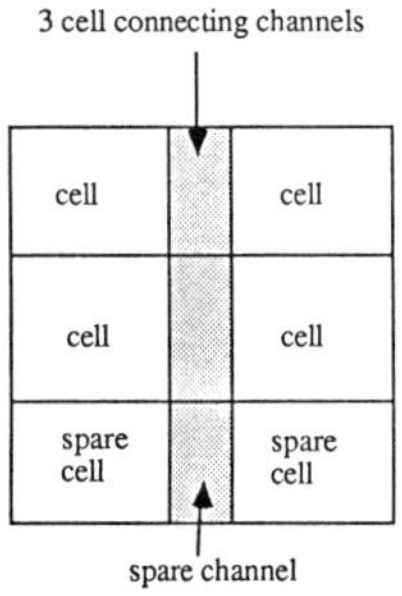

Fig. 1a A 2 by 2 block

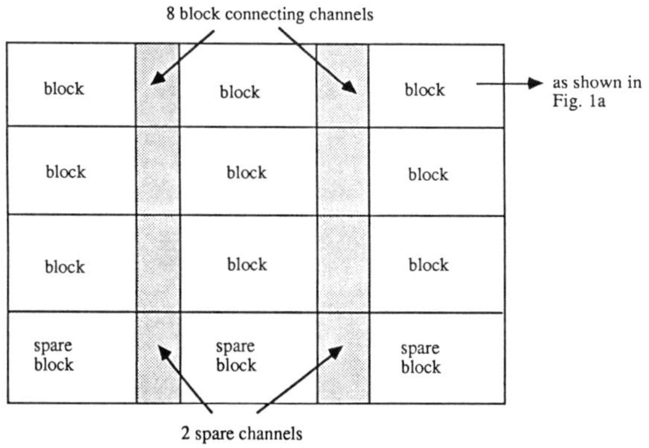

Fig. 1b A 6 by 6 array with two-level redundancy

The advantages of two-level redundancy scheme over one-level redundancy are as follows.

(1) The local redundancy is effective for distributed defects and the global redundancy is effective for clustered defects.
(2) Length of connecting wires after reconfiguration is shorter.
(3) Number of required tracks between cells for effective reconfiguration are reduced.
(4) Self-reconfiguration control circuit is simpler, hence increasing the speed of reconfiguration.
(5) Failure of reconfiguration control circuits at local level can be isolated without causing the system failure.

There are various redundancy and reconfiguration schemes which can be employed in both levels. Among them, we shall explain the column-bypass and whole-column/row-bypass scheme as examples. For *the column-bypass scheme,* there are one or more spare cells in each column. Faulty cells are bypassed and replaced only by spares in the same column provided that there are enough spares. Fig. 2 shows a 4 by 4 mesh array with two-level redundancy in which the column-bypass scheme is employed at both levels. For *the whole-column/ row-bypass scheme,* there are one or more spare columns and rows. A whole column (or row) containing faulty cell(s) is bypassed and replaced by a redundant column (or row) provided that there are enough spare columns or rows. Fig. 3 shows the reconfiguration result of a 4 by 4 block with two spare columns and two spare rows.

We shall present a yield estimation model of WSI systolic arrays with two-level redundancy in which the column-bypass or the whole-column/row-bypass reconfiguration scheme is employed.

DEFECT MODELS FOR YIELD ESTIMATION

Our yield estimation model is based on the combinatorial analysis result of an occupancy problem which randomly distributes r indistinguishable balls into n distinguishable boxes and finds the probability Q(r, n, i) of exactly i boxes being occupied. A box may contain more than one ball. According to (Feller 1967),

$$Q(r, n, i) = C(n, i) \sum_{v=0}^{i} (-1)^v C(i, v) [(i-v) / n]^r \qquad (2)$$

where C(n, i) = n! / [i! (n-i)!] is the total number of combinations of i elements out of n.

Another useful result from combinatorial theory is the probability $P_d(r, k, t)$ of distributing k out of r indistinguishable balls into a box, given that the probability of a ball being distributed into the box is t.

$$P_d(r, k, t) = C(r, k)\ t^k\ (1-t)^{r-k} \qquad (3)$$

We assume that an element (e.g. a cell, a wire or a switch) of a WSI systolic array is faulty if there is at least one defect in it. We treat defects on a wafer as indistinguishable balls, and elements as distinguishable boxes. The yield of a systolic array is the probability that defects are distributed into the array in such a way that the array can be successfully reconfigured with the current redundancy scheme.

In current manufacturing processes, the number of defects on a wafer have been shown to follow the general negative binomial distribution (or Polya-Eggenberger distribution) as shown in equation (1).

The distribution of distributed defects is modeled by uniform distribution. In other words,

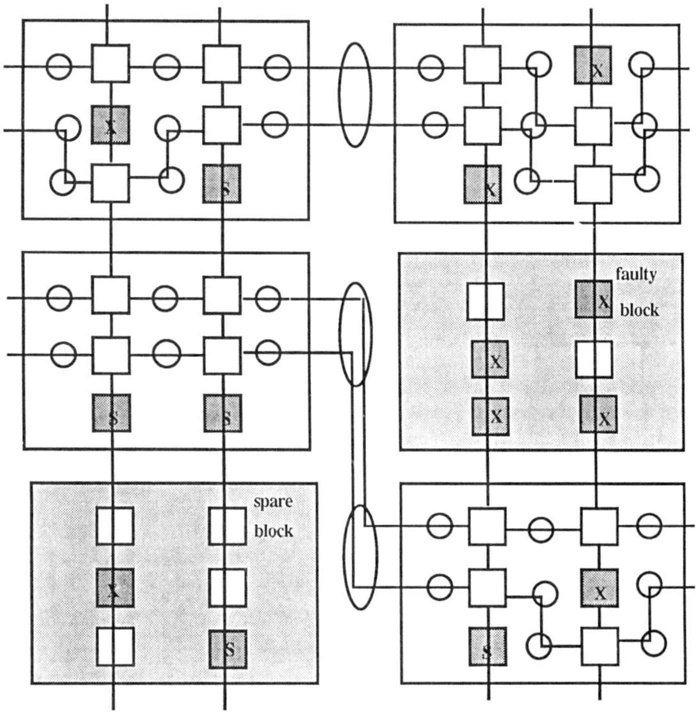

Fig. 2 A 4 by 4 mesh array with two-level redundancy and column-bypass scheme after reconfiguration

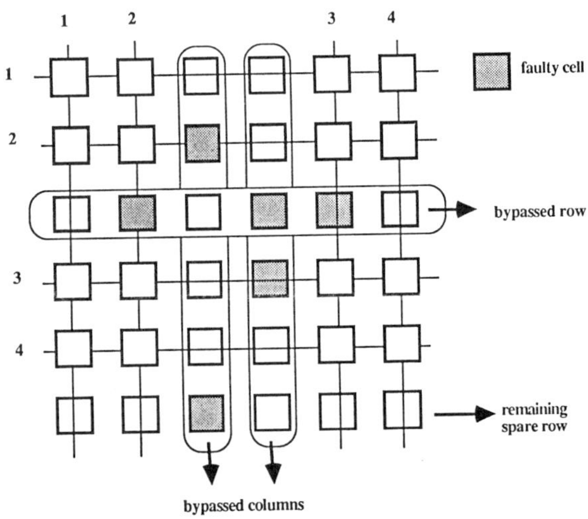

Fig. 3 A 4 x 4 arrary with 2 spare rows/columns and whole column/row bypass scheme

a distributed defect can be distributed into any point in a wafer with the same probability. The probability density function of the distributed defects is

$$f_u(x, y) = 1 / A, \tag{4}$$

where A is the area of the wafer. Previous research considered only uniformly distributed defects. We added into the calculations clustered defects. The distribution of clustered defects is modeled by a *truncated normal distribution*. The probability density function of clustered defects on a wafer is

$$f_c(x, y) = C \exp\left(-1/2 \{ [(x-u_x) / \sigma]^2 + [(y-u_y) / \sigma]^2 \} \right), \tag{5}$$

where (u_x, u_y) is the central location of the clustered faults, and σ is the standard deviation.

The coefficient C can be calculated by solving the following equation

$$\int_0^a \int_0^b f_c(x, y)\, dx\, dy = 1, \quad \text{where } a * b \text{ is the area of the wafer.}$$

The center of the clustered defects and the deviation can be obtained by analyzing the manufacturing data.

YIELD ESTIMATION

To calculate the yield of an array, Y_r, we take into account the yield of each block Y_{bj} (i.e. the probability of the jth block being reconfigurable), the yield of each connecting channel between adjacent blocks Y_{wi}, and the reconfiguration scheme. After the yield of each block and block connecting channel has been obtained, the yield of the array can be calculated according to the reconfiguration scheme.

Yield of a Block

We shall present here our yield model to calculate the yield of the jth block, Y_{bj}. Assume that the wafer has k defects and among them, c_1k defects are uniformly distributed and c_2k defects are clustered, where $c_1 + c_2 = 1$, as shown in Fig. 4. Note that c_1 and c_2 can be also obtained from the manufacturing data. Assume that r_1 defects out of c_1k distributed defects and r_2 out of c_2k clustered defects are distributed into the jth block. Among those r_1+r_2 defects in the block, r defects are in cells and the rest of r_1+r_2-r defects are in cell connecting channels . Then the yield of the jth block can be expressed as

$$Y_{bj} = \sum_{k=0}^{\infty} \sum_{r_1=0}^{c_1k} \sum_{r_2=0}^{c_2k} \sum_{r=0}^{r_1+r_2} P_w(k)\ P_d(c_1k, r_1, t_d)\ P_d(c_2k, r_2, t_c)\ P_d(r_1+r_2, r, t_e)\ Q_c(r)\ Q_w(r_1+r_2-r) \tag{6}$$

where

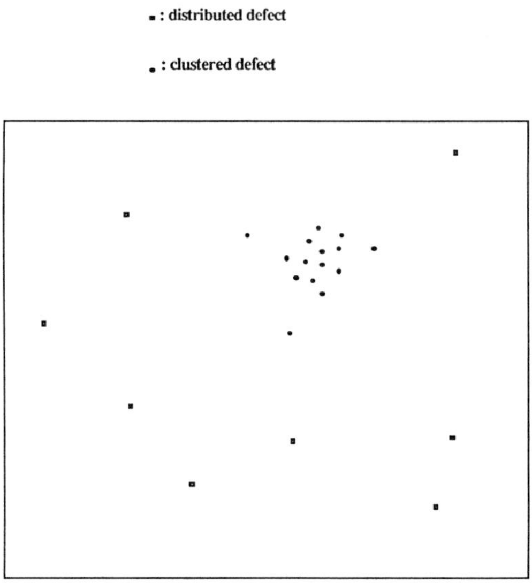

Total number of defects: k=23
Number of distributed defects: 8 (c1=0.35)
Number of clustered defects: 15 (c2=0.65)

Fig. 4 A wafer with distributed and clustered defects

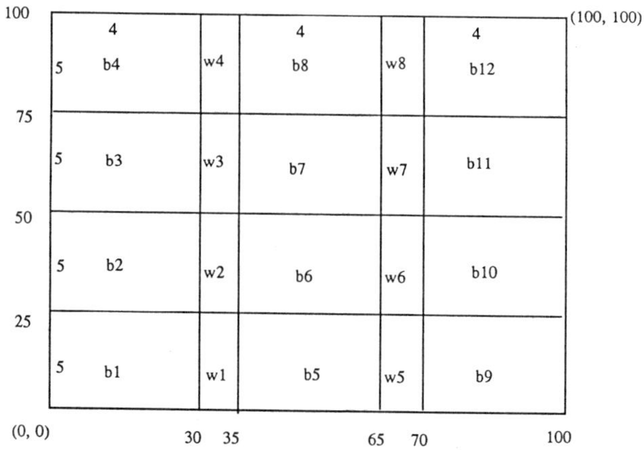

Fig. 5 A 12 by 12 mesh array with two-level redundancy

$P_w(k)$ is the probability of the array having k defects as shown in equation (1).

$P_d(c_1k, r_1, t_d)$ is the probability of exactly r_1 defects out of c_1k distributed defects being located on the jth block, and t_d is the probability of one distributed defect being located on this block. t_d is obtained by integrating equation (4) on the area of the block,

$$t_d = \int_{x_1}^{x_2}\int_{y_1}^{y_2} f_u(x, y)\,dx\,dy,$$ where (x_1, y_1) and (x_2, y_2) are the bottom left and top right corners of the block respectively.

$P_d(c_2k, r_2, t_c)$ is the probability of exactly r_2 defects out of c_2k clustered defects being located on block j, and t_c is the probability of a clustered defect being located on this block. t_c is obtained by integrating equation (5) on the area of the block,

$$t_c = \int_{x_1}^{x_2}\int_{y_1}^{y_2} f_c(x, y)\,dx\,dy,$$ where (x_1, y_1) and (x_2, y_2) are the bottom left and top right corners of the block respectively.

$P_d(r_1 \mid r_2, r, t_e)$ is the probability of exactly r defects out of r_1+r_2 defects being located on cells of the block j, and t_e is the probability of a single defect being located in cells of the block, i.e.

t_e = (area of cells of the block) / (area of the block) .

$Q_c(r)$ is the probability that by randomly distributing r defects into cells of block , all defective cells can be replaced by spares.

$Q_c(r)$ can be calculated as $$Q_c(r) = \sum_{i=0}^{n_{cb}} Q(r, n_{cb}, i)\,R(i),$$ where

$Q(r, n_{cb}, i)$ is the probability of randomly distributing r defects into exactly i out of n_{cb} cells in the block (see equation 1),

$R(i)$ is the probability that the block is reconfigurable given that there are exactly i defective cells on the block.

Assume that a block of size (m+1) by m uses the column-bypass reconfiguration scheme with one spare row, $R(i)$ of the block is determined as follows.

1. Let $n_{cb} = m(m+1)$.
2. If $i \le m$, $R(i) = \prod_{j=0}^{i-1} (n_{cb} - j(m+1)) / (n_{cb} - j)$.
3. If $i > m$, $R(i) = 0$.

If a block of size (m+1) by (m+1) uses the whole-column/row-bypass scheme with one spare row and one spare column, $R(i)$ is as follows.

1. Let $n_{cb} = (m+1)^2$.
2. If $i > 2m+1$, $R(i) = 0$.
3. If $2m+1 \ge i > m+1$, $R(i) = n_{cb}\, C(2m+1, i) / C(n_{cb}, i)$.

4. If $i \leq m+1$, $R(i) = \{ [2(m+1)] C(m+1,i) + n_{cb} [C(2m+1,i) - 2 C(m+1, i)] \} / C(n_{cb}, i)$

$Q_w(r_1+r_2-r)$ is the probability that by randomly distributing r_1+r_2-r defects into cell connecting channels, all defective channels can be replaced by spares.

$Q_w(r_1+r_2-r)$ can be determined in a similar way to $Q_c(r)$.

Yield of a Block Connecting Channel

Assume that a block connecting channel consists of m connecting wires and t spare wires. A block connecting wire is reconfigurable (i.e. can provide the useful service) if there are m defect-free wires among m+t wires. The yield of a block connecting channel, Y_{wi}, is the probability that at least m out of m+t wires are defect-free. Therefore, similar to equation (6),

$$Y_{wi} = \sum_{k=0}^{\infty} \sum_{r_1=0}^{c_1 k} \sum_{r_2=0}^{c_2 k} \sum_{x=0}^{t} P_w(k)\ P_d(c_1 k, r_1, s_d)\ P_d(c_2 k, r_2, s_c)\ Q(r_1+r_2, m+t, x)$$

where

$P_w(k)$ is the probability of the array having k defects .

$P_d(c_1 k, r_1, s_d)$ is the probability of exactly r_1 defects out of $c_1 k$ distributed defects being located on the ith block connecting channel, and s_d is the probability of one distributed defect being located on this block. s_d is obtained by integrating equation (4) on the area of the channel,

$$s_d = \int_{x_1}^{x_2} \int_{y_1}^{y_2} f_u (x, y)\, dx\, dy ,$$ where (x_1, y_1) and (x_2, y_2) are the bottom left and top right corners of the channel respectively.

$P_d(c_2 k, r_2, s_c)$ is the probability of exactly r_2 defects out of $c_2 k$ clustered defects being located on channel i, and s_c is the probability of a clustered defect being located on this channel. s_c is obtained by integrating equation (5) on the area of the channel,

$$s_c = \int_{x_1}^{x_2} \int_{y_1}^{y_2} f_c (x, y)\, dx\, dy ,$$ where (x_1, y_1) and (x_2, y_2) are the bottom left and top right corners of the channel respectively.

$Q(r_1+r_2, m+t, x)$ is the probability that r_1+r_2 defects are distributed into the channel and x out of m+t wires in the channel contain defects (see equation 2).

Yield of an Array

Assume that the column-bypass scheme with one spare row of blocks is also used at the array level. To simplify the calculation of the yield, in the above equation, we neglect the possibility of the array failure due to the relative position of defective blocks and defective block connecting channels. The yield of the array can be approximated by the following equation.

Yield = $\prod_{\text{all } i}$ (probability that the ith column of blocks has at most one defective block) x

$\prod_{\text{all } j}$ (probability that the jth column of block connecting channel has at most one defective channel)

[Example] We show the yield calculation for a 12 by 12 mesh array in Fig. 5. The block size is 5 by 4 and there are 4 * 3 blocks. The coordinates of each block and interconnection wire are marked on the figure. The WSI has a length and width of 100. Each block has a physical size of 25 x 30 and each block connecting channel has a size of 5 x 25.

Assume that :

a. The average number of defects on a wafer $\lambda = 40$. The variation parameter $\alpha = 2$.
b. 60% (c1) of defects are distributed defects and 40% (c2) are clustered defects.
c. The clustered defects are centered at (ux=30, uy=20). The standard deviation $\sigma = 5$.
d. The column-bypass reconfiguration scheme is employed at both level. A block contains one spare row. The array contains one spare row of blocks.

The coefficient C of truncated normal distribution is

$$C = 1 / \left(\int_0^{100} \int_0^{100} \exp\{ -1/2 \{[(x-30) / 5]^2 + [(y-20) / 5]^2\} \}dx\, dy \right) = 0.0063663.$$

Area of wafer = 100 x 100 = 10000
Area of a block = 25 x30 = 750.
Area of a block connecting channel = 25 x 5 = 125.

$$t_d = \int_0^{30} \int_0^{25} 1/10000\, dx\, dy = 750/10000 = 0.075.$$

t_c for each block and channel are calculated as follows by using the equation

$$t_c = \int_0^{30} \int_0^{25} f_c (x, y)\, dx\, dy \qquad \text{(see equation 5 for } f_c (x, y) \text{)}$$

t_c (b1) = 0.3651	t_c (b2) = 0.17235	t_c (b3) = 0.00083	t_c (b4) = 0
t_c (b5) = 0.2294	t_c (b6) = 0.108	t_c (b7) = 0.00005	t_c (b8) = 0
t_c (b9) = 0.000002	t_c (b10) = 0	t_c (b11) = 0	t_c (b12) = 0

t_c (w1) = 0.16	t_c (w2) = 0.076	t_c (w3) = 0.00003	t_c (w4) = 0
t_c (w5) = 0.00001	t_c (w6) = 0	t_c (w7) = 0	t_c (w8) = 0

Yields of each block and wire are calculated as follows.

Y_{b1} = 0.1225	Y_{b2} = 0.425	Y_{b3} = 0.845	Y_{b4} = 0.851
Y_{b5} = 0.326	Y_{b6} = 0.593	Y_{b7} = 0.840	Y_{b8} = 0.842
Y_{b9} = 0.844	Y_{b10} = 0.845	Y_{b11} = 0.846	Y_{b12} = 0.846

Assume that each block connecting channel has one spare wire and therefore can tolerate one defect.

$Y_{w1} = 0.3557$ $Y_{w2} = 0.698$ $Y_{w3} = 0.983$ $Y_{w4} = 0.988$
$Y_{w5} = 0.983$ $Y_{w6} = 0.987$ $Y_{w7} = 0.988$ $Y_{w8} = 0.988$

Since the column-bypass scheme with one spare block per column is also used in the array level, yield of the array is

$$\text{Yield} = \prod_{\text{all } i} (\text{probability that the ith column of blocks has at most one defective block}) \times \prod_{\text{all } j} (\text{probability that the jth column of block connecting channel has at most one defective channel}) = 0.172$$

CONCLUSION

We have presented an analytical method for estimating the yield of fault-tolerant WSI systolic arrays with two-level redundancy. Distributed and clustered defects are modeled by the uniform distribution and the truncated normal distribution respectively. Defects on connecting channels are taken into consideration.

This analytical method can be used to evaluate the effectiveness of various fault-tolerant schemes. Another application of this yield estimation method is to determine the optimal allocation of redundant components in both levels.

REFERENCES

Hofstein S R and Heiman F D 1963 "The silicon insulated-gate field-effect transistor", Proc. IEEE, vol. 51, 1963, 1190-1202.

Murphy B T 1963 "Cost-size optima of monolithic integrated circuits", Proc. IEEE vol 52, 1964, 1537-1545.

Stapper C H 1973 "Defect density distribution for LSI yield calculations", IEEE Trans. Electronic Devices, July, 1973, 655-657.

Mangir T and Avizienis A 1982 " Fault-tolerant design for VLSI: effect of interconnect requirements on yield improvement of VLSI designs", IEEE Trans. Compt. July 1982, 609-615.

Koren I and Pradhan D K 1987 "Modeling the effect of redundancy on yield and performance of VLSI systems", IEEE Trans. Compt. March 1987, 344-355.

Wang M, Cutler M and Su S Y H 1987 "On-line error detection and reconfiguration of array processors with two-level redundancy", Proceedings of the 1987 CompEuro Conference on VLSI and Computers, Hamburg, West Germany, May 11-15, 1987.

Feller W 1967 "An introduction to probability theory and its applications," John Wiley & Sons, New York, 1967.

6.4 A Cost Orientated Redundancy Model for Defect-Tolerant VLSI/WSI Systems

Ulrich Ramacher

INTRODUCTION

Redundancy may be used to improve a system's defect-tolerance or failure-tolerance, i.e. its yield or reliability figure. Irrespective of the quality chosen, it is desirable to know how to determine the degree of redundancy useful for implementation. This is easily accomplished only if one deals with a system of highly irregular architecture for which **T**riple **M**odular **R**edundancy is most appropriate. For systems with inherent regularity (for example, a 2 dimensional systolic array of identical processor elements) TMR is a very unfavourable concept to use, and one is forced to start a systematic investigation by means of probability theory and defect statistics in order to bring to light the functional dependence of yield on redundancy (Mangir, 1981; Ketchen, 1985; Franzon, 1986) .

Because costs (for processing, testing, packaging, etc.) are area dependent and redundancy requires extra area, redundancy influences sytem costs. Thus, one has to care for a cost optimal value of redundancy, too, and to extend common yield models accordingly.

Besides, a redundancy model should answer 2 questions which are of central importance to the designer of WSI systems:

> With which system size should redundant implementation begin and end respectively ?
> How seriously should the computed value of redundancy be taken if fluctuations of wafer defect density and cluster coefficient are accounted for?

In the first section, parameters and statistics relevant to the redundancy model are described and the extension of yield modeling to cost modeling is presented. The second section concentrates on design issues and proposes a test method which allows the designer to split the overhead into a small non-defect-tolerant part and a much bigger defect-tolerant part. It will be shown that the non-defect-tolerant part of the overhead comprises mainly the switches for test and configuration, but not the switch setting logic. In the third section, the influence of overhead and costs on redundancy is discussed for a 2-dimensional systolic array, the fourth section discusses word slice partitioning and "local versus global redundancy". Finally, the model dependence of the computed value of redundancy is investigated, and reasonable system sizes for Wafer Scale Integration are determined. The last section considers the example of a matrix multiplier which is under design and comprises 10^6 transistors.

INPUTS OF THE REDUNDANCY MODEL

Assume N·L identical modules to be implemented (L indicates redundancy) and the yield of each given by generalized negative binomial statistics:

$$Y(A) = \left[1 + D \cdot \frac{A}{E}\right]^{-E} \quad . \tag{1}$$

Here, D is the mean wafer defect density, A the module area and E the cluster coefficient (Bräckelmann, 1985) . The probability of getting **N** good modules from N·L implemented modules can be calculated according to the formula

$$Y(N,L,A) = \sum_{j=0}^{(L-1)N} \binom{NL}{j} \cdot Y(A)^{NL-j} \cdot [1 - Y(A)]^{j} \quad , \tag{2}$$

(Stapper, 1983) . The yield of the system consisting of modules, wiring and overhead for test & configuration is the product of corresponding yields :

$$Y_r = Y(N,L,A) \cdot Y_{Overhead} \cdot Y_{Wiring} \quad . \tag{3}$$

To account for the reduced number of masks needed for wiring, and for the decreased sensibility to particles interrupting alu lines of width $\geq 5\mu m$, an additional exponent is introduced into (1) if interconnection yield is computed (typically, the exponent is less than 0.5 for a common CMOS process) .
The following two quantities prove to be useful for comparison of redundant Wafer Scale Integration with some more conventional discrete realizations :

$$figure\ of\ merit\ \ FM := \frac{(yield/area)_{redundant\ realization}}{(yield/area)_{discrete\ realization}} \tag{4}$$

$$costs\ ratio := \frac{costs\ with\ discrete\ realization}{costs\ with\ redundant\ realization} =: Kd/Kr \tag{5}$$

Redundant WSI stands for monolithic large area integration by means of redundant design, as opposed to discrete WSI which accomplishes integration by mounting pretested chips on a prewired wafer. If discrete Wafer Scale Integration is taken for comparison (Ramacher, 1987a), the expression for the costs ratio reads (Ramacher, 1987b) :

$$Kd/Kr = \frac{1}{L} \cdot \frac{\left(\frac{q_p}{q_c} + \frac{q_m}{q_c}\frac{1}{Y_m}\right) \cdot \frac{A_{fp,d}}{A_{fp,r}} + \frac{1}{Y_m \cdot Y(A)} \cdot \frac{A_p}{A_{fp,r}}}{\frac{q_p}{q_c} + \frac{1}{Y_r}} \quad . \tag{6}$$

Here, costs are assumed to be proportional to area, and

A_p is the module area with pads

$A_{fp,d}$ is the footprint of a module with discrete WSI comprising chip- and wafer pads and average wafer wiring

$A_{fp,r}$ is the footprint of a module with redundant WSI comprising overhead and wiring

Y_m is the yield of chip mounting and wafer wiring with discrete WSI

q_m is the costs per unit area with discrete **WSI** for chip mounting and wafer wiring inclusive test

q_p is the costs per unit area for packaging the WSI system and eventual assembly at a PCB or elsewhere

q_c is the costs per unit area for chip production including design, simulation, processing, test, etc. .

(6) may be rewritten as

$$Kd/Kr \;=\; FM \cdot \frac{\left(Q_p \cdot\; Y_m + \; Q_m\right) \cdot\; Y(A) + \dfrac{A_p}{A_{fp,d}}}{Q_p \cdot\; Y_r + \; 1} \quad , \tag{7}$$

with normalized costs $Q_p := q_p/q_c$ $\quad Q_m := q_m/q_c$.

Thus, the costs ratio appears as a simple continuation of the figure of merit FM.

Formulae similiar to (6), (7) result if instead of discrete WSI a conventional multilayer PCB realization of the system is considered. In this case, the FM value of discrete WSI must be multiplied by

$$\frac{A_{PCB}}{A_W} \cdot \frac{Y_m}{Y_{m,PCB}} \quad ,$$

with

A_{PCB} the area of the system using PCB
A_W the area of the system using discrete WSI
$Y_{m,PCB}$ the mounting yield using PCB

and in the quotient of (7), Y_m and $A_{fp,d}$ must be replaced by $Y_{m,PCB}$ and by the module's PCB footprint $A_{fp,PCB}$ resp..

Because mounting yield should reach values close to 1 , the formula (7) has simply to be multiplied by the factor A_{PCB}/A_W to give a first order correct expression for the comparison of redundant WSI with PCB.

In order to evaluate (6), (7) one must compute (3) , i.e. find a set of parameters with which modeling of overhead and wiring is described. They are listed in table 1.

discrete & redundant WSI	discrete WSI	redundant WSI
wafer defect density	interconnection area	transistors per module
cluster coefficient	mounting yield	transistors for global test logic
# of blocks	mounting costs	wiring area for global test logic
# of identical modules	pad pitch	mean length of a reconfig.line
normalized packaging costs		mean width of a reconfig. line
module area		vertical & horizontal word width

Table 1 : Parameters of the redundancy model

OVERHEAD DESIGN FOR HIGH YIELD

To illustrate the constituents of the overhead, a systolic array is considered, say, with 2x4 modules and one extra module per row. Its reconfiguration scheme is shown in figure 1.

As the number of switches and reconfiguration lines for test and configuration is a simple function of redundancy and vertical & horizontal word width, the switch and wiring overhead is easily evaluated in terms of the number of transistors and wiring area. Sometimes, this part of the total overhead may amount to several 10000 transistors (as is the case for the matrix multiplier considered below) and, thus, it is important to design the global test- and local switch setting logic for high yield. Figure 2 illustrates the test strategy chosen for this purpose.

The array is tested row by row, C_j is a test selector bit which takes the value 1 if the module in the j-th column is to take part in a configuration of 4 out of 5 modules, and 0 if this is not the case. After selecting a set of C_j and enabling the row, the test is run and the output fed into the global evaluation logic . After a negative test result, a new set of C_j is created and the test repeated until a working configuration is found in which case the coefficients C_j are loaded into local latches. Otherwise, the second row is tested by broadcasting new coefficients C_j to its modules.

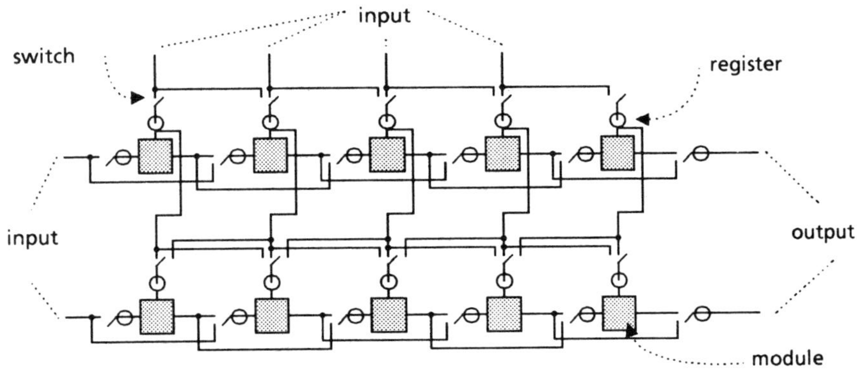

Figure 1 : Systolic array with one one extra module per row

Basic to this test strategy is the assumption that the set of C_j is enough to set the switches for each module. Indeed, needed for fixing the position of the switches of a module M_j, say, in the second row and j-th column is

information about the state of the selector bits of the neighbours of M_j, i.e. about C_j from the first row and about C_{j-1}, C_j, from the second row

information about a zero coefficient C eventually occurred in the first or second row up to the (j-1)-th column

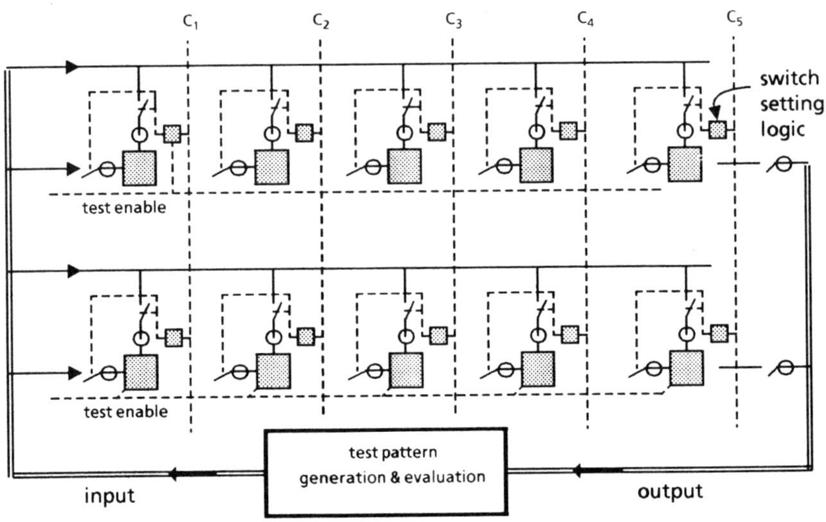

Figure 2 : Test strategy for the array mentioned in figure 1

It is easy to propagate information concerning the occurrence of a zero coefficient along a row. Thus, all the information required is locally transferred. The switch setting logic may be split into two parts : one part shares the benefit of defect-tolerant design, the other part lacks such a quality (figure 3) . The defect-tolerant part consists of the switch setting logic where evaluation of a set of equations (one for each of the 5 switches) with 5 logical variables (Z_{j-1},C_{j-1} , C_j from row 2, C_j, Z_{j-1} from row 1) is done and eventual malfunction would affect only the module. The zero propagation logic (which is just an AND gate in case of 1 extra module per row) constitutes the non-defect-tolerant part because a defect, happening there, would affect all the subsequent modules of the row and could make the global test logic decide that no working configuration

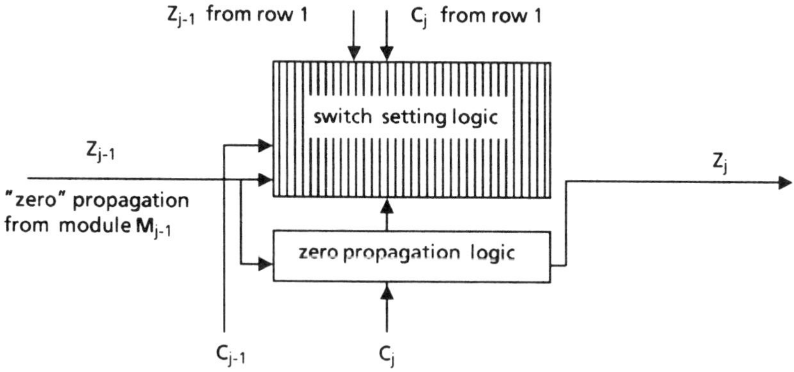

Figure 3 : Defect-tolerant design of local switch logic

could be found although there might have been one. Clearly, the zero propagation logic lends itself to implementation in triple modular redundancy because of its very low hardware amount (not only measured with respect to the switch setting logic, but also to the absolute number of transistors required even for two defects to repair in a row).

As a result, the overhead is determined mainly by the number of switches required for test & configuration of the modules (provided that the global test logic is implemented with triple modular redundancy,too, which always is affordable).

With these provisions made, the evaluation of (3), (6), (7) can be undertaken by fixing a set of realistic parameters (table 2) which is understood to hold for all the coming figures unless other values are specified, explicitly. Redundancy is understood to be provided row-wise (see figure 1).

discrete & redundant WSI		discrete WSI		redundant WSI	
defect density	1cm^{-2}	interconnect. area	300 mm^2	transistor density	2000
cluster coefficient	2.5	mounting yield	0.85	transistors for global test logic	10000
# of blocks	8	rel.mounting costs	1	wiring area for "	30mm^2
# of modules	8	pad pitch	140 µm	mean length of a reconfig.line	5mm
rel. packaging costs	1			mean width of a reconfig. line	5µm
module area	5mm^2			vertical & horizontal word width	(8,8)

Table 2 : Parameters for the redundancy model

INFLUENCE OF OVERHEAD AND COSTS ON THE OPTIMAL VALUE OF REDUNDANCY

Figure 4 shows the various different yields. It demonstrates the need for exact computation of redundancy. Furthermore, the maxima of yield, FM and costs ratio are reached at different optimal values of redundancy, namely 1.375, 1.25 and 1.125 (see figure 5).

As explained before, the overhead is determined by the number of vertical & horizontal bits per module. Figure 6 proves the expected decrease of yield etc. with increasing word width.

Also, the module area recommendable for redundant implementation is limited , as figure 7 points out.

If packaging and mounting costs are taken to be negligible it follows from (7) that the costs ratio is equal to the FM value, up to a constant factor. Figure 8 shows that inclusion of costs supports redundant WSI. Particularly, the mounting costs help redundant implementation. Remember also that the axis of Kd/Kr must be up-scaled by a factor reflecting the loss in packaging density when turning from discrete WSI to PCB realization.

However, figure 8 shouldn't be interpreted without a look at figure 9. For, it may happen that the costs ratio is still greater than 1 , but redundancy reaches 1 . In this case, non-redundant one-

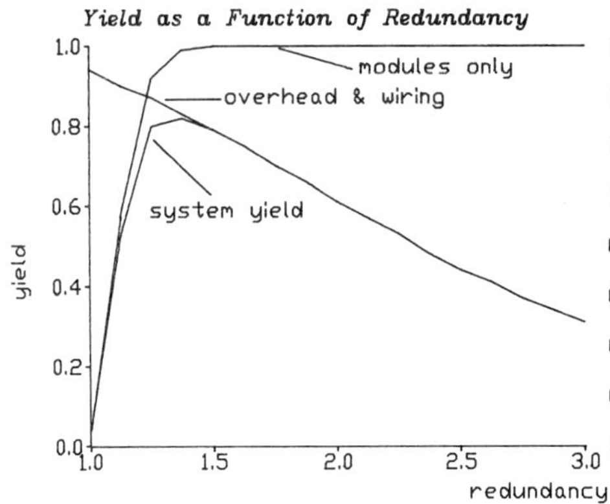

Figure 4

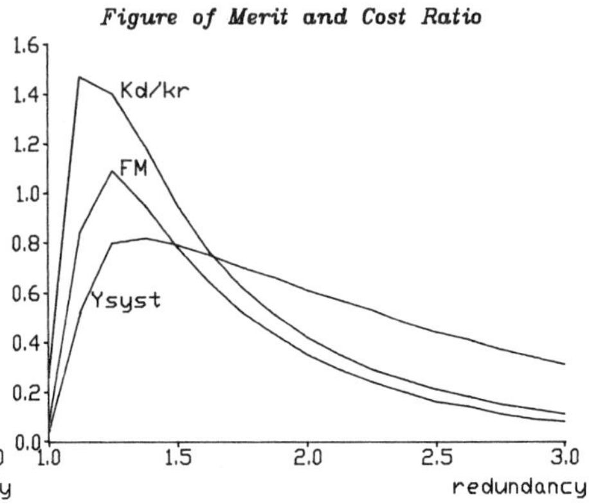

Figure 5

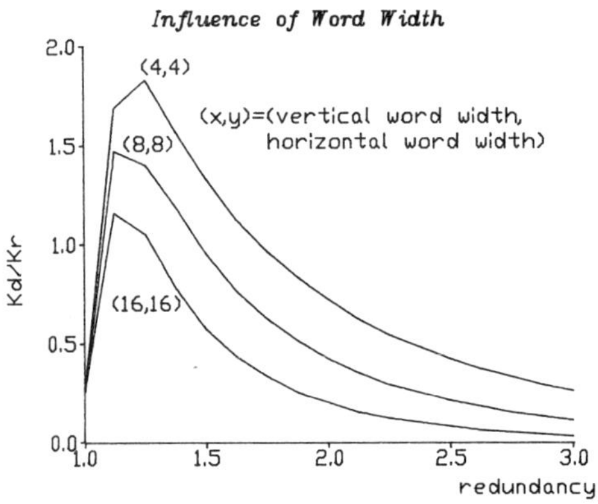

Figure 6

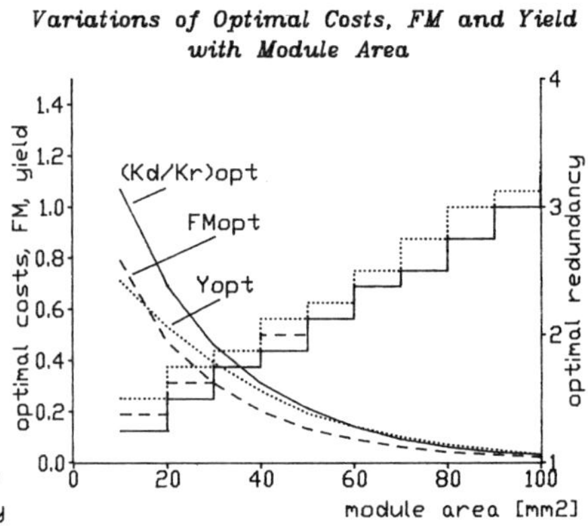

Figure 7

chip implementation is preferable. Figure 9 shows that optimal redundancy is constant over a fairly wide range of packaging and mounting costs. This property is quite welcome because it is hard, if not impossible, to get exact data. On the other side, approximate data for packaging and mounting costs are available (Carlson and Neugebauer, 1986) .

REDUNDANCY AND ARCHITECTURE

So far, redundancy is provided for each row of the 8x8 array. A dramatic decrease in the number of "vertical" switches is obtained, however, if a full column (or parts of it) is taken as the unit which is redundantly implemented (figure 10) . This has to be contrasted with the increased module area (it grows to 8x5 mm^2 for a full column) . Thus, a decrease in overhead must be balanced against an increase in module area (remember figure 7). Figure 11 shows that for the array under consideration redundancy should be provided for each row separately (the curves' indices refer to the module size measured in parts of a full column) .

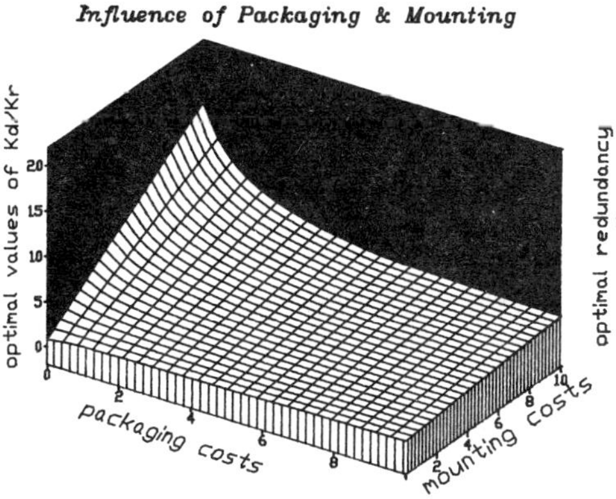

Figure 8 (wafer defect density = 0.5)

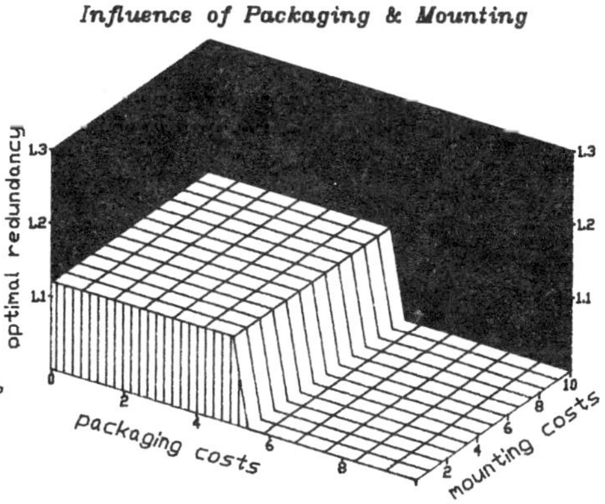

Figure 9 (wafer defect density = 0.5)

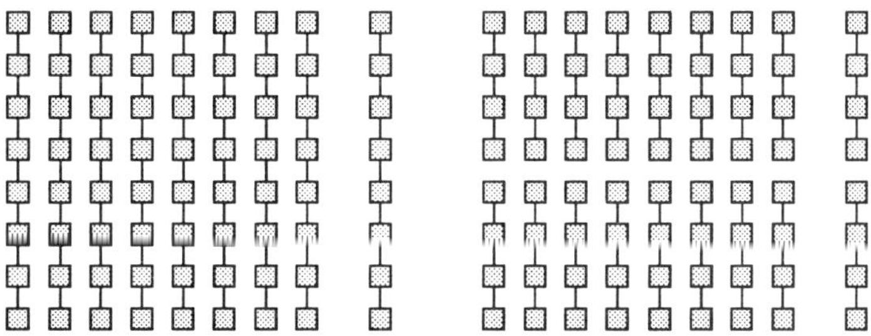

Figure 10 : 2 partitionings of a column of the8x8 array

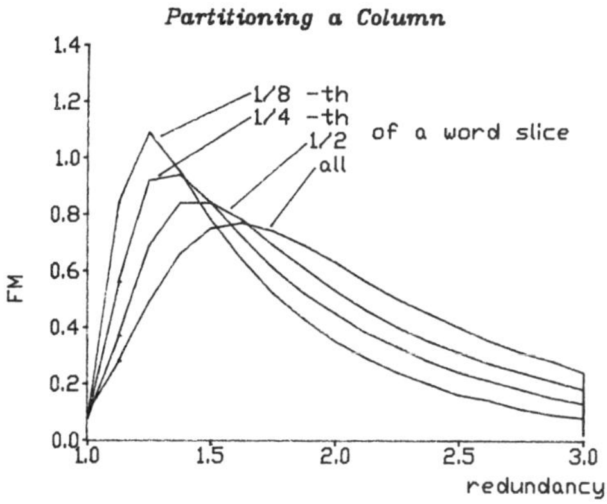

Figure 11

Next, a one dimensional array of modules is considered, and redundancy provided to all 64 modules, or to 2 groups of 32 modules separately, to 4 groups of 16 modules separately, and so on (see figure 13) . Clearly, this chain of 64 modules is not meant to replace the 8x8 array. It turns out that global redundancy is not optimal although it needs only 5 extra modules to reach the FM maximum whereas the partition into 8 groups of 8 modules needs 16 extra ones (figure 13).

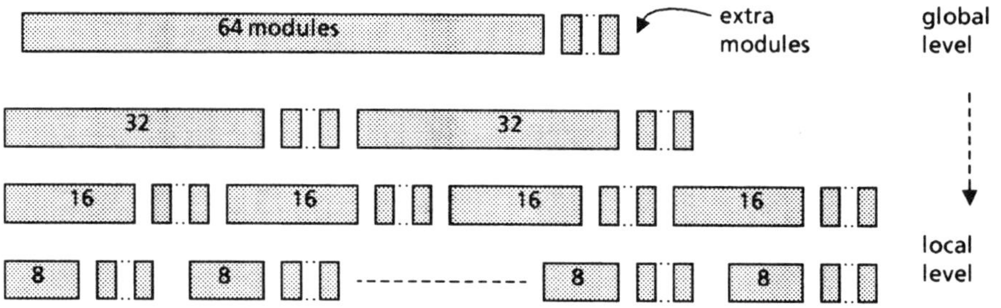

Figure 12 : Providing redundancy at local to global level

Similiar curves hold for yield and costs ratio. The explanation is found in the much higher overhead for repairing 5 defects within 64 modules than 8-times 2 defects within 8 modules (figure 14) .

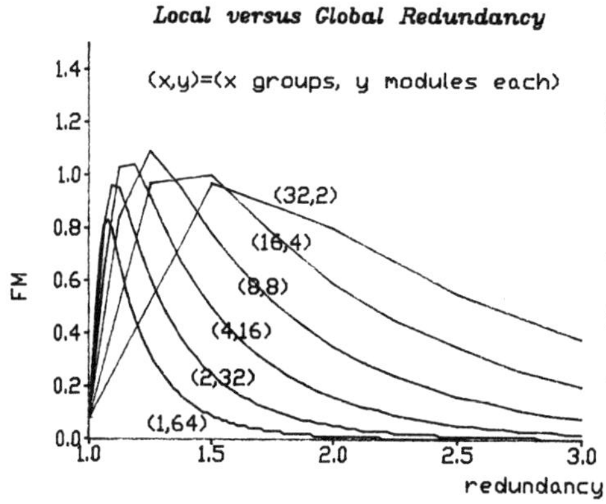

Figure 13

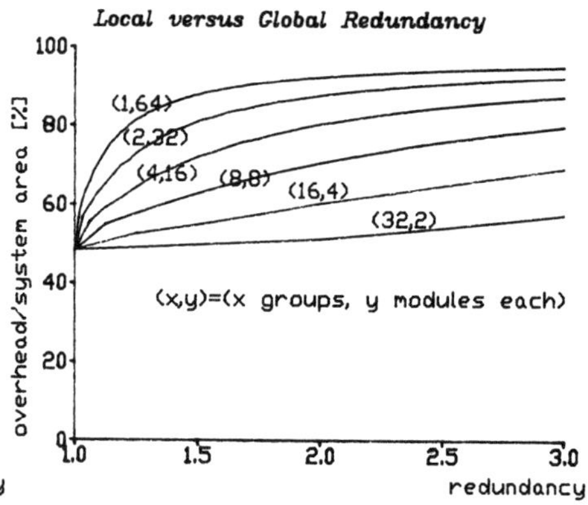

Figure 14

MODEL DEPENDENCE OF THE VALUE OF OPTIMAL REDUNDANCY

For the designer of a WSI system it is crucial to know how much he can trust the value for optimal redundancy when the basic statistical quantities, namely wafer defect density and cluster coefficient, are known to fluctuate considerably. Figure 15 shows that today's inexactness of the knowledge of the data for cluster coefficients hardly influences the decision which value of optimal redundancy is to be chosen.

A quantity of greater influence is wafer defect density (figure 16) . For example, if values for wafer defect density vary between 0.5 and 5 defects per cm^2 the corresponding optimal redundancy shifts from 1.25 to 1.625 . Which one of the values should be chosen? As figure 17 shows, it is advantageous to choose 1.5 or 1.625 if yield is a concern and knowledge about the distribution function of the wafer defect density not available . Thus, with the proposed redundancy model, the designer of a WSI system can decide which value of redundancy should be used for implementation. When it comes to system production, i.e. to costs, this decision is more delicate and the reliability of statistical data is of higher importance (figure 18) .

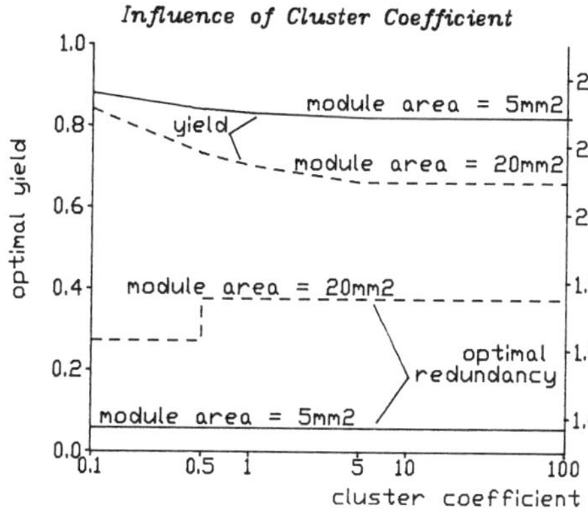

Figure 15

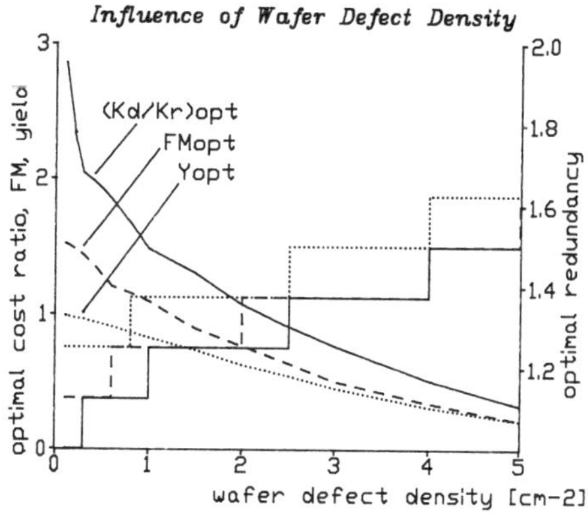

Figure 16

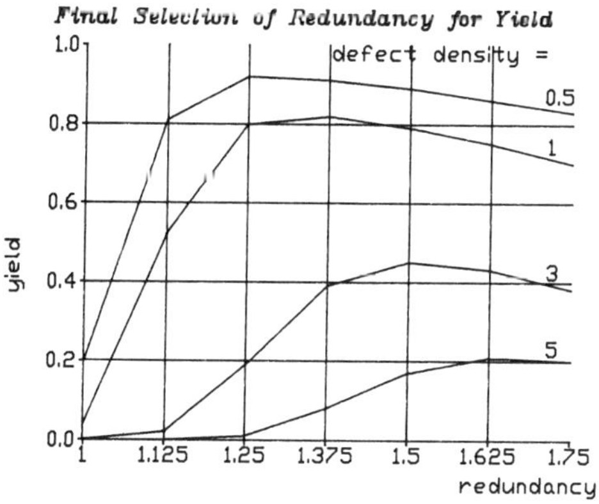

Figure 17

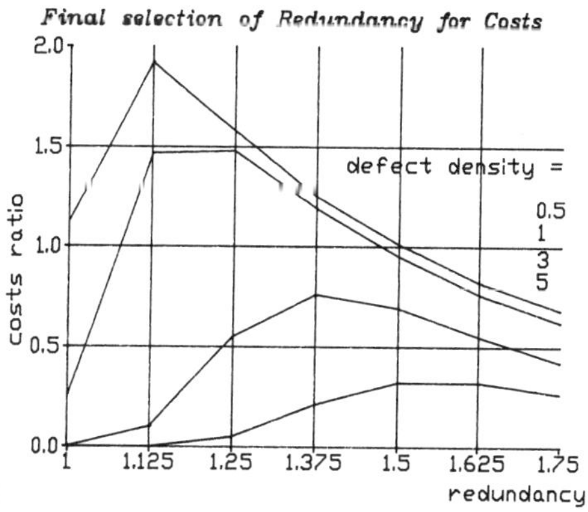

Figure 18

REASONABLE MODULE AREAS FOR WSI

In figures 19-20, minimum and maximum module and system size respectively are shown as a function of wafer defect density. Figure 19 is obtained for the FM criterion, and figure 20 for the costs ratio criterion. The region between upper and lower curves represents the module and system size respectively with which redundant implementation is superior to non-redundant one-chip realization or to discrete multi-chip WSI or PCB realizations. The region below the lower curves comprises systems which are best realized by non-redundant one-chip implementation, and the region above the upper curves indicates that discrete multi-chip WSI or PCB realization is preferable. It is obvious that with today's wafer defect density interval (1..5 cm^{-2}) redundant WSI isn't very appealing <u>if costs are not a concern</u>. Not before wafer defect densities of less than 0.5 cm^{-2} were attained could a sytem designer feel invited by redundant WSI. The situation changes quite dramatically when costs become important. Furthermore, if one takes a comparison between a redundant WSI system and an equivalent PCB system - instead of discrete WSI system -- the situation improves even more in favour of redundant WSI.

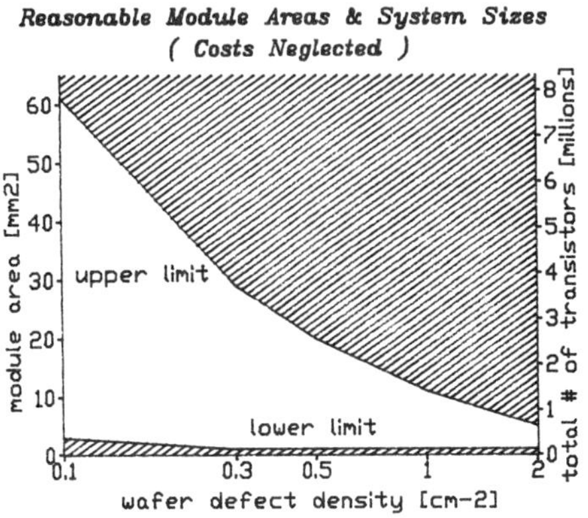

Figure 19

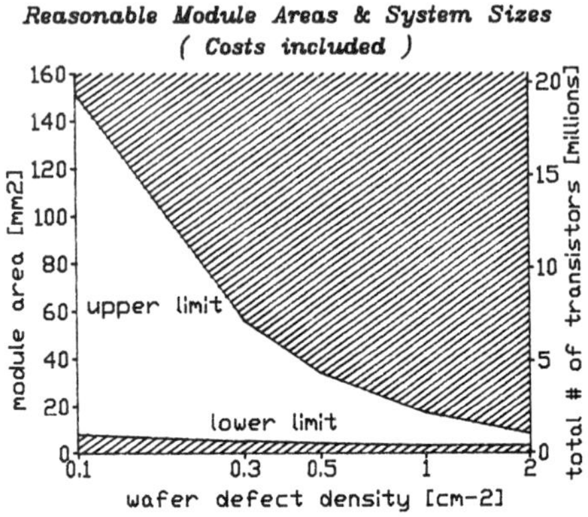

Figure 20

MODELING A WSI MATRIX MULTIPLIER

A suitable object to start redundant WSI is a matrix-matrix multiplier consisting of an array of 64 multiplier & accumulator modules, each of 16x16 bit input width and 35 bit output width. A pipelined multiplier & accumulator module (ca. 10000 transistors) and a matrix multiplier architecture with two level pipelining (see figure 21) were used. The architecture shown below guarantees low pad density and simple configuration&test logic as compared with contra data flow arrangement. Redundancy turned out to be most efficient if provided separately for each row, requiring 2 extra modules. The configuration of the array is shown in figure 22 . Ca. 17000 switches are needed for test and configuration. The local switch control logic is split into a defect-tolerant part and a non-defect-tolerant part, the latter accounting for just 4 gates per module. As global test logic is implemented with triple modular redundancy, the total overhead amounts to less than 40000 transistors, i.e less than 3 % .The parameters relevant for the computation of optimal redundancy are the same as in table 2, except the values for vertical and horizontal word width. For more details on the WSI matrix multiplier see Beichter and Ramacher (1987) . Figure 23 shows the expected yield.

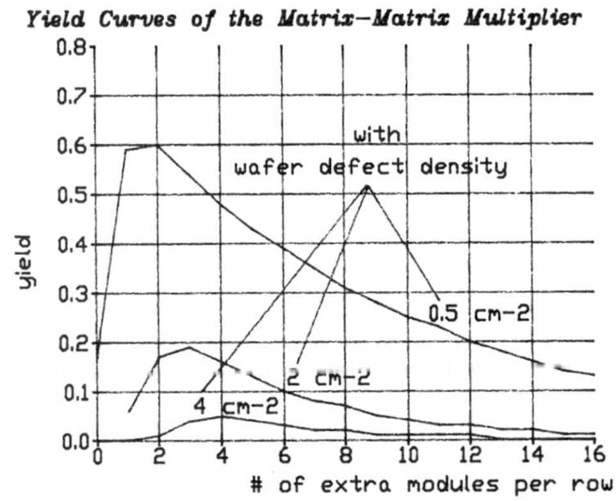

Figure 23

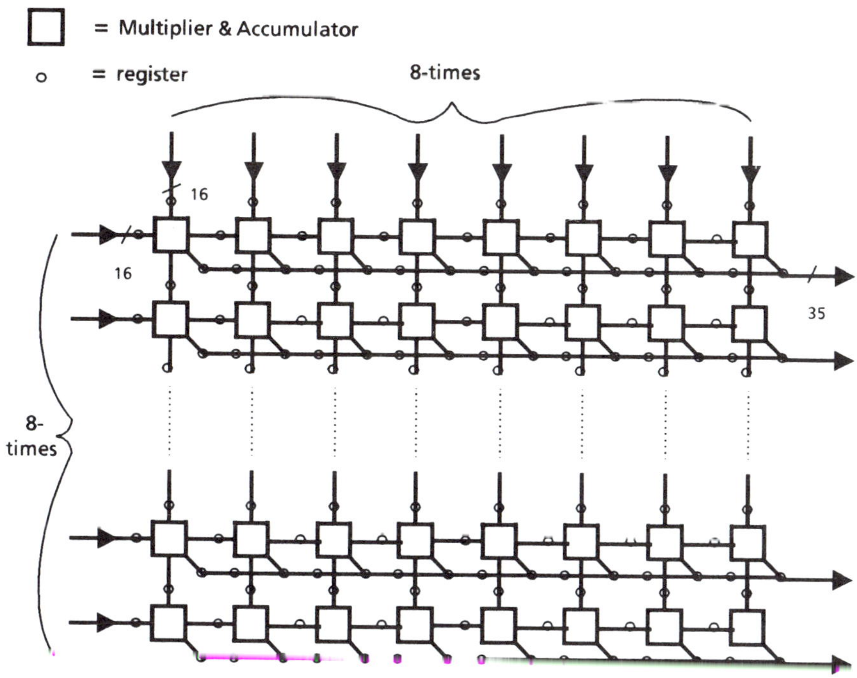

Figure 21 : Architecture of the matrix multiplier

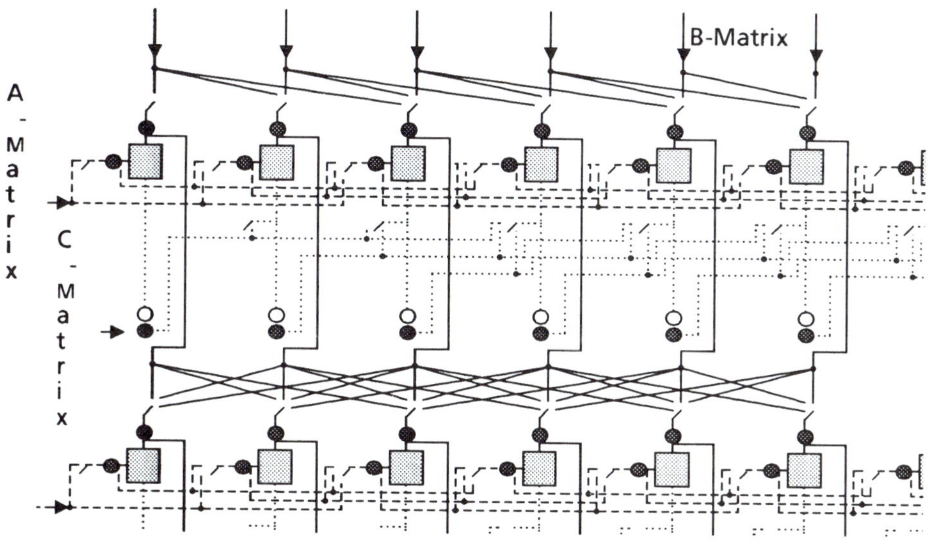

Figure 22 : Switch configuration of the matrix multiplier repairing 2 defects per row

CONCLUSIONS

The proposed redundancy model is a necessary tool for the designer of Wafer Scale Integrated systems. It helps him

1. to plan the realization of highly complex systems (> 10^6 transistors)
2. to partition the system in such a way that redundancy is placed most favourable
3. to determine the amount of redundancy needed for implementation

Ad1. The introduction of cost factors for mounting and housing chips supports redundant Wafer Scale Integration and dramatically enlarges the field of application where redundant implementation is superior to more conventional realizations (figure 8; figure 19-20). Besides, incorporation of costs into the redundancy model allows a decision to be made as to whether a system under consideration should be realized in a redundant or non-redundant fashion. This feature is even more important with future wafer defect densities (< 0.5 cm^{-2}) . It is remarkable that the *system sizes* which can be covered by redundant Wafer Scale Integration *grow with future wafer defect densities* (this tendency is maintained with the use of submicron structure sizes). With 0.5 defects per cm^2 and 2µm minimal structure size, redundant implementation of typical WSI-systems ranging from ca. 250 thousand to 2.5 million transistors is possible *and* preferable in terms of yield (and of costs , figure 19-20). *Thus, redundant Wafer Scale Integration will establish a new way to monolithic system integration which is, by no way, restricted by future submicron technology and whose outstanding complexity can not be reached other than by large area integration.*

Ad2. Providing redundancy for a system of 10^6 transistors and more is quite a complex task that needs systematic treatment and support by a redundancy model. It is not the designer's free will to decide on what is considered to be the elemental unit for redundant implementation. *The correct placement of redundancy is as important as the skills of the architect who must get acquainted with algorithms of highly complex systems and map these into systolic architectures featuring a sufficient degree of regularity and parallel computation.*

Ad3. An important feature of the redundancy model is that it provides the designer with a safe value of redundancy. For, redundant implementation would be uneconomic if redundancy were to react sensitively to new values of the wafer defect density, typically encountered with a more mature state of line engineering. Fortunately, redundancy is quite tolerant to improvements of line parameters and cost factors (figure15-17) . Besides, the redundancy model doesn't need exact values of the parameters (mentioned in table 1), only the range of parameter variation needs to be specified. Clearly, this requires a greater measure of communication between line technologyst, designer and salesman than practiced today. It is remarkable that the amount of extra modules needed for implementation is rarely more than 1.5 times the number of modules required to be functional.

REFERENCES

Bräckelmann 1986 Bipolar gate arrays in central processing units,
Siemens Res. & Devel. Reports, Vol.15 No.2, p.56

Beichter,Ramacher 1987 A self testing WSI matrix-matrix multiplier, accepted for publication at the Workshop on Wafer Scale Integration , Brunel University

Carlson,Neugebauer 1986 Future trends in Wafer Scale Integration,
Proc. IEEE, Vol. 74 No.12 , p. 1741-1752

Franzon 1986 Yield modeling for fault-tolerant VLSI arrays,
International Workshop on Systolic Arrays, University of Oxford, N4 p.1-10

Hilberg 1986 Mikroelektronik, Physik in unserer Zeit, 17. Jg. No.1, p. 18-28

Ketchen 1985 Point defect yield model for Wafer Scale Integration,
IEEE Circuits and Devices Magazine, p.24-34

Mangir 1981 Use of on-chip redundancy for fault-tolerant very large scale integrated circuit design, dissertation, Univ.of California, Los Angeles

Ramacher 1987a Wege zur Großintegration -- diskrete und redundante Wafer Scale Integration,
ITG-Fachberichte 98 Großintegration, Baden-Baden, p.81-91

Ramacher 1987b Internal report, ZFE ME 22 Siemens, p.1-17

Stapper, Armstrong and Saji 1983 Integrated circuit yield statistics,
Proc. IEEE, Vol. 71 No.4 , p. 453-470

7 TESTING

This last chapter touches on the vital problem of testing VLSI circuits and illustrates some areas that require more information on defect models.

Transistor level fault models are known to present more difficult testing problems than higher-level 'stuck-at' models, and Sastry and Breuer (§7.1) analyse the effectiveness of built-in-test techniques to cover such faults.

Marnane *et al.* (§7.2) look at a structured approach to testing systolic arrays and show that this can be very effective and also that it can permit any desired fault model to be incorporated at the lower level. In view of the interest in designing fault-tolerant systolic arrays (see e.g. §6.2) these results will also be useful to the diagnosis of defective modules.

In the final contribution, Kim and Reddy (§7.3) consider the problem of diagnosing the defective cells in more general defect-tolerant processor arrays and show the kind of hardware overhead which may be necessary.

7.1 Analysis of BIST Techniques for CMOS Stuck-Open Faults

Sarma Sastry and Melvin Breuer

Introduction

Built-in self test (BIST) (McCluskey 1985a, McCluskey 1985b) is a testing methodology where some or all of the test circuitry is included on the chip. One class of BIST schemes generates random or pseudo-random test vectors on-line, i.e., as they are being applied. Random testing involves generating a *random* sample of test vectors and applying them to the circuit under test. One common technique to generate such vectors is via a linear feedback shift register (LFSR), which consists of a string of D flip flops and one or more feedback lines passing through XOR gates (McCluskey 1985a, McCluskey 1985b). An N bit LFSR has a period of at most $2^N - 1$ and can easily be modified to generate all 2^N input patterns in order to exhaustively test an N-input combinational circuit. Exhaustive testing often has the advantage of not requiring expensive fault simulation. The modified LFSR technique used for performing exhaustive testing is an attractive BIST scheme as it provides for 100% fault coverage of all detectable stuck-at (s-a) faults in combinational circuits which do not produce sequential behavior under the presence of a fault.

The situation where a combinational circuit behaves like a sequential circuit in the presence of a fault causes unique testing problems. CMOS circuits, being a three state logic technology, exhibit this unique failure mode for certain faults such as *stuck-open* (s-op) faults which can be caused by missing connections either at the gate, drain or source terminal of a FET (El-Ziq 1981, Reddy 1986). Some of the s-op faults require only a single test vector to be detected while others require an ordered pair of test vectors. If exhaustive testing is performed then all s-op faults that require only a single test vector can be detected. However, this is not the case for s-op faults that require an ordered pair of test vectors.

In this paper we focus our attention on the probabilistic analysis of s-op faults when tested with a *pseudo-random exhaustive test sequence* (PRETS), i.e. a test *sequence* consisting of all 2^N test patterns generated by a modified N-bit LFSR. Such a scheme can be viewed as *testing without replacement*. The results when testing using a PRETS are compared with the results obtained when testing is done with *pure random sampling with replacement*. For the sake of conciseness the proofs of all the theorems are omitted. The proofs appear in (Sastry, et al. 1987a).

Stuck-Open Faults in CMOS circuits

Consider the CMOS inverter shown in Figure 1. Assume that the pFET is s-op (fault f). The truth table for this circuit is shown in Table 1, where Z denotes a high impedance state and F^{-1} indicates that the present output is the same as the previous output. From the truth

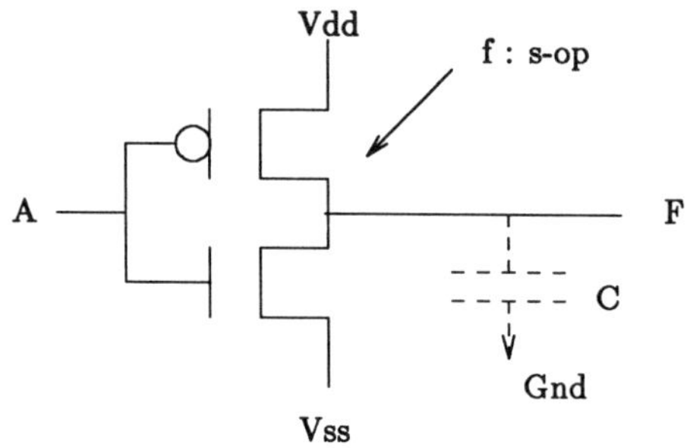

Figure 1: CMOS Inverter

Table 1: Truth Table for CMOS Inverter

Input A	F Fault Free	F Fault f
0	1	Z or F^{-1}
1	0	0

table we see that it is impossible to set $F = 1$ in the faulty circuit and hence this fault can be modeled as F s-a-0.

Now consider the CMOS gate shown in Figure 2 which is assumed to be embedded in some larger network $\mathcal{C}$. For the input condition ABCD = 1x00 both the series path A-B and the pull-down network are open. The series path C-D is closed. Hence $F = 1$. If the fault f is present then $F = Z$ or $F = F^{-1}$ because both the pull-up and pull-down networks are open. To test for this fault, the output F must first be set to 0 and then followed by a test vector that sets $F = 1$ by closing only the path C-D and leaving the path A-B open. Thus to detect f requires an ordered pair of test vectors, (t_0, t_1), satisfying the following conditions : **(1.)** t_0 sets $F = 0$. **(2.)** t_1 sets $F = 1$, t_1 closes the path C-D and t_1 sets up a sensitized path from F to a primary output in the circuit $\mathcal{C}$.

Condition 2 states that t_1 is a test for F s-a-0 and t_1 sets up the condition such that if the fault exists then the pull-up network is open. Hence t_1 must set up the condition which results in all closed paths in the pull-up network between VDD and F passing through the p-FET being tested. The test vectors t_0 and t_1 are referred to as the *initialization* and *sensitization* vectors respectively. Tests invalidated by hazard conditions are not be considered in this paper.

Definition 1 *A two terminal pull-up or pull-down network is said to contain a series subnetwork, refered to as an s-type network, if all paths between the two terminals must go through the subnetwork. A network which is not an s-type network is refered to as a non s-type network.*

Theorem 1 *In an s-type subnetwork within a pull-up network a s-op fault in a p-FET can be modeled by F s-a-0. Similarly in an s-type subnetwork within a pull-down network a s-op fault in an n-FET can be modeled as F s-a-1. Thus any s-op fault in an s-type CMOS subnetwork can be modeled by a s-a gate fault.*

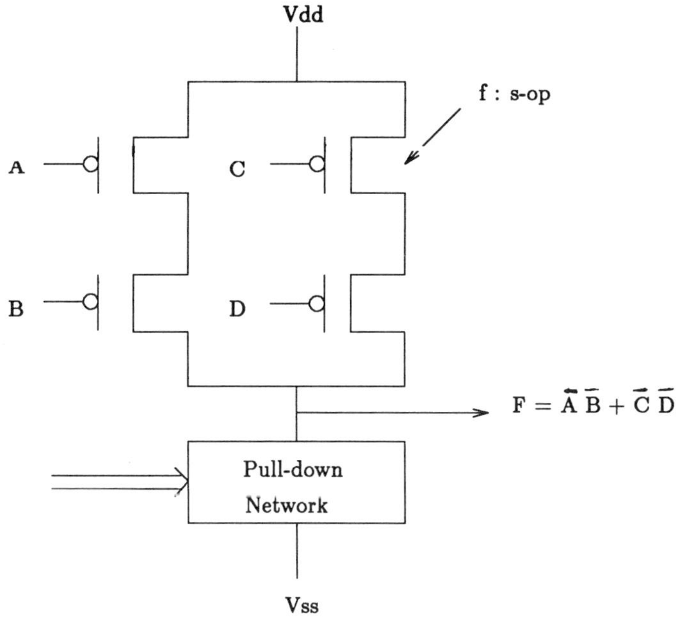

Figure 2: CMOS Gate

Theorem 2 *To test for a s-op fault in a non s-type pull-up (pull-down) network associated with a gate F within some circuit C, a two vector sequence t_0, t_1 is required, where t_0 initializes the output of the gate F to 0(1) and t_1 is a test for F s-a-0 (s-a-1) and sensitizes a path from the site of the fault to F.*

Probabilistic Analysis of s-op Faults

In this section we analyze the detectability of s-op faults in non s-type CMOS combinational networks when tested using a modified N-bit LFSR to generate all $m = 2^N$ input combinations of a N input network. Any one of these $m!$ possible sequences of test vectors is referred to as a *pseudo random exhaustive test sequence* (PRETS). Testing a circuit using a PRETS may be viewed as testing *without replacement*, i.e., selecting one of the m possible test vectors at random and applying it to the circuit under test, never repeating the selection. The other choice is random testing *with replacement*, which refers to selecting one of the m test vectors and replacing the selected vector in the sample. The following notation will be used throughout the remainder of this paper.

N - number of inputs of a combinational circuit

m - total number of input combinations $= 2^N$

f_i - denotes some s-op fault

$\mathcal{T}_{0i}, \mathcal{T}_{1i}$ - non-empty sets of initialization and sensitization vectors associated with fault f_i respectively.

k_{0i}, k_{1i} - cardinality of sets $\mathcal{T}_{0i}$ and $\mathcal{T}_{1i}$ with $2 \leq k_{0i} + k_{1i} \leq m$

$P^{w/o}, P^{w}$ - probability expressions when sampling without replacement and with replacement, respectively

$(x)_j \; = x(x-1)\cdots(x-j+1)$

$[x]_j \; = x(x+1)\cdots(x+j-1)$

Probability of Detection using PRETS

For a s-op fault in a non s-type network the vectors in the initialization set are those that result in $F = 0$ and this set can be quite large. However, in general the sensitizaiton set is relatively much smaller since these vectors must result in $F = 1$ and sensitize a path involving the p-FET being tested to the primary output. The following theorems provide general formulas for the probability of detecting a s-op fault in a non s-type network under random testing. For simplicity we will consider pull-up networks. Similar analysis holds for pull-down networks.

Theorem 3 *Let f denote a s-op fault in an N-input CMOS combinational network that requires a pair of vectors (t_0, t_1) to be detected. Then the probability of detecting f using n test vectors selected without replacement is given by*

$$P_d^{w/o}(k_0, k_1, m, n) = 1 - \sum_{j=0}^{\lfloor \frac{n}{2} \rfloor} \frac{[-k_0]_j[-k_1]_j[-n]_{2j}}{[-m]_{2j}[-n]_j[1]_j} \tag{1}$$

For $n < m$ Equation 1 does not have a closed form solution. However for $n = m$, a simpler closed form is obtained. This is stated in the following corollary.

Corollary 1 *If testing is done using a PRETS, i.e., $n = m$ in the statement of Theorem 3, then*

$$P_d^{w/o}(k_0, k_1, m) = 1 - \frac{\binom{m-k_1}{k_0}}{\binom{m}{k_0}} \tag{2}$$

The parameters k_0 and k_1 are functions of the circuit structure and the fault under consideration. In general, the typical sizes of circuits being tested with PRETS will range from $N = 10$ to $N = 20$. For such large values of m ($m = 2^N$) with $k_1 \ll m$, a simple and excellent approximation to $P_d^{w/o}(k_0, k_1, m)$ is given by

$$P_d^{w/o}(k_0, k_1, m) = 1 - \frac{\binom{m-k_1}{k_0}}{\binom{m}{k_0}} \approx 1 - (1 - k_0/m)^{k_1} \tag{3}$$

As an example, consider the two level NAND circuit shown in Figure 3. Details of gate F are shown in Figure 3a. A test for one of the s-op faults in the pull-up network of gate H requires $\mathcal{T}_0 = \{\tilde{x} | H = 0\}$ and $\mathcal{T}_1 = \{\tilde{x} | F = 0, G = 1\}$. Thus $k_0 = 49$ and $k_1 = 7$. Using Equation 2 we obtain $P_d^{w/o}(k_0, k_1, m) = 0.99998$, while the approximation given by Equation 3 yields 0.99996.

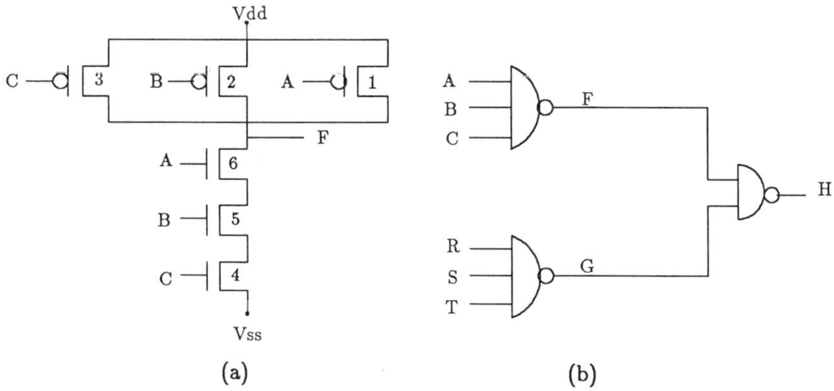

Figure 3: Two Level NAND network

Table 2: Values of k_0/m and k_1 for probability of detection ≥ 0.99 using a PRETS

k_0/m	k_1
.01	≥ 460
.05	≥ 92
.1	≥ 46
.5	≥ 9
.7	≥ 6

Equation 3 can be used to determine the values of k_0 and k_1 necessary for achieving a predefined probability of detection using a PRETS. Using the fact that for $0 < x < 1$, $-\ln(1-x)$ lies between x and $x/(1-x)$, we can solve the inequality

$$P_d^{w/o}(k_0, k_1, m) \approx 1 - (1 - k_0/m)^{k_1} \geq \alpha \tag{4}$$

to obtain

$$\frac{k_0 k_1}{m} \geq \ln\left(\frac{1}{1-\alpha}\right). \tag{5}$$

Table 2 shows values of k_0/m and k_1 for $\alpha = .99$. Thus, if it is known that $k_0/m = 0.5$ then $k_1 = 9$ is sufficient to achieve a probability of detection of 0.99.

Probability of Detection when testing with replacement

Consider now the situation where n test vectors are selected randomly, one at a time, applied to the circuit under test and replaced back in the population. In this section we present expressions for the probability of detecting a s-op fault when testing is done with replacement.

Theorem 4 *Let $q = (k_0 k_1)/m^2$. If testing is done with replacement then the probability of detecting f using n test vectors is given by*

$$P_d^w(k_0, k_1, m, n) = \begin{cases} 1 - \frac{1}{\sqrt{1-4q}}\left[\left(\frac{1+\sqrt{1-4q}}{2}\right)^{n+1} - \left(\frac{1-\sqrt{1-4q}}{2}\right)^{n+1}\right] & q < \frac{1}{4} \\ \\ 1 - \frac{n+1}{2^n} & q = \frac{1}{4} \end{cases} \tag{6}$$

When testing with a PRETS the number of test vectors is m. The number of test vectors when testing is done with replacement is n, and n can be less than, equal to or greater than m. Clearly, as $n \to \infty$, $P_d^w(k_0, k_1, m, n) \to 1$. However, for $n \leq m$, the probability of detection of a given s-op fault using PRETS will be greater than the probability of detection when testing with replacement. Furthermore, when m is large the probability of detection of a s-op fault when testing with replacement will be close to the probability of detection using a PRETS. That is, Equation 6 should approach Equation 1 from below. These observations are summarized in the following two theorems.

Theorem 5 *For $n \leq m$, $P_d^{w/o}(k_0, k_1, m) \geq P_d^w(k_0, k_1, m, n)$.*

Theorem 6 *Let $\lim_{m\to\infty} k_0/m = \lambda_0$ and $\lim_{m\to\infty} k_1/m = \lambda_1$. Then*

$$\lim_{m\to\infty} P_d^{w/o}(k_0, k_1, m, n) = P_d^w(k_0, k_1, m, n) \tag{7}$$

where $P_d^w(k_0, k_1, m, n)$ is given by Equation 3 with $q = \lambda_0\lambda_1$.

To compare testing with a PRETS and testing with replacement, consider again the previous example of a two level NAND circuit shown in Figure 3, where $m = 64$. Consider the faults f_1 and f_2 defined by p-FET # 1 being s-op and one of the p-FETs in the pull-up network of gate H being s-op, respectively. For fault f_1 we find $k_{01} = 8$ and $k_{11} = 7$ and for fault f_2 we find $k_{02} = 49$ and $k_{12} = 7$. The probability of detecting these two faults as a function of the number of samples (n) when testing without replacement ($P_d^{w/o}(k_{01}, k_{11}, m, n)$) and when testing with replacement ($P_d^w(k_{01}, k_{11}, m, n)$ is shown in Figure 4. Figure 4 also shows similar plots for other values of k_0 and k_1.

We see that for a fixed m, the difference between testing with and without replacement become negligible when the sizes of the initialization (k_0) and sensitization (k_1) sets are relatively large. The term k_0/m (k_1/m) represents the fraction of of the test vectors that satisfy the conditions for an initialization (sensitization) vector and their product $q = k_0k_1/m^2$ measures the *detectability* of a given s-op fault. Figure 5 shows a plot of the probability of detection versus q. For faults of high detectability (i.e. q large) there seems to be little or no difference between testing with and without replacement. However, for small values of q, analysis based on testing with replacement does not provide an accurate approximation to testing without replacement.

Error Latency

The number of times that a s-op fault f is detected by a given PRETS $s = (s_1, \ldots, s_m)$ is ≥ 0. Since a s-op fault f need not be detected even when all the test vectors have been applied, test length is not well defined. In this case an equivalent measure is the *duration of test*, denoted by τ_f. For a given PRETS $s = (s_1, \ldots, s_m)$, τ_f has the value t if f is detected by s for the first time by test vectors s_{t-1} and s_t, for $2 \leq t \leq m$. On the other hand if f is not detected by s then τ_f has the value m.

To determine the probability that τ_f has a value t, let ξ_t denote the event that the s-op fault f is detected by test vectors at positions $t-1$ and t and let ξ_t' denote the complement of ξ_t. Then the probability that τ_f has the value t, when testing without replacement, is given by

$$P^{w/o}(\tau_f = t) = \begin{cases} \text{Prob}(\xi_2' \cdots \xi_{t-1}'\xi_t) & 2 \leq t \leq m-1 \\ \\ \text{Prob}(\xi_2' \cdots \xi_{m-1}'\xi_m) + \text{Prob}(\xi_2' \cdots \xi_{m-1}'\xi_m') & t = m \end{cases} \tag{8}$$

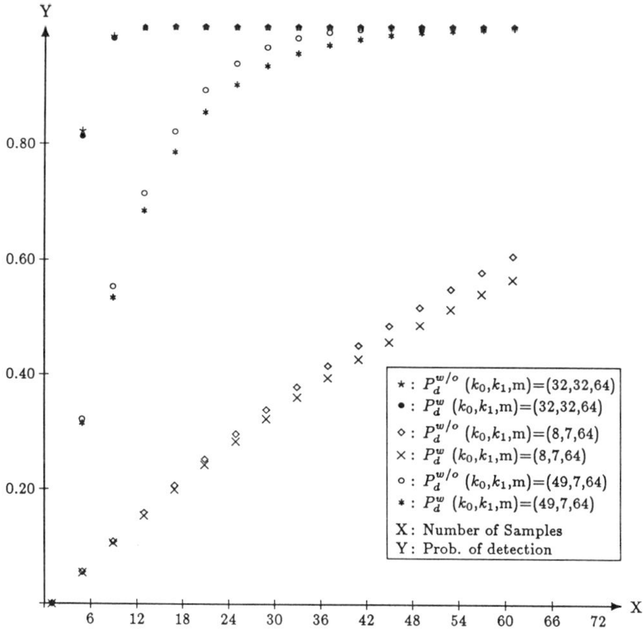

Figure 4: Probability of detection vs. Number of Samples

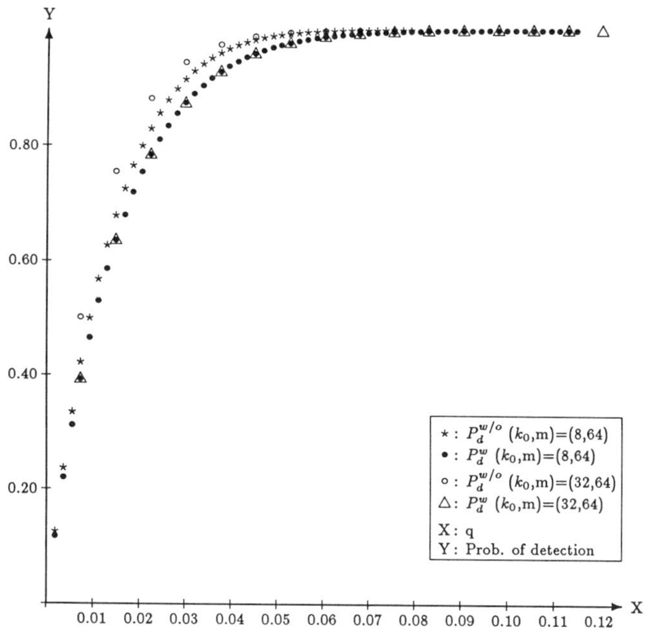

Figure 5: Probability of detection vs. q

This leads to the following theorem.

Theorem 7 *Let k_0 and k_1 be the sizes of the initialization and sensitization sets of a s-op fault f. If testing is done without replacement then*

$$P^{w/o}(\tau_f = t) =$$

$$\begin{cases} \frac{k_0 k_1}{m(m-1)} \sum_{j=0}^{\lfloor \frac{t-2}{2} \rfloor} \frac{[-k_0+1]_j[-k_1+1]_j[(-t+2)/2]_j[(-t+3)/2]_j}{[(-m+2)/2]_j[(-m+3)/2]_j[-t+2]_j[1]_j} & 2 \le t \le m-1 \\ \frac{[-m+k_1]_{k_0}}{[-m]_{k_0}} \left\{ \frac{k_0 k_1}{(m-k_0)(m-k_1)} + 1 \right\} & t = m \end{cases} \tag{9}$$

The *expected* error latency when testing without replacement is given by

$$\mathrm{E}^{w/o}(\tau_f) = \sum_{t=2}^{m} t P^{w/o}(\tau_f = t) \tag{10}$$

When testing is done with replacement the testing need not terminate. Thus the range of τ_f is $2 \le \tau_f \le \infty$. In this case *test length* and *duration of test* are synonymous.

Theorem 8 *If testing is done with replacement then*

$$P^w(\tau_f = t) = \begin{cases} \frac{q}{\sqrt{1-4q}} \left[\left(\frac{1+\sqrt{1-4q}}{2} \right)^{t-1} - \left(\frac{1-\sqrt{1-4q}}{2} \right)^{t-1} \right] & q < \frac{1}{4} \\ \frac{t-1}{2^{t-2}} & q = \frac{1}{4} \end{cases} \tag{11}$$

Equation 11 yields the probability that the test length is equal to t. When $q < 1/4$, the first term in Equation 11 is the dominant term. We use this fact to bound the tail of the distribution. That is, for a given ϵ, we wish to find a value of t such that

$$P^w(\tau_f \ge t) \le \epsilon \tag{12}$$

For $q < 1/4$, $P^w(\tau_f \ge t)$ is given by

$$P^w(\tau_f \ge t) = \frac{1}{\sqrt{1-4q}} \left[\left(\frac{1+\sqrt{1-4q}}{2} \right)^t - \left(\frac{1-\sqrt{1-4q}}{2} \right)^t \right] \tag{13}$$

Discarding the second term in Equation 13, we find that if

$$t \ge \frac{\ln(\epsilon\sqrt{1-4q})}{\ln((1+\sqrt{1-4q})/2)} \tag{14}$$

then the inequality given in Equation 12 will be satisfied.

As an example, consider a s-op fault in the pull-up network of gate H shown in Figure 3, where $m = 64$, $k_0 = 49$, $k_1 = 7$ and $q = 0.0837$. If $\epsilon = 0.01$ then the right hand side of Equation 14 evaluates to 50. Thus to detect a s-op fault in the pullup network of gate H, the probability that the test length will exceed 50 will be less than 0.01.

The expectation of τ_f when testing with replacement is $1/q$. Figure 6 shows a plot of the expected error latency versus q. We find that expected error latency when testing without replacement is always smaller then when testing with replacement. That is, analysis made assuming sampling with replacement provides a conservative bound for a PRETS. Referring to Figure 6, where $m = 64$ and $k_1 = 8$, if $q \ge .02$, then the expected number of test vectors needed to detect a given fault is ≤ 30. The probability of detecting a fault for $q \ge .02$ is $\approx .80$.

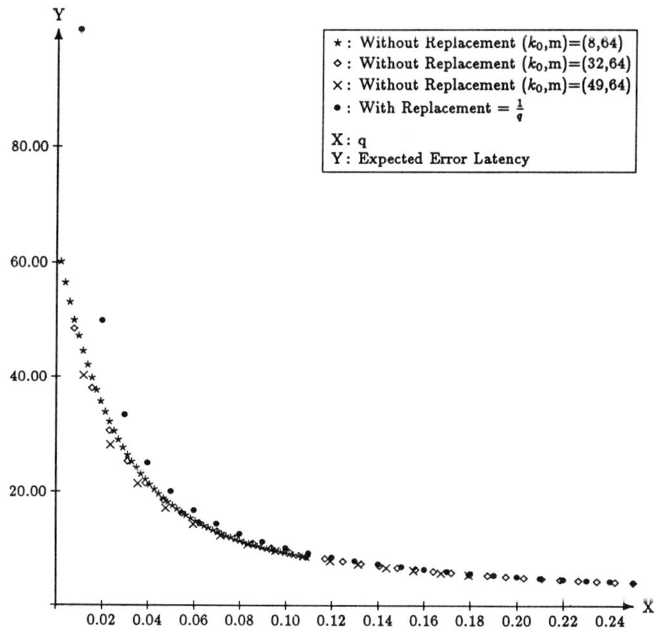

Figure 6: Expected Error Latency vs. q

Fault Coverage

The effectiveness of any test procedure is measured by the percentage of faults that are detected. When testing is randomized the number of faults detected is a random variable and one is naturally interested in the expected number of faults detected by the application of a random sample of test vectors. Determining the fault coverage when testing is done using a PRETS poses an extremely difficult combinatorial problem since the initialization and sensitization set of different faults may have common elements. We are therefore led to the task of obtaining *good*, *conservative* approximations. The exact analysis would require knowledge of all possible pairwise intersections of the initialization and sensitization sets of each fault.

We can arrive at a simple lower bound on the expected fault coverage when testing with a PRETS by first considering the expected fault coverage when testing with replacement. To do this, assume that each trial consists of a random selection of a *pair* of test vectors and ignore the possibility that the second vector of some trial i and the first vector of trial $i+1$ may detect a fault. Let $\mathcal{F} = \{f_i\}$, for $i = 1, \ldots, r$, denote the set of all possible s-op faults and ξ^i denote the event that fault f_i is detected on any single trial. Thus on any trial one of the $r+1$ possible events $\{\xi^1, \ldots, \xi^r, \xi^{r+1}\}$ may occur, where ξ^{r+1} denotes the event that none of the r faults in $\mathcal{F}$ were detected.

In a sequence of n trials conducted with replacement, if ξ^i occurs one or more times then the fault f_i is detected. Thus the sum of the number of times each of the events $\xi^1, \ldots, \xi^r$ occurs is represented in the sequence of n trials is the total number of faults detected by the sequence.

Let Z_i denote the number of occurences of event ξ^i in n trials with replacement and q_i be the probability of event ξ^i on any trial. Then

$$P\{Z_1 = i_1, \ldots, Z_r = i_r\} = \frac{n!}{i_1! \cdots i_r! i_{r+1}!} q_1^{i_1} \cdots q_r^{i_r} q_{r+1}^{i_{r+1}} \tag{15}$$

where $q_i = k_{0i}k_{1i}/m^2$.

Let Y denote the number of faults detected in a sequence of n trials. Then the expected number of faults detected is given by

$$E(Y) = \sum_{i=1}^{r} P\{Z_i \geq 1\} = \sum_{i=1}^{r} [1 - (1 - q_i)^n] \tag{16}$$

This leads directly to the following theorem.

Theorem 9 *The expected number of faults, ENF, detected by a PRETS satisfies*

$$ENF \geq r - \sum_{i=1}^{r} (1 - q_i)^m \tag{17}$$

where $q_i = k_{0i}k_{1i}/m^2$

Consider again the two-level NAND network shown in Figure 3. The total number of s-op faults in the pull-up network of each gate is 8. For each such fault the values of q_i were computed and the bound, given by Equation 17, on the expected number of faults detected was found to be 4.72. Thus, the expected fault coverage is $\geq$ 59%.

Now consider the more general structure of a p-input NAND gate Q embedded in some large circuit shown in Figure 7. The *cone of logic L* shown in the figure defines the set of j inputs that effect the output of Q. Of the 2^p possible input combinations to Q only one will produce an output of 0. Assume now that this fraction is maintained at the j primary inputs of the cone L. Thus the number of primary input combinations that produce a 0 at the output of Q is $k_0(Q) = 2^{i+j-p} = m2^{-p}$. A typical value of p lies between 2 and 3. In general, $p \leq 5$. For $p = 5$ and $m = 2^{20}$ (i.e. a 20 input circuit), $k_0(Q) = 2^{15}$. Hence, one may assume that there are numerous initialization vectors. However, the number of sensitization vectors, $k_1(Q)$, will be far fewer. For the analysis here we have assumed a rather pessimistic situation. An optimistic model would assume that all lines are equally likely to be 0 or 1, and hence $k_0(Q) = 2^{N-1}$.

We can use Equation 17 to obtain a *conservative* bound on the number of sensitization vectors needed to achieve a fault coverage $\geq \alpha$, for a given α. From Equation 17 we find this value to be

$$k_1(Q) \geq 2^p \ln(\frac{1}{1-\alpha}) \tag{18}$$

Thus using a PRETS to achieve a fault coverage of $\geq$ .95 for a 20 input circuit with $p \leq 5$ the number of sensitization vectors for any fault need not be greater than 94, which is a relatively small fraction of the total number of input combinations.

Closing Remarks

All the formulas presented in this paper are functions of the sizes of the initialization and sensitization sets. The sizes of these sets can be estimated by using one of several techniques such as the statistical fault coverage estimator known as STAFAN (Jain, et al. 1984) or by computing approximations to the signal probabilities using the *cutting algorithm* (Savir, et al. 1984a).

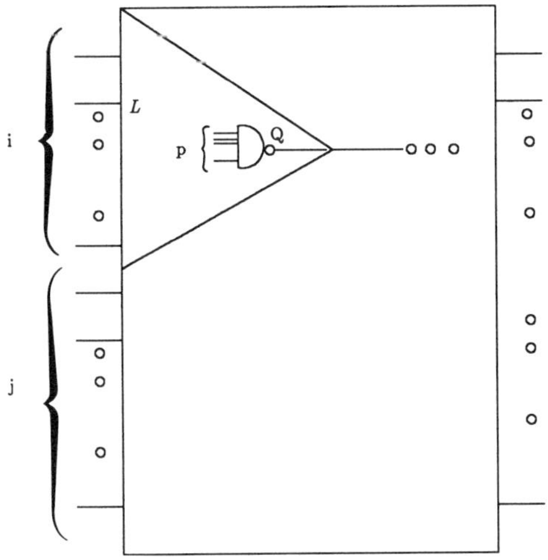

Figure 7: Gate Q and its cone of logic L

References

El-Ziq Y. M. 1981 Automatic Test Generation for Stuck Open Faults in CMOS VLSI. In *Proceedings of the 18th Design Automation Conference*, pages 347–354, IEEE, June 1981.

Jain S. K. and Agrawal V. D. 1984 Statistical fault analysis. *IEEE Design and Test of Computers*, 2:38–44, February 1984.

McCluskey E. J. 1985a Built-In Self-Test Techniques. *Design and Test of Computers*, 2(2):21–28, April 1985.

McCluskey E. J. 1985b Built-In Self-Test Structures. *Design and Test of Computers*, 2(2):29–36, April 1985.

Reddy S. M., Agrawal V. D. and Jain S. K. 1984 A Gate Level Model for CMOS Combinational Logic Circuits with Application to Fault Detection. In *Proceedings of the 21st Design Automation Conference*, pages 504–509, IEEE, June 1984.

Reddy S. M. and M. K. Reddy S. M. 1986 Testable Realizations for FET Stuck-Open faults in CMOS Combinational Logic Circuits. *IEEE Transactions on Computers*, C-35(8):742–754, August 1986.

Sastry S. and Breuer M. 1987a *Detectability of CMOS Stuck Open Faults using Random and Pseudo Random Test Sequences*. Technical Report CRI-87-30, Computer Research Institute, Dept. of EE-Systems, University of Southern California, June 1987.

Savir J, Ditlow G. S. and Bardell P. H. 1984a Random Pattern Testability. *IEEE Transactions on Computers*, C-33(1):79–90, January 1984.

Wagner K. D., Chin C. K. and McCluskey E. J. 1987 Pseudorandom Testing. *IEEE Transactions on Computers*, C-36(3):332–343, March 1987.

7.2 Structured Test of Bit Level Systolic Arrays

William Marnane, Will Moore and Eric Gautrin

INTRODUCTION

Bit level systolic arrays have a highly regular structure and any testing strategy for them should exploit this regularity. Our test strategy (Marnane et al 1987) is based on the principle of divide and conquer. Instead of generating test vectors for the whole array the fault analysis is carried out on one cell of the array only. The test vectors are then propagated throughout the array such that the array is completely tested. This test strategy is not restricted to any specific fault model; a fault model can be chosen to suit the particular technology. In this paper two different fault models are applied to a cell and the required test vectors generated. These test vectors are combined into test sets which are propagated throughout the array testing every cell.

TEST STRATEGY

Our proposed test strategy divides the problem of testing the array into testing a cell and then determining how to propagate the required test vectors throughout the array. The testing scheme consists of three steps:

1. Generating the test vectors for a single cell using an appropriate fault model.
2. Propagating the test vectors to every cell and to the outputs of the array. Thus having generated the required test vectors for one cell, every cell in the array is tested by the same test vectors.
3. Detecting any faults at the outputs of the array.

The set of test vectors produced by the fault analysis is dependent on the fault model chosen. The propagation of these test vectors is independent of the fault model but is determined by the propagation properties of the array. The results presented in this paper assume that there is at most one fault in each row and column of the array and that it is only necessary to detect the fault and not to locate it.

TEST GENERATION

The test strategy is illustrated for a bit level systolic correlator array (McCanny et al 1986) with a unidirectional structure shown in Figure 1. The signal A is the reference coefficient, the signal X is the input data, the signal Y is the partial result and the signal C is the carry. Two different fault models are applied to the processing element namely the *"stuck-at" fault model* and the *MOS switch level model.*

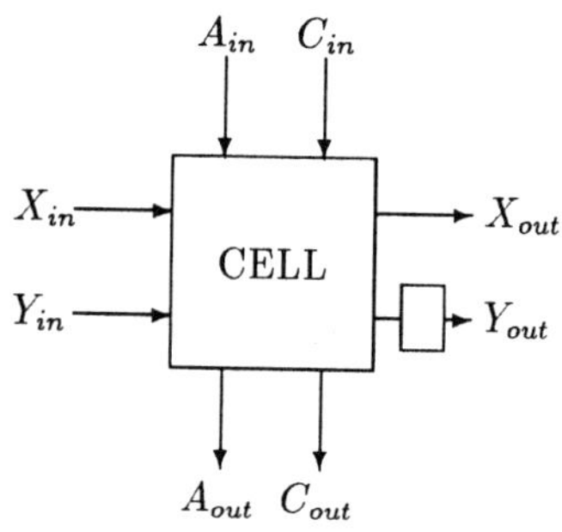

$$X_{out} = \triangle X_{in}$$
$$A_{out} = \triangle A_{in}$$
$$Y_{out} = \triangle^2\{Y_{in} \oplus C_{in} \oplus (A_{in}.X_{in})\}$$
$$C_{out} = \triangle\{[C_{in}.Y_{in}] + [C_{in}.(A_{in}.X_{in})] + [Y_{in}.(A_{in}.X_{in})]\}$$
$$\triangle = 1 \text{ Clock Delay}$$

Figure 1. Correlator Cell

"Stuck-at" Fault Model

The stuck-at fault model (Eldred 1959) postulates that all physical defects in logic circuits will manifest themselves as nodes in the circuit stuck-at logic 1 or logic 0. In this example for simplicity, the circuit considered for the fault analysis is the block diagram of Figure 1, and stuck-at faults are considered at the inputs X_{in}, A_{in}, C_{in}, Y_{in} and outputs X_{out}, A_{out}, C_{out}, Y_{out} of the cell.

To test a supposed faulty node, it is necessary to determine a test vector which will sensitize the fault and allow it to be detected. For example if the node Y_{in} is stuck-at 1, then one such test which will detect the fault is the vector $\underline{C}$ ($\underline{C}$ is the hexadecimal notation for $X_{in} = 1, A_{in} = 1, C_{in} = 0, Y_{in} = 0$). Applying $\underline{C}$ to a cell gives the fault free response of $Y_{out} = 1$ and $C_{out} = 0$. If Y_{in} is stuck-at 1, then the response is $Y_{out} = 0$ and $C_{out} = 1$. Note that the fault effect can be propagated in the row direction (signal Y), and the column direction (signal C). Thus the fault can be detected at two observable outputs.

Table 1 shows possible test vectors which can be used to detect all these stuck-at faults. From this it can be seen that the application of the set of five vectors ($\underline{E}$, $\underline{C}$, $\underline{9}$, $\underline{7}$, $\underline{5}$) to a cell is sufficient to test for the above list of stuck-at faults.

MOS Switch Level Fault Model

A more realistic fault model for MOS integrated circuits is to consider the circuit at the transistor switch level (Galiay et al 1980). In this analysis transistor stuck-on and stuck-open faults are considered. Figure 2 shows a possible CMOS implementation of the correlator cell (excluding the latches on the outputs).

TABLE 1.
Test Vectors for Stuck-at Faults

Test Vector	Fault
E	X_{in} S@0, A_{in} S@0
7	C_{in} S@0, Y_{in} S@0, C_{out} S@0, Y_{out} S@1
9	X_{out} S@0, A_{in} S@1, A_{out} S@1
5	A_{out} S@0 ,Y_{out} S@0, X_{in} S@1, X_{out} S@1, C_{out} S@0
C	C_{in} S@1, Y_{in} S@1

In a CMOS circuit, a stuck-open fault can lead to a state where a node is left open circuit and its value is determined by its charge storage. Thus the fault manifests itself as a parasitic latch (Starke 1984). To detect such a fault a sequence of two test vectors is required. For example considering transistor p1 with a stuck-open fault, a first test vector D sets the outputs $C_{out} = 1$ and $Y_{out} = 0$. Then by applying the vector 6 the parasitic capacitance manifest itself by retaining the previous values of $C_{out} = 1$ and $Y_{out} = 0$. In the fault-free case, the correct values would be $C_{out} = 0$ and $Y_{out} = 1$.

A set of the sequence of vectors required to detect each transistor stuck-open fault is shown in Table 2. This shows that all such stuck-open faults can be tested with 10 sequences of 2 test vectors, which contain 11 distinct vectors.

The detection of transistor stuck-on faults requires only the application of single test vector, similar to the stuck-at case. To detect n1 stuck-on, one such test vector is C which will sensitize the fault. In the presence of a stuck-on fault a CMOS gate acts as a voltage divider with the output voltage level a function of the on-resistances of the transistors in the gate and their topology. Therefore one method used to detect transistor stuck-on faults is to monitor the power supply current, since the activation of a stuck-on fault will produce a temporary increase of several orders of magnitude in the supply current (Levi 1981).

The vectors required to detect all transistor stuck-on faults are shown in Table 3. From this it can be seen that all stuck-on faults are detected with the vectors that are required to test for stuck-open faults.

TABLE 2.
Test Sequences for Transistor Stuck-open Faults.

Initialization Vector	Sensitization Vector	Stuck-open Transistors
D	6	p1-p6-n13
E	9	p2-p5-n12-n16
7	C	p3-p4-p17-n7-n14-n15
6	D	p7-p10-n1
9	7	p8-p16-n3-n4
C	B	p9
5	E	p11-n2-n5-n6
9	4	p12-p13-p15
F	8	p14-n17
4	F	n8-n9-n10-n11

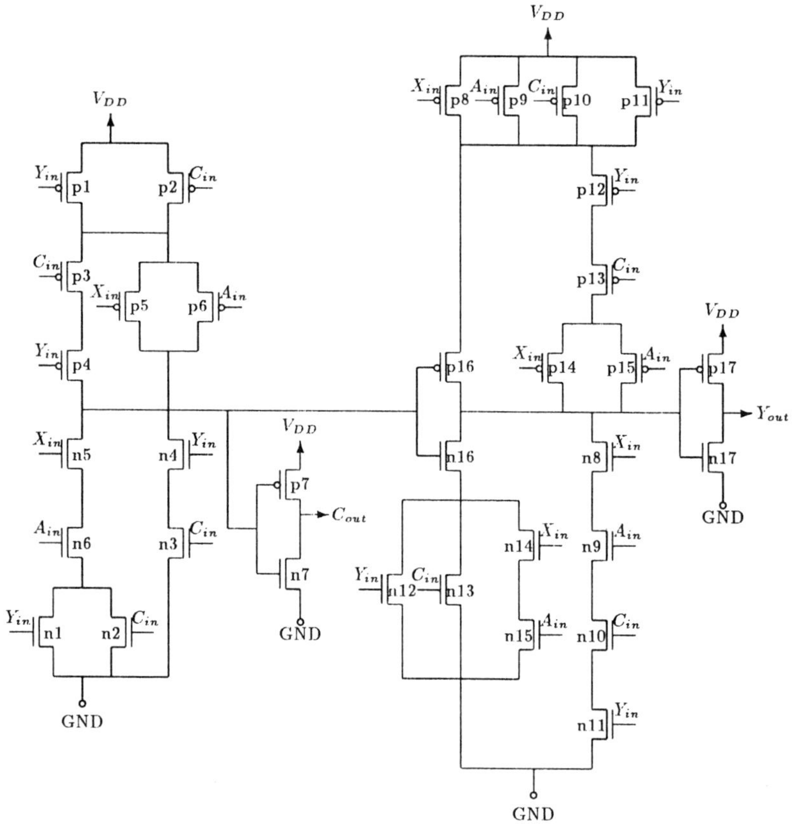

Figure 2. CMOS Circuit Diagram of Correlator Cell

TABLE 3.
Test Vectors for Transistor Stuck-on Faults.

Test Vector	Stuck-on Transistor
B	p1-p2-n9
E	p3-p5-n11-n16
D	p4-p6-p17-n10
C	p7-p14-p15-n1-n2
F	p8-p9-p10-p11-n17
9	p12-p16-n3-n6
6	p13-n4
5	n5
7	n7-n8
8	n12-n13-n14-n15

TEST VECTOR PROPAGATION

Testing the whole array consists of propagating the test vectors to every cell. The problem is to determine particular sequences of inputs to the array which will propagate the required test vectors of a specific fault model to every cell. In this paper we present test vectors for both the stuck-at fault model and the MOS switch level fault model. Of course such input sequences, if they exist, can be determined from the test vector propagation properties of the array. These properties are defined by the data communication between cells and by the algorithm performed by each cell.

Data Communication

In systolic arrays the data flow can be in different directions and at different speeds. Therefore an approach is required to describe these dependencies and relate the data in time. The method adopted here is the Data Dependence Graph (DDG) (Kung 1986).

The correlator of Figure 1 has interesting properties because of the different speeds of the signals X and Y in the row direction. For a 2-dimensional array of these cells the DDG has three dimensions. The main components are shown in Figure 3. Note that since the signals A and C move in the same direction with the same speed, they are represented together in the DDG. The DDG shows the movement of the X, Y, A and C data relative to each other in time and space. The **i** direction represents the movement of X data across the row in time. The **j** direction represents the movement of A and C down the column in time. The **k** direction represents the sequence of data in the array. Therefore since Y has a slower speed than X its movement is represented by the **(i+k)** diagonal.

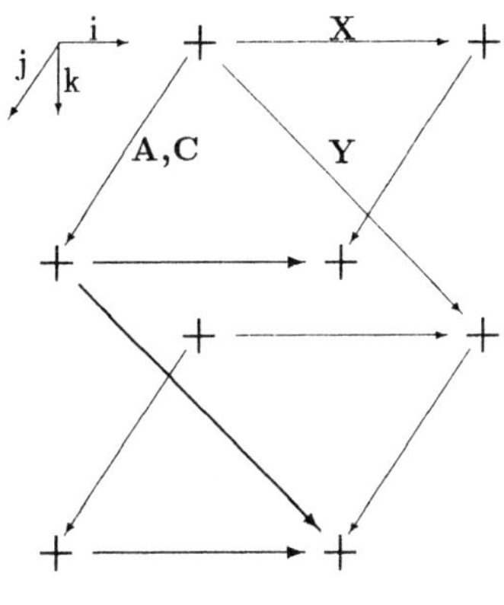

Figure 3. Data Dependence Graph

Cell Logic

For each distinct direction in the DDG the precise nature of the dependence can be determined by examining the logic of a cell. This may be described by its input-output function f_{cell}:

$$V_{out} = f_{cell}(V_{in})$$

The vector V_{in} represents the set of input signals and the vector V_{out} the set of output signals of the cell:

$$V_{in} = (S_{in}^1, S_{in}^2, S_{in}^3, \ldots)$$
$$V_{out} = (S_{out}^1, S_{out}^2, S_{out}^3, \ldots)$$

To determine the effect of f_{cell} on each signal a new operator called the **propagation function** is presented. Applying the propagation function f_{Sj} to an input vector V_{in} produces a set of output vectors which are determined only by the output S_{out}^j. The definition of the propagation function is:

$$f_{Sj}(V_{in}) = \{(\ldots, \bullet, S_{out}^j, \bullet, \ldots)\}$$

where the notation $\bullet$ means "don't care".

In the correlator cell shown in Figure 1 there are 4 signals, X, A, C, Y, so the input and output vectors are given by $V = (X, A, C, Y)$. The signals A and C have the same direction and speed and hence can be considered together. Therefore there are three distinct propagation functions for this cell:

$$f_X(V_{in}) = \{(X_{out}, \bullet, \bullet, \bullet)\}$$
$$f_{AC}(V_{in}) = \{(\bullet, A_{out}, C_{out}, \bullet)\}$$
$$f_Y(V_{in}) = \{(\bullet, \bullet, \bullet, Y_{out})\}$$

These propagation functions must be applied to all the vectors of the truth table. For example applying input vector 3 to a cell gives $Y_{out} = 0$, and the propagation function f_Y produces the result:

$$f_Y(3) = \{(\bullet, \bullet, \bullet, 0)\}$$
$$= \{\underline{0},\underline{2},\underline{4},\underline{6},\underline{8},\underline{A},\underline{C},\underline{E}\}$$

The results of f_Y applied to all the vectors in the truth table can be represented by the propagation graph of Figure 4(a). Similarly Figures 4(b) and 4(c) are the propagation graphs of f_X and f_{AC}.

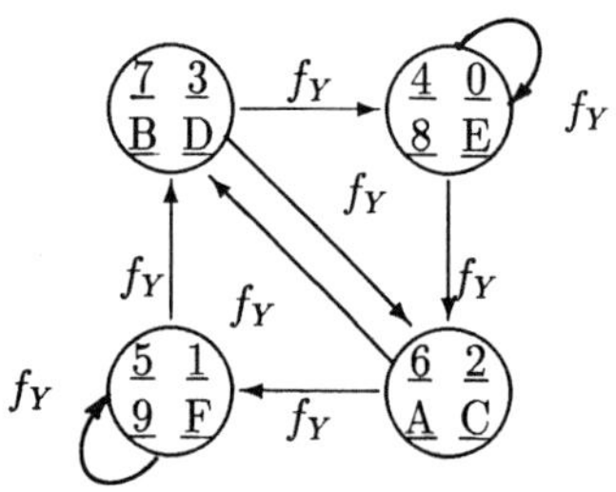

Figure 4(a). Graph of f_Y

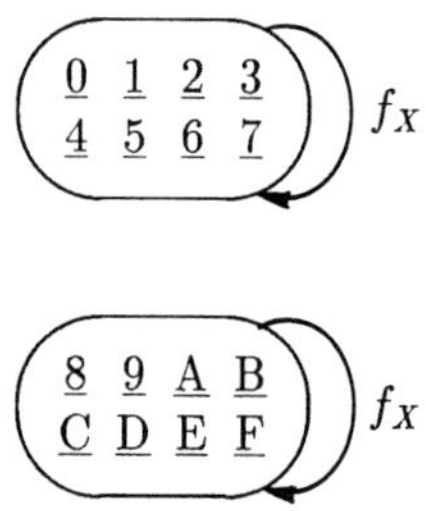

Figure 4(b). Graph of f_X

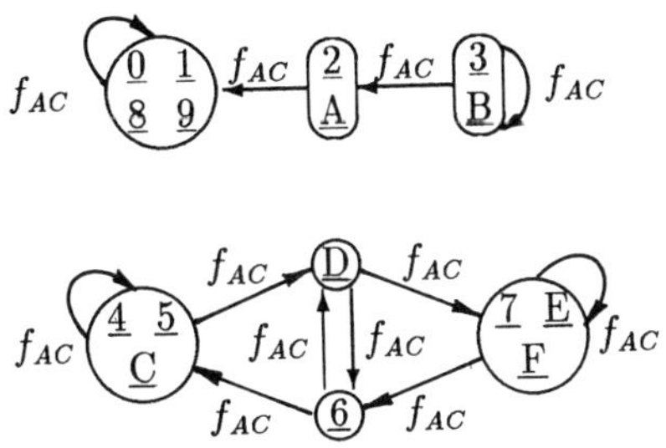

Figure 4(c). Graph of f_{AC}

Test Vector Dependence Graphs

The DDG of an array, and the propagation function together define how a test vector can be propagated throughout an array. For example in the correlator systolic array, the DDG (see Figure 3) represents the data communications, and the propagation functions f_X, f_Y and f_{AC} (see Figure 4) represent the effect of the cell logic. The propagation of the test vectors through the array can be easily defined by using a Test Vector Dependence Graph (TVDG) which is a DDG with the test vectors mapped onto the nodes and the choice in vector propagations are constrained by the propagation functions. For example consider the TVDG of Figure 5. In this TVDG the vectors $\underline{P}$, $\underline{N}$, and $\underline{M}$ are previously defined and we wish to choose $\underline{L}$. The effect of the cell logic on the other vectors and the exact nature of the dependence of $\underline{L}$ on these vectors must be considered. From the TVDG, $\underline{L}$ is determined by the propagation functions $f_Y(\underline{P})$, $f_X(\underline{N})$ and $f_{AC}(\underline{M})$. Thus $\underline{L}$ is a member of the set of vectors

$$\underline{L} \in \{f_Y(\underline{P}) \cap f_X(\underline{N}) \cap f_{AC}(\underline{M})\}$$

In other words the vector $\underline{L}$ is affected by all the propagation functions.

Thus by mapping the required test vectors onto the array DDG to give a TVDG, we can describe how test vectors can be propagated throughout the array. From the TVDG the test patterns for the whole array can be derived by the test vectors applied at the boundary of the array.

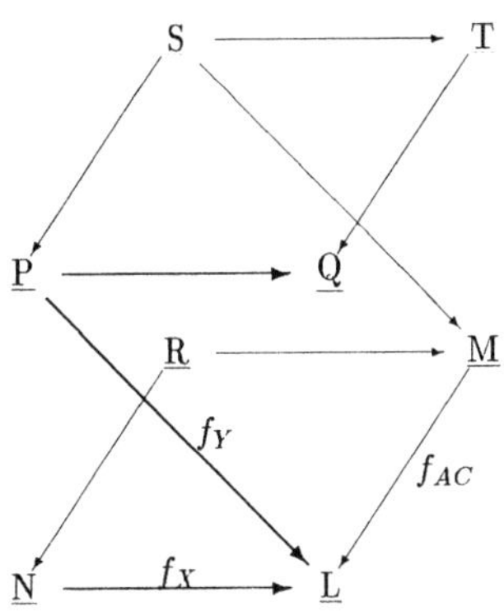

Figure 5. Test Vector Dependence Graph

CYCLIC TVDG

A method is required for mapping vectors onto TVDG's, such that they propagate in an efficient manner. For preference this means it should be independent of the size of the array and propagate at maximum speed through the array. Such a solution is a mapping of test vectors that will propagate in the array in a repeating fashion called cycles. Cycles are defined as closed paths around the propagation graphs. In a cyclic TVDG we only require cycles in the directions, **i** and **j**. To illustrate this concept consider the TVDG for the row direction with the signals X and Y only. In this case an example of a cyclic TVDG for a row of the correlator cells is:

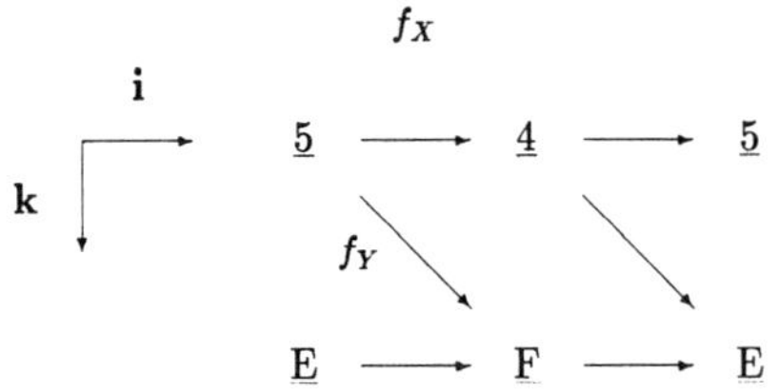

The application of this cyclic TVDG to the row of correlator cells is shown in Figure 6. The **i** direction represents the vector propagation along the row, and contains the cycles such that the vectors 5, 4, E and F are propagated to every second cell in the array independent of the array size. The *sequence* of vectors applied to the cells is represented by the **k** direction where cycles are not required.

Cyclic TVDG's are defined by a mapping of vectors onto the DDG. The test vectors required by a fault model must be combined into one or more cyclic TVDG's such that these test vectors can be propagated efficiently to every cell of the array.

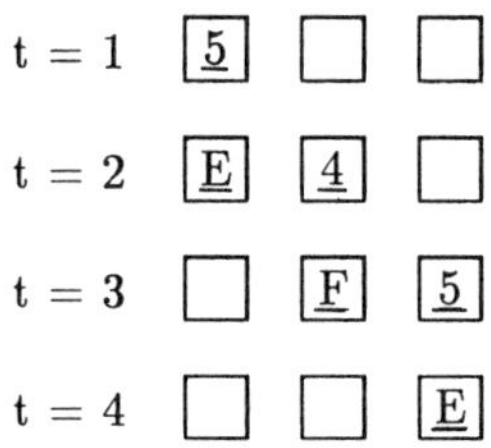

Figure 6.
Application of a Cyclic Test Set to a Row of the Correlator

Stuck-at Cyclic TVDG's

The 5 vectors required for the stuck-at fault model are 5, 7, 9, C and E. These can be mapped onto a cyclic TVDG. One such cyclic TVDG is shown in figure 7. This TVDG has cycles in the row **i** and the column **j** directions. These cycles are of length 1, such that each vector can propagate to itself in both directions.

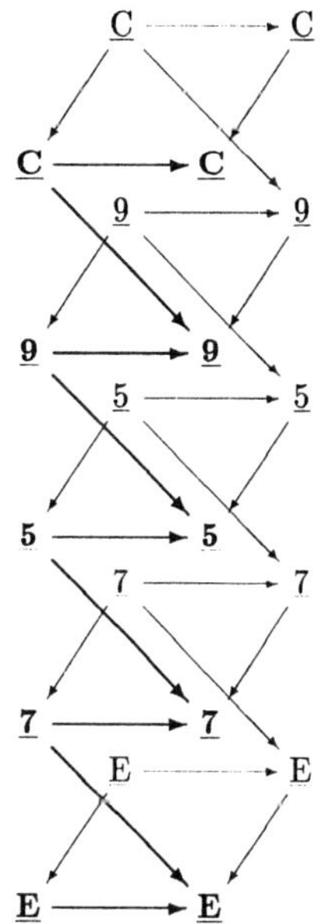

Figure 7. Stuck-at Cyclic TVDG

MOS Switch Level Cyclic TVDG's

The MOS switch level fault model requires the 10 test vectors: $\underline{4}$, $\underline{6}$, $\underline{7}$, $\underline{8}$, $\underline{9}$, $\underline{B}$, $\underline{C}$, $\underline{D}$, $\underline{E}$ and $\underline{F}$; and the 10 test sequences (setup-test): $\underline{D}$-$\underline{6}$, $\underline{6}$-$\underline{D}$, $\underline{9}$-$\underline{7}$, $\underline{7}$-$\underline{C}$, $\underline{C}$-$\underline{B}$, $\underline{5}$-$\underline{E}$, $\underline{E}$-$\underline{9}$, $\underline{9}$-$\underline{4}$, $\underline{4}$-$\underline{F}$, $\underline{F}$-$\underline{8}$. These are mapped onto cyclic TVDG's such that the sequence of setup vector followed by test vector for the stuck-open faults is applied to every cell in the array. Three such TVDG's are required to propagate all the test sequences and these are shown in figure 8.

In the TVDG of figure 8(a). the cycles in the column (**j**) direction are of length 2 and in the row (**i**) are of length 1. In figure 8(b). the cycles are of length 1 in both directions. In figure 8(c). the cycles are of length 1 in the column direction and are of length 2 in the row.

FAULT DETECTION

The detection of a fault is dependent on the propagation of the fault effect to an observable output, i.e. the outputs of the array. For example with the fault Y_{in} stuck-at 1 the test vector is $\underline{C}$. The fault free response is $Y_{out} = 1$, $C_{out} = 0$ and the faulty response is $Y_{out} = 0$, $C_{out} = 1$. The fault affects signals in both the row and column directions. Test vector propagation in the array is dependent on the data flow. Therefore the effect of faults will be to disrupt this propagation of vectors. Thus if the stuck-at fault test set is applied to a cell with Y_{in} S@ 1, the fault effect is:

Input sequence	E 7 5 9 C
Expected output sequence	E 7 5 9 C
Faulty output sequence detected at signal Y	E 7 5 8 C
Faulty output sequence detected at signal C	F 7 5 9 E

The vector sequence is affected in both directions and this fault effect will propagate to the outputs of the array. Then a single comparison between the outputs and the expected outputs will detect any fault.

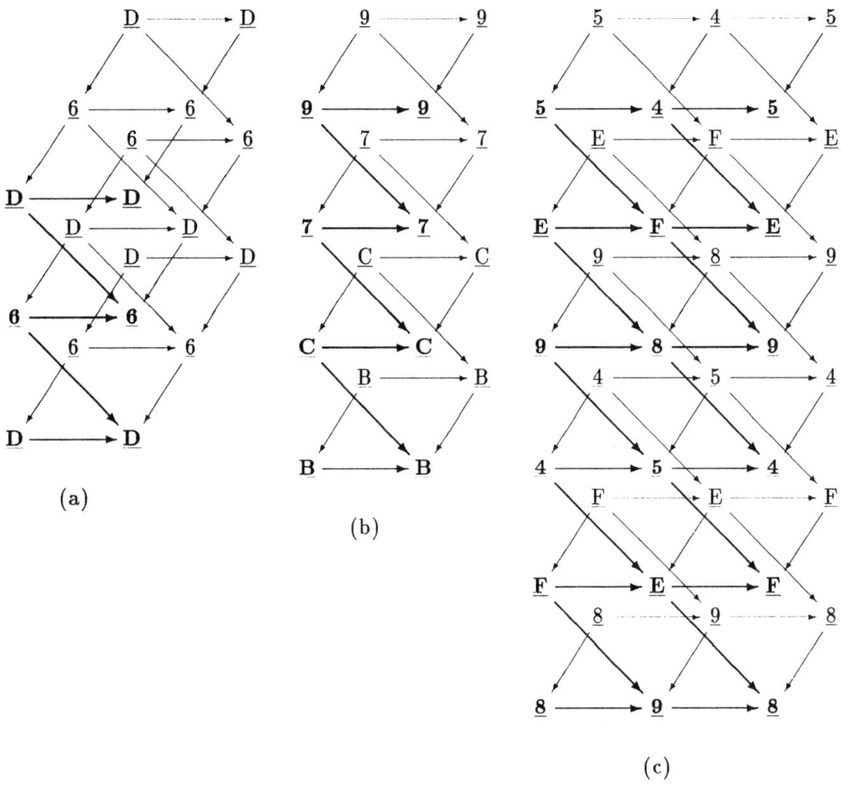

Figure 8. MOS Switch Level TVDG

CONCLUSION

This paper presents a structured test strategy for bit level systolic arrays. First the test generation for a specific cell is performed using the most appropriate fault model. Two different fault models have been used to generate test vectors for a systolic correlator. To test the whole array, we propose a strategy which consists of propagating the required test vectors to every cell in the array. This method is based on the Data Dependence Graph and the cell logic decomposed into propagation functions. A test vector sequence is found by mapping the test vectors on the Data Dependence Graph labelled by the propagation functions,which is called the Test Vector Dependence Graph. Moreover we search for efficient cyclic tests which include, if possible, all the vectors and thus is independent of the size of the array.

ACKNOWLEDGEMENT

This work has been undertaken as part of the UK Alvey Programme (Project ARCH013) whose support is gratefully acknowledged. We are also grateful to GEC and Plessey for permission to publish this material.

References

Eldred R D 1959, Test routines based on symbolic logic statements, J. ACM 6 33-36

Galiay J, Crouzet Y and Vergnault M 1980, Physical versus logical fault models MOS LSI circuits : Impact on their testability, IEEE Transactions on Computers C-29 (June) 527-531

Kung S Y 1986, VLSI array processors, in Systolic Arrays, Moore W R et al (Eds), (Bristol Adam Hilger) 7-24

Levi M 1981, CMOS is more testable, IEEE International Test Conference 217-220

Marnane W P, Moore W R, Yassine H M, Gautrin E, Burgess N and McCabe A 1987, Testing bit level systolic arrays, IEEE International Test Conference 906-914

McCanny J V, Evans R A and McWhirter J G 1986, Use of unidirectional data flow in bit level systolic array chips, Electronic Letters 22 540-541

Starke C W 1984, Built in self test for CMOS circuits, IEEE Test Conference 309-314

7.3 On the Design of Easily Testable and Reconfigurable VLSI Processor Arrays

Jung Hwa Kim and Sudhakar Reddy

INTRODUCTION

To achieve high speed computing, concurrently operating arrays of identical processors have been designed and proposed (Mead C. 1980). Processor arrays are gaining in popularity in real time and signal processing applications (Kung H.T. 1979). The continuing growth of interest in processor arrays poses new problems in ensuring acceptable yield and reliability of computations performed by such systems (Kung H.T. 1984, Kim J.H. 1985). To increase the yield of such systems, several approaches to easily reconfigurable arrays have been proposed recently (Leighton T. 1985, Kung H.T. 1984, Sami M. 1986). Easy reconfiguration is also essential to improve availability of the computational capacity during the operational use of the array. The main emphasis in this paper will be on the achievement of yield enhancement through a proposed general reconfiguration algorithm. The time complexity of the proposed reconfiguration algorithm is analysed.

PRELIMINARIES

Earlier we proposed methods to design Easily Testable And Reconfigurable (ETAR) systolic arrays (Kim J.H. 1987a). We also described new 2-D systolic arrays which lend themselves to easy reconfiguration as well as efficient implementation of algorithms (Kim J.H. 1987a). The proposed 2-D unidirectional array is built from linear unidirectional arrays by using linear unidirectional arrays as rows of 2-D unidirectional arrays. Figure 2 shows a 2-D unidirectional array to perform multiplication of the band matrix with the full matrix given in Figure 1. The 2-D unidirectional array requires O(2N) time units to perform the desired matrix operation. The PEs of the proposed 2-D arrays are numbered according to a matrix ordering: the PE on row i, column j is designated as PE[i,j]. The PE used in the 2-D unidirectional array is shown in Figure 3. Details of the modification to the PEs of the proposed 2-D arrays to render them easily testable and reconfigurable and also the test procedures are given in (Kim J.H. 1987b). The test procedures can detect faults and if faults are not fatal, locate the faulty PEs in the proposed 2-D ETAR arrays. A main feature of the proposed ETAR designs is that the PEs in the 2-D arrays can all be tested simultaneously, thus reducing the testing time significantly.

In this paper, we propose a general reconfiguration algorithm for 2-D processor arrays to achieve yield enhancement. The time complexity of the proposed reconfiguration algorithm is

† The reaserch reported has been supported in part by SDIO/IST Contract No. N00014-87-K-0419 managed by U.S. Office of Naval Research.

$$\begin{pmatrix} a_{11} & a_{12} & & 0 \\ a_{21} & a_{22} & a_{23} & \\ & a_{32} & a_{33} & a_{34} \\ 0 & & a_{43} & a_{44} \end{pmatrix} \times \begin{pmatrix} b_{11} & b_{12} & b_{13} & b_{14} \\ b_{21} & b_{22} & b_{23} & b_{24} \\ b_{31} & b_{32} & b_{33} & b_{34} \\ b_{41} & b_{42} & b_{43} & b_{44} \end{pmatrix} = \begin{pmatrix} c_{11} & c_{12} & c_{13} & c_{14} \\ c_{21} & c_{22} & c_{23} & c_{24} \\ c_{31} & c_{32} & c_{33} & c_{34} \\ c_{41} & c_{42} & c_{43} & c_{44} \end{pmatrix}$$

Fig. 1. Multiplication of a full-matrix by a band-matrix.

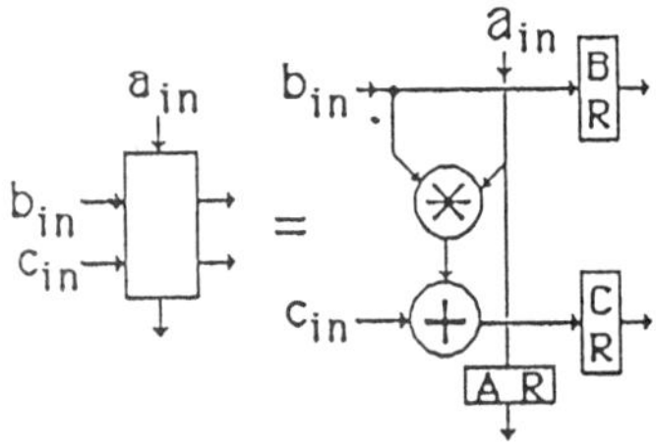

Fig. 3. The PE of a 2-D unidirectional array.

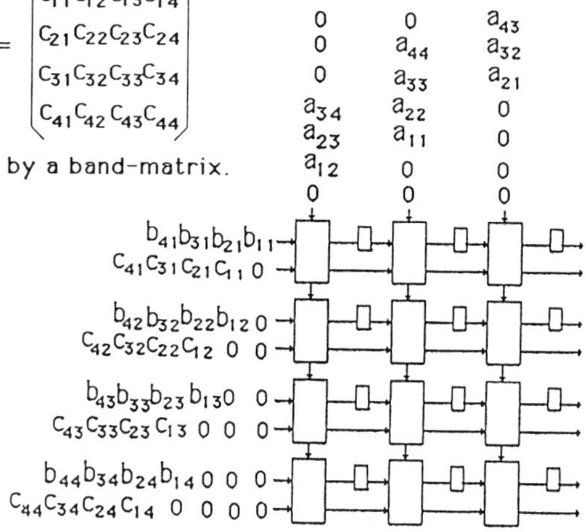

Fig. 2. The 2-D unidirectional array to solve the problem of Fig. 1.

O(N^2) to reconfigure a potentially faulty Nx(N+SC) physical array into a fault-free NxN logical array, where SC is the number of spare column. For the purpose of completeness of presentation we include the augmentation to the PEs of the proposed 2-D arrays and testing procedures for the proposed 2-D arrays in this paper.

PES FOR 2-D ETAR ARRAYS AND FAULT MODEL

Augmented PEs

In this subsection, the augmentation to the PEs of the proposed 2-D arrays to achieve simplicity in testing and reconfiguration are given. In the sequel we will call the data paths in the 2-D arrays A-data path (or simply A-path) that corresponds to the vertical input to a PE, B-data path and C-data path (these paths correspond to the horizontal inputs to PEs). The proposed augmented PE for 2-D unidirectional arrays is shown in Figure 4. The augmented PEs shown can be used to reconfigure the array in the presence of no more than one faulty PE in a row. Extension to arbitrary number of faulty PEs in a row is possible and given in (Kim J.H. 1987b). Two clocks are assumed to be available for the array. Clock TCK, called test clock, is used to scan-in and scan-out test inputs and test results and clock SCK is the system clock.

For 2-D unidirectional arrays, the extra hardware in the augmented PE, added to achieve easy testing and reconfiguration, is as follows (cf. Figure 3); bypass registers: BP_b and BP_c, which are used to store data when the PE is bypassed, thus allowing reconfiguration without increasing inter-PE wire length, multiplexers: MUX1 and MUX2 to bypass or not to bypass the PE, and a control flip-flop CM to control MUX1 and MUX2, a delay register DR_a which is used to resynchronize meeting of three data at a PE, a multiplexer MUX3 to introduce or not to introduce one time delay into A-data path (i.e. the vertical input to a PE) of the multiplier for resynchronization, a control flip-flop CD to control MUX3, a switch SW to connect PEs located in neighboring rows, and control flip-flops CS to control SW. The switch SW is connected only to three neighboring PEs in the preceeding row, that is, SW in PE[i,j] can receive input data only from PE[i-1,j-1], PE[i-1,j] and PE[i-1,j+1]. Depending on fault distribution, the switch SW has three states as shown in Figure 5. The control flip-flops could come from a shift register stage whose individual stages are in different PEs of the array. Additionally the delay register shown between two PEs of the unaugmented array (cf. Figure 2) is now incorporated as DR_b into the augmented PEs. Finally the registers in the original PEs (BR and CR) are modified to two-port, two-clock registers. Note that all registers except BR and CR are clocked by CK, which is active if TCK or SCK is active. TCK is used in testing mode only.

Fault Model

Functional faults are the basic feature of the fault model used. A cell is defined as faulty if it produces an incorrect functional result which is different from the expected result. Since a fault in a switch SW connected to the vertical (A) input of a PE may not allow the use of the vertical output of a PE, a failure in switch SW[i,j] is treated as equivalent to both PE[i-1,j] and PE[i,j] being faulty and hence not available for use in a reconfigured array.

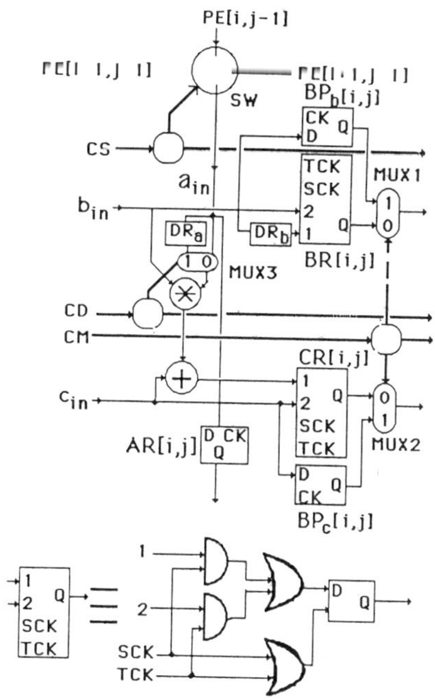

Fig. 4. The PE of a 2-D unidirectional ETAR systolic array.

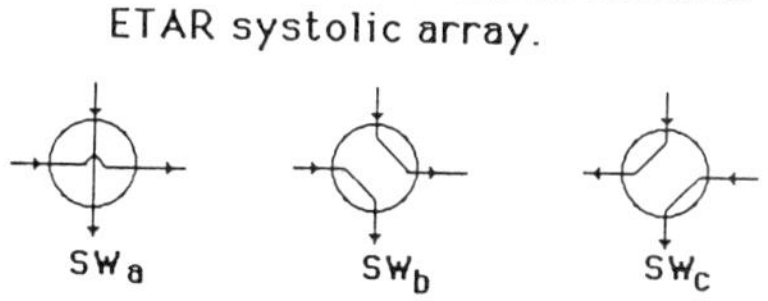

Fig. 5. The function of a switch SW.

TEST PROCEDURES FOR 2-D ETAR ARRAYS

This section describes how to detect faults and if faults are not fatal, how to locate faults in 2-D arrays. Most of the existing literature dealing with fault-tolerant array architectures consider only the reconfiguration phase. The problem of error detection and location of the faulty PEs, which is the prerequisite for reconfiguration, is not treated adequately. In the proposed test procedure the basic strategy is to use the scan path to scan-in test patterns and scan-out responses. By conducting tests in every row at the same time, all PEs in the 2-D array are tested simultaneously. Test procedures for a 2-D unidirectional ETAR array are given next.

For the purpose of testing, each PE is divided into four functional units : control unit, reconfiguration unit (RU), scan unit (SU), and computation unit (COMU). The elements of the four units of the PE are as follows : the control unit consists of the control flip-flops and related links, the scan unit SU consists of port 2 of BR and CR, port 0 of MUX1 and MUX2, AR, SW, and related links, the reconfiguration unit RU consists of BP_b,BP_c, port 1 of MUX1 and MUX2, and related links, and computation unit COMU consisting of the rest of the PE. *Scan path (SP)* is defined as a linear array of SUs, and ***reconfiguration path (RP)*** as a linear array of RUs.

Test procedure for 2-D unidirectional ETAR arrays is divided into four steps. Details of the test procedure are given in Figure 6. In the first step, the control units are tested. If there is a fault in a control flip-flop CS, the array needs no more testing since the fault is fatal. If there is a fault in other parts of control units, the row is regarded as faulty and is bypassed. In the second step the scan path SP is tested to detect faults in the horizontal and vertical data paths, and the faults in SWs or vertical data paths are located with proper setting of the control register CS. To locate faults in SWs or in vertical data paths, vertical data paths and rows of inter-PE buses are tested with all SWs set to the SW_a state by controlling CS. If rows of inter-PE buses are all fault-free and one vertical data path is faulty, a fault is in the vertical data path of one PE, and the exact location of a faulty PE can be found by testing a reconstructed vertical data path such that some PEs in faulty vertical path replace those in the neighboring fault-free vertical path with proper setting of SWs. If vertical data paths are all fault-free and one row of inter-PE buses is faulty, the faulty inter-PE bus can be located by testing vertical paths with SW in the faulty row set to the SW_b state. The fault in a inter-PE bus is regarded as a fault in SW to avoid using this bus in reconfiguration. Next, the reconfiguration path RP is tested as the third step to detect

Fig. 6. Testing procedures in a 2-D unidirectional array.

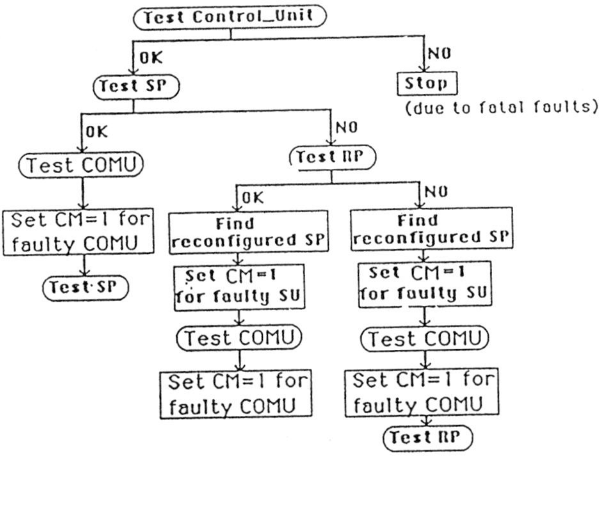

faults. Even if both RP and SP are faulty, the array can be still testable and reconfigurable as long as one of RU and SU in a PE is fault-free. The process to find a reconfigured SP which combines the good part of the faulty RUs and faulty SUs, is to connect RU and SU together by proper setting of control flip-flop CM. Only when there exists a fault-free SP or reconfigured SP, can we test the COMU of PEs as the fourth step. Once the locations of faulty PEs are determined, they can be bypassed by setting CM to 1 permanently, if RUs of faulty PEs are fault-free. If RP is faulty, we apply Procedure Test_SP with the control registers of faulty PEs set to 1 permanently to check whether RUs of faulty PEs function correctly.

It is helpful to understand how an identical test pattern can be brought into corresponding registers of all PEs simultaneously. For the sake of simplicity, assume that SP is fault-free. If CM for all PEs is held at 0 and test clock TCK is clocked an adequate number of times, while holding A, B and C inputs constant, all corresponding registers (cf. Figure 4) in each PE will have identical contents. Thus identical test inputs can be scanned into the registers of PEs through the fault-free SP. By pulsing SCK (system clock) once the test result produced by computation units of all PEs are put into CR register simultaneously. Clearly the test results can be similarly scanned out through the fault-free SP. By noting the positions of incorrect outputs in the scanned-out test results, the location of the faulty COMUs can be easily determined. The number of tests applied to locate all faulty COMUs (or PEs) in a 2-D array can be seen to be *equal to* the number required to detect faults in a single COMU (or PE).

Details of testing procedures to detect and locate all faulty cells in 2-D processor arrays are given in (Kim J.H. 1987b). The testing procedures can detect faults and if faults are not fatal, faulty PEs can be located and the faulty unit of the faulty PEs in 2-D ETAR arrays can be identified.

RECONFIGURATION ALGORITHM FOR THE 2-D PROCESSOR ARRAYS

In this section, we describe how reconfiguration is achieved in the 2-D processor arrays, after faulty PEs have been located. The reconfiguration strategy proposed utilizes the diagnostic information provided by the testing procedure outlined in the preceeding section. The proposed reconfiguration scheme is a row-oriented scheme with the proviso that rows can be bypassed. That is, columns are organized through the rows with local connections only, taking one and only one column element from each and every row except the bypassed rows.

An example of a 2-D unidirectional array containing one faulty PE in a row is shown in Figure 7. Figure 7 shows the reconfiguration scheme from 4x4 physical array to 4x3 logical array, implying SC=1 and SR=0. In Figure 7 the delay register shown between two PEs (cf. Figure 2) is now incorporated into the PE. After faulty PEs are located, they are bypassed by setting control flip-flop CM of faulty PEs to 1. When bypassed, the horizontal data (i.e. B and C data) go through bypass registers. Therefore, the horizontal data are delayed by one time unit in the PEs located on the right side of the faulty PE. To accommodate this time skew in the data to the PEs right of the bypassed PE, a delay is introduced into the vertical data (i.e. A-data) paths to the multiplier of those PEs, by setting each control flip-flop CD of these PEs to 1. That is, the delay in B and C data arrival at the PEs to the right of the faulty PE can be accommodated by simply delaying, by one time unit, A-data applied to these PEs. However, this delay does not affect the transmission of A data from PE to PE (cf. Figure 7), hence the throughput does not decrease. In Figure 7 the PEs labeled D have been reconfigured, by setting CD=1, to introduce an extra delay into A data path to the corresponding multiplier.

A sketch of the algorithm to reconfigure a faulty array and two examples to illustrate the reconfiguration strategy are given below. Our aim is to reconfigure a potentially faulty (N+SR) x (M+SC) physical array into a fault-free N x M logical array, where *SR* and *SC* are the number of

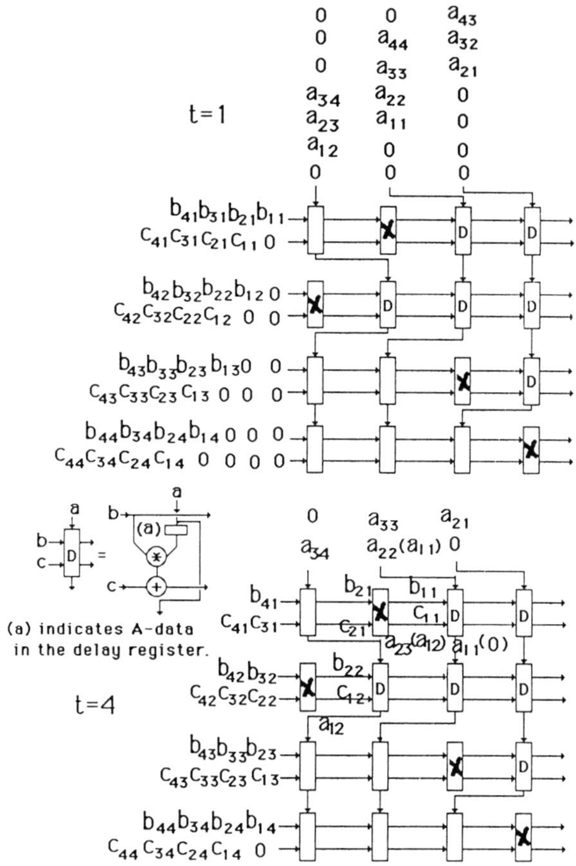

Fig. 7. The proposed reconfiguration scheme of a 2-D unidirectional ETAR array.

spare rows and spare columns, respectively. For the sake of simplicity we consider only the case of N=M.

Case 1 : There is no spare row; that is, SR=0.

The proposed reconfiguration algorithm is given in Figure 8. If there exists a row containing more than *SC* faulty PEs we cannot get an N x N logical array, and the algorithm is aborted. A column of the desired fault-free array is obtained by attempting to connect a PE, say PE[i,j], to a cell in row[i+1] to its left first, below next, and finally to its right. That is, the algorithm attempts to connect PE[i,j] to PE[i+1,j-1], PE[i+1,j] and PE[i+1,j+1], in that order. Once a PE is visited, the PE is marked as ***used*** and will not be visited again from a preceeding row. If the attempt fails due to one of the following three reasons : (1) no PE has been used and all three PEs are faulty, (2) PE[i+1,j-1] has already been used and both PE[i+1,j] and PE[i+1,j+1] are both faulty, and (3) both PE[i+1,j-1] and PE[i+1,j] have already been used and PE[i+1,j+1] is faulty, then the algorithm backtracks to the (i-1)th row, continues to connect next available good PE. The reconfiguration algorithm continues until N columns connecting N good PEs are obtained or it is determined that a fault-free N x N array is not possible.

An example of a 2-D processor array is shown in Figure 9. Figure 9 shows the reconfiguration scheme from a faulty 4x7 physical array to a fault-free 4x4 logical array, implying

Figure 8. The proposed reconfiguration algorithm.

```
Algorithm Reconfiguration
* This algorithm reconfigure  a faulty Nx(N+SC) array  *
* into a fault-free NxN array,                          *
* where  SC = number of spare columns.                  *
BEGIN
  If there is a row containing more than SC faulty cells
  then  return failed
  else
  { successcol=0;
    for col=1 to N+SC  do
    {if PE[1,col]= good
     then { i=1;  i=i+1;path=true; j=col;
          while (path and i<=N)
           { if PE[i,j-1] is good and not used
             then {define the PE used; i=i+1;j=j-1;}
             else if PE[i,j] is good and not used
                  then {define the PE used;i=i+1;}
                  else if PE[i,j+1] is good and not used
                       then {define PE used;i=i+1;j=j+1;}
                       else if i=1 then path=false
                                    else i=i-1; *backtrack*
           } *while*
          if path=true then successcol=successcol+1;
          if successcol=N then return succeed;
    } *if good*
   return  failed;
  } *if *
END.
```

SC=3 and SR=0. In Figure 9 the third column is connected as follows. PE[1,4] is first connected to PE[2,4] since PE[2,3] has already been used. From PE[2,4] no further connection can be made since PE[3,2], PE[3,3] and PE[3,4] are all faulty. So the algorithm has to backtrack to PE[1,4] and PE[1,4] is connected to PE[2,5] and so on.

Next we compute the time complexity of the proposed reconfiguration algorithm to reconfigure a potentially faulty N x (N+SC) array into a fault-free N x N array. In the proposed algorithm, once a PE is visited the PE is marked as *used* and the PE will not be visited again from a preceeding row, but may be revisited up to three times from a succeeding row through three different paths when backtrack occurs. Since no PEs will be visited more than four times (forward once and backward three times) and there are ($N^2 + N*SC$) PEs in the array, the time complexity of the proposed reconfiguration algorithm is, in the worst case, $O(N^2 + N*SC)$. Our objective is to improve yield, that is, to maximize the number of acceptable chips from a given wafer, not to increase simply the probability that a single chip is acceptable. The addition of spare columns and the introduction of fault-tolerance into the array increase the chip area and as a result a smaller number of chips will now fit into the same wafer area. Therefore, the number of spare columns needed is $\leq O(N)$. Hence, the time complexity of the proposed rconfiguration algorithm to reconfigure a potentially faulty Nx(N+SC) array into a fault-free NxN array is $O(N^2)$, For example, the best yield of a chip consisting of 15X15 PEs is obtained with 1 spare row and 1 spare column, according to the yield calculation done in (Kim J.H. 1987a).

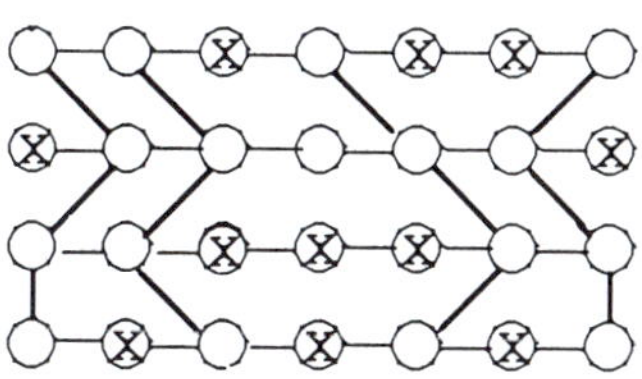

Fig. 9. Reconfiguration from 4X7 to 4X4

We also propose a statistical measure for the algorithm complexity as follows:

Statistical complexity of the algorithm to configure a column = C_s
= [number of PEs visited until backtrack occurs or a column is complete] + Pr[backtrack occurs]([number of PEs visited until backtrack occurs or a column is complete] + Pr[backtrack occurs]([number of PEs visited until backtrack occurs or a column is complete] + Pr[backtrack occurs]([number of PEs visited until backtrack occurs or a column is complete] + ...)))

$$< N + P_1 (N + P_1 (N + P_1 (N .....))..)))$$

$$= N + P_1 N + P_1^2 N + P_1^3 N + ... + P_1^s N$$

$$= N \frac{(1-P_1^{s+1})}{(1-P_1)} \quad (A)$$

where N is the number of rows, P_1 is the probability of backtrack occuring and s is the number of backtracks attempted.

Clearly an upper bound for the expression in (A) is obtained by substituting an upper bound for P_1. If P_c is the probability of a cell failure, then P_c can be seen to be an upper bound on P_1, since each new backtrack requires that an additional cell, not visited earlier, is faulty. Hence $C_s < \frac{N(1-P_c)^{s+1}}{(1-P_c)} < \frac{N}{(1-P_c)}$.

For a given P_c clearly C_s is of O(N). In an analogous manner the statistical time complexity to arrive at an N x N array can be shown to be $O(N^2)$.

Theorem 1 : The reconfiguration algorithm given in Figure 8 reconfigures a potentially faulty N x (N+SC) physical array into a fault-free N x N logical array in $O(N^2)$ time, where SC is the number of spare columns and $\leq O(N)$, or determines the irreconfigurability of the array under given conditions in the same period of time.

Proof: The proof is given separately for the two possible situations noted below:

Situation 1: There exists a row containing more than SC faulty PEs.

Since the column is organized taking one and only one good cell from each and every row and the row containing more than SC faulty cells has less than N good cells. one cannot get N columns. Therefore, the Nx(N+SC) faulty array cannot be reconfigured into the NXN fault-free array. Since each row consists of O(N) PEs counting the number of faulty PEs in each row has a time complexity of O(N). Since there are N rows the overall time complexity to determine irreconfigurability in this situation is clearly $O(N^2)$.

Situation 2: There exists no row containing more than SC faulty PEs.

The PE, say PE[i,j], visited from a preceeding row for the first time results in one of the following 4 situations in the worst case:

(1) A column is successfully formed during the first visit of PE[i,j].

(2) A column is not successfully formed during the first visit of PE[i,j], but successfully formed during the second visit of PE[i,j] from a succeeding row.

(3) A column is not successfully formed during the first and the second visit of PE[i,j], but successfully formed during the third visit of PE[i,j] from a succeeding row through the path which is different from that taken during the second visit.

(4) A column cannot be successfully formed through PE[i,j] and the algorithm visits PE[i,j] the fourth time from a succeeding row through the path which is different from those taken during the second and third visit and backtracks to a preceeding row.

If a column is successfully formed, PE[i,j] should not be used again, and hence will not be visited again. If a column cannot be successfully formed through PE[i,j], PE[i,j] need not be visited again since the column reconfiguration through PE[i,j] was found to be unsuccessful. Therefore, a PE will be visited at most four times, once from a preceeding row and three times through each of three different path from a succeeding row. Since there are $O(N^2)$ PEs in the array, the time complexity of the proposed reconfiguration algorithm is, in the worst case, clearly $O(N^2)$. If N columns connecting N good cells are not obtained, the algorithm determines the irreconfigurability of the array under given conditions. <Q.E.D.>

Case 2 : There are SR spare rows.

The following rows must be regarded as faulty rows and bypassed because they can not be used in the logical array :

(1) rows which contain more than *SC* faulty PEs,

(2) rows which contain faults in the control flip-flop CM or CD,

(3) rows which can not provide fault-free scan paths (SP) or reconstructed SP.

If there exist K must-be-bypassed rows and K>SR we cannot get an N x N logical array, and the algorithm is aborted. If there exist K must-be-bypassed rows and K<SR there are still (SR-K) spare rows after bypassing K rows. In case that T rows contain more than one faulty PE but up to maximum of SC faulty PEs, the reconfiguration algorithm can find a fault-free N x N logical array trying as many times as all combinations of (SR-K) out of T with bypassing (SR-K) rows at a time in the decending order of number of faulty cells if the reconfiguration from the faulty (N+SR) x (N+SC) array into the fault-free N x N array is possible. The time complexity of the proposed reconfiguration scheme reconfiguring a potentially faulty (N+SR) x (N+SC) physical array into a fault-free N x N logical array is, in the worst case, $O(N^{2+SR})$, since there are $O(N^{SR})$ combinations each of which require, in the worst case, $O(N^2)$.

Figure 10 shows the reconfiguration scheme from a faulty 6x7 physical array into a fault-free 4x4 logical array, implying SC=3 and SR=2. Let us assume that testing procedures indicate that PE[2,1] has a fault in vertical data path and the rest faulty PEs have faults in computation units. Since the 4th row contains more than three faults the 4th row has to be bypassed. After bypassing the 4th row there is one more row to be bypassed. The only way to get a fault-free 4x4 array is to bypass the 2nd row, even though the 2nd row contains the smallest number of faulty PEs. When the 2nd row is bypassed the 1st logical column is constructed by connecting PE[1,1] to PE[2,2] to avoid the use of the faulty data path in PE[2,1]. Since the number of spare rows needed is usually small to improve yield the time complexity to choose rows to bypass is expected to be low.

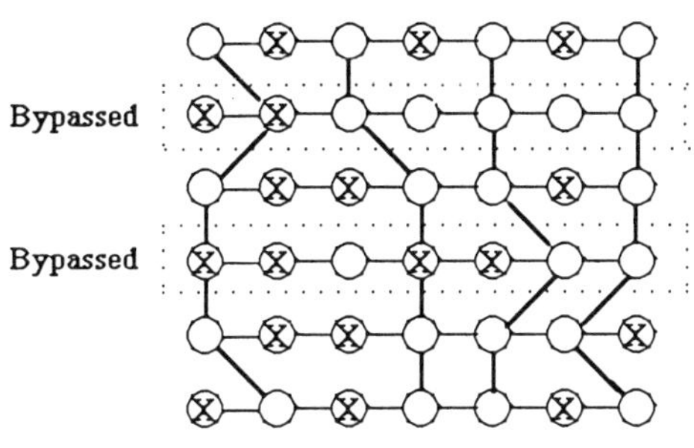

Fig. 10. Reconfiguration from 6X7 to 4X4

Next we discuss a more general case that a PE is connected to (2K+1) PEs of the next row, that is, PE[i,j] is connected to PE[i+1,l] where (j-k)$\leq$ l $\leq$(j+k). Since no PEs will be visited more than (2k+2) times i.e. once from the preceeding row and (2k+1) times from the succeeding row through (2k+1) different paths in the proposed algorithm, and there are O(N^2) PEs in the array, the time complexity of the proposed reconfiguration algorithm to reconfigure a potentially faulty N x (N+SC) array into a fault-free N x N array is clearly O($N^{2)}$.

CONCLUDING REMARKS

In this paper, We have proposed the general reconfiguration algorithm for 2-D processor arrays to achieve yield enhancement. The time complexity of the proposed reconfiguration algorithm is, in the worst case, O(N^2) to reconfigure a potentially faulty N x (N+SC) physical array into a fault-free N x N logical array, where SC is the number of spare column.

REFERENCES

Leighton T. and Leiserson C.E. 1985, "Wafer Scale Integration of Systolic Arrays," IEEE TC, May 1985 pp.448-461

Kim J.H. and Reddy S.M. 1985, "A Fault-Tolerant Systolic Array Design Using TMR Method," ICCD 85 pp.769-773

Kim J.H. and Reddy S.M. 1987a, "On Easily Testable and Reconfigurable Two-Dimensional Systolic Arrays" Proc. 1987 Intl.Conf. on Parallel Processing, August 1987 pp.101-109

Kim J.H. 1987b, "On the Design of Easily Testable And Reconfigurable Systolic arrays", Ph.D. Thesis, Dec. 1987

Kung H.T. 1979, "Let's design algorithms for VLSI systems," Proc. Caltech. Conf. on VLSI, Jan. 1979, pp.65-90.

Kung H.T. and Lam M.S. 1984, "Fault-tolerance and two level pipelining in VLSI systolic arrays," MIT Conference on ADV. Research in VLSI, Jan. 1984, pp.74-83.

Mead C. and Conway L. 1980, Introduction to VLSI systems, chapter 8, Addison-Wesley, 1980.

Sami M. and Stefanelli R. 1986, "Reconfiguration Architectures for VLSI Processing Arrays," Proc. of the IEEE, May 1986, pp.712-722

Epilogue

Our purpose in this book has been to show the vital importance to the IC designer of a clear understanding of defect mechanisms and of yield modelling in order to achieve economic fabrication of microelectronic chips and effective testing. The pressures towards ever more complex circuits make the increasing use of defect-tolerant designs inevitable and these can only be exploited in the future through a proper understanding of the underlying issues.

It is also evident that there is much basic research and practical development to be undertaken, but we hope that this collection of papers gives a good picture of current and ongoing work and a pointer to future possibilities.

Index